Faxed

Johns Hopkins Studies in the History of Technology
Merritt Roe Smith, Series Editor

Faxed

The Rise and Fall of the Fax Machine

JONATHAN COOPERSMITH

Johns Hopkins University Press

Baltimore

© 2015 Johns Hopkins University Press
All rights reserved. Published 2015
Printed in the United States of America on acid-free paper

Johns Hopkins Paperback edition, 2016
2 4 6 8 9 7 5 3 1

Johns Hopkins University Press
2715 North Charles Street
Baltimore, Maryland 21218-4363
www.press.jhu.edu

*The Library of Congress has cataloged the hardcover edition of this book
as follows:*
Coopersmith, Jonathan, 1955–.
Faxed : the rise and fall of the fax machine / Jonathan Coopersmith.
pages cm. — (Johns Hopkins studies in the history of technology)
ISBN 978-1-4214-1591-8 (hardcover : alk. paper) — ISBN 1-4214-1591-7
(hardcover : alk. paper) — ISBN 978-1-4214-1592-5 (electronic) —
ISBN 1-4214-1592-5 (electronic) 1. Fax machines—History. 2. Facsimile
transmission—Technological innovations. I. Title.
TK6710.C65 2015
621.382'35—dc23
2014016734

A catalog record for this book is available from the British Library.

ISBN-13: 978-1-4214-2123-0
ISBN-10: 1-4214-2123-2

*Special discounts are available for bulk purchases of this book.
For more information, please contact Special Sales at 410-516-6936 or
specialsales@press.jhu.edu.*

Johns Hopkins University Press uses environmentally friendly book
materials, including recycled text paper that is composed of at least
30 percent post-consumer waste, whenever possible.

Contents

Preface

My interest in faxing began when my mother purchased a fax machine to communicate with friends in Russia and Thailand. Our family is not the most technically competent (after my father died, we found a closet full of analog answering machines in his office; apparently, when the tape filled, he simply bought a new machine). So my mother's enthusiastic acquisition of a trendy new technology intrigued me. I investigated and found a machine so capable it could communicate worldwide yet so easy that a Coopersmith could operate it. Even more exciting for a historian of technology, the fax machine had a long, if not always commercially distinguished, pedigree. Yet strangely, only a few books existed about faxing and none of them were historical. I had found a topic that fulfilled the two basic rules for doing research: study the unstudied to make a contribution, and study the interesting to keep your sanity.

Like its story, the research has taken me literally around the globe to work in archives and libraries. Fellowships from the National Science Foundation and the Center for International Exchange of Scholars and a grant from the Texas A&M Military Studies Institute made my travel possible. A Fulbright research-teaching fellowship enabled me to spend a year at Tokyo Institute of Technology. A Faculty Development Leave from Texas A&M University provided the opportunity to spend a year at the National Museum of American History, one of America's great treasures, with Barnie Finn and his colleagues. I am grateful to the American and Japanese taxpayers who made this travel and research possible.

Many archivists, librarians, engineers, managers, administrators, historians, fax users and others in the United States, Japan, France, Germany, Britain, Italy and Switzerland generously gave their time and shared their expertise to help me. This book would not have been possible without them. I am especially indebted for their thoughts and comments to Roger Beaumont, Andy Butrica, Paul Ceruzzi, Ed Constant, Thomas Haigh, Dave Hochfelder, Sheldon Hochheiser, David Hochman, Richard John, Marcel Lafollette, Harold Livesay, Alex Magoun,

Peter Meyers, Steve Myers, James Rafferty, Geoff Thompson, Jim Weisser, Audra Wolfe, and JoAnne Yates. Jean Marie Linhart provided statistical support, while Denise Wernikoff wandered down some unproductive alleys so I did not have to.

Many people in Japan enhanced the value of my visits there by graciously arranging appointments, translating, accompanying me, and otherwise kindly going out of their way to help me. In particular, I want to acknowledge the invaluable assistance of Yakup Bektas, Masanori Kaji, Eiichi Matsumoto, Hideto Nakajima, Hiroshi Oda, Eiichi Ohno, Hitoshi Omori, and Yuzo Takehashi.

When listening to me describe my research on Russian electrification, people interrupted to ask my thoughts on contemporary Russia. When describing this research, however, listeners often interrupted me with "Let me tell you my fax story." Even decades later, many people still vividly remembered their first encounter with a fax machine, a device that seemed almost miraculous to them. Not surprisingly, interviews provided a personal window on users that documents rarely recorded and access to the insights of engineers, managers, and industry analysts. As well as being generous with their time, many people lent me reports, newsletters, sales manuals, and other materials they had written or acquired in their fax careers.

I am very grateful to the following who discussed faxing or otherwise assisted me: Kazunori Ariki, Takayuki Arimura, Kunihiko Asakura, Oskar Blumtritt, Dennis Bodsen, Tony Borg, William Caughlin, Tom Colbert, Matthew Connally, Steve Cotler, Leonard David, Frank DiSanto, David Duehre, Robert Dujerric, Matthew Edmund, Jeff Eller, Tadashi Endo, Toshiaki Endoh, Larry Farrington, Richard Forman, Jun Fukami, Kazuhiro Fukuda, Eric Gan, Eric Giler, Greg Glassman, Norihiro Hagita, Takehiko Hashimoto, Yoshiaki Hayashi, Hiroshi Hirai, Hiro Hiranandani, S. K. Ho, Jim Imamzade, Satoshi Ishibashi, Ayuko Ishigawa, Akio Ishikawa, Sadayuki Ito, Mitsuo Kaji, Osamu Kamei, Hajime Karatsu, Koichi Karube, Sadamoto Kato, Maury S. Kauffman, Takashi Kawade, Mitsuru Kawasaki, Nick D. Kenyon, Iida Kiyoshi, Kazuo Kobayashi, Noboru Kobayashi, Kohshi Komatsuzaki, Kazumi Komiya, Mindy Kotler, Denis Krusos, Keijiro Kubota, Mitsuo Kurachi, Ryoko Kurata, Terrie Lloyd, Jim MacKintosh, Masahiro Maejima, Hank G. Magnuski, Tsuguhiro Matsuda, Akio Matsui, Mitsuji Matsumoto, Ken McConnell, Kaji Mitsuo, Yukikazu Mori, Tetsuro Morita, John Morrison, Chiaki Motegi, Takayuki Naitoh, Fumiharu Nakayama, Toshiya Nakazawa, Shinitsu Narazaki, Jeff Neff, Hisatomo Negishi, Mitsuru Nishiyama, Mutsuo Ogawa, Kunio Ohtake, Hiroshi Okazaki, Yuji Okita, Kenji Onishi, Masayoshi Orii, Skipp Orr, Judy Pisani, Alan Pugh, James Rafferty, Dave Rampe, T. R. Reid, Garvice H. Ridings, Laurel Rodd, Lisa Rosenberg, Masahiro Sakai, Syuzo Sasaki, Masaki Satsuka, Genbei Sawada, Toshiya Sayama, Roger Sherman, Keiichi Shimizu, I. Shirotani, John Shonnard, Ludwell Sibley, Elliot Sivowitch, Michael Smitka, George Stamps, Hideo Suzuki, Taiji Suzuki, Kimihiro Tajima, Masanori Tanaka, Shigeo Tanaka, Toshikazu Tanida, Eloise Tholen, Keiichi Torimitsu, Kazuo Usui, Eric von Hippel, Kazuo Wada, Kaoru Wakabayashi, Terry White, Gavin Whitehead, Akihiko Yamada, Naoya 'Rocky' Yamada, Toyomichi Yamada, Hajime Yamamoto, Yasuhiro Yamazaki, Daqing Yang, Yasuhiko Yasuda, Emi Yasukawa, and Takeshi Yuzawa. Any misinterpretations and mistakes, however, are mine alone.

Curators at the Deutches Museum in Munich, the Science Museum in London, the Museo Nazionale della Scienza e della Tecnica in Milan, the NTT museum in Tokyo, and the Smithsonian National Museum of American History in Washington, D.C., generously provided the opportunity to study their wonderful collections of fax machines. One particular joy during my year at the National Museum of American History was acting as its representative when GammaLink donated its first fax modem to the museum.

This book also benefitted from comments at conferences and meetings (especially at the Society for the History of Technology, the Business History Conference, and the Asian-Pacific Business and Economic History Conference) in North America, Asia, and Europe and the works-in-progress colloquium at the Texas A&M History Department organized by April Hatfield and Adam Seipp. Indeed, the History Department has been a patient and stimulating academic home with colleagues (especially Walter Kamphoefner, Hoi-eun Kim, and Cynthia Bouton) generously sharing their expertise and competent staff, particularly Kelly Cook, who helped immensely in this project's material production. Joel Kitchens, Jared Hoppenfeld, and the Interlibrary Loan staff at the Evans Library of Texas A&M University proved patient providers of many documents and books. Donald Link, Hiroshi Ono, and Guillema Syon helped with translations, and Alan Anderson assisted with some of the research. Turning the manuscript into a book required the patient, mature guidance of Bob Brugger and his colleagues at Johns Hopkins University Press. Elizabeth Yoder similarly proved to be a delightful wordsmith who improved my writing.

I want to thank the publishers of the following articles for allowing me to draw from them for this book:

"A Florentine in Paris: The Caselli Pantelegraph and Its Successors, 1859–1871," Third IEEE History of Electro-Technology Conference (HISTELCON), Pavia, Italy, Sept. 5–7, 2012, 1–6.

"Creating Fax Standards: Technology Red in Tooth and Claw?" *Kagakugi-*

jutsushi. The Japanese Journal for the History of Science and Technology 11 (July 2010), 37–66.

"Old Technologies Never Die, They Just Never Get Updated," *International Journal for the History of Engineering and Technology* 80, 2 (July 2010), 166–82.

"The Changing Picture of Fax," in Bernard Finn, principal editor, *Presenting Pictures: Artefacts* 4 (2004), 116–28.

"From Lemons to Lemonade: The Development of AP Wirephoto," *American Journalism* 17, 4 (Fall 2000), 55–72.

"Creating the Commons: Establishing a Civic Space for a New Form of Communications," *Business and Economic History* 28, 1 (Fall 1999), 115–24.

"Texas Politics and the Fax Revolution," *Information Systems Research* 7, 1 (March 1996), 37–51; reprinted with a coda in JoAnne Yates and John Van Maanen, eds., *Information Technology and Organizational Transformation: History, Rhetoric, and Practice* (Thousand Oaks, CA: Sage Publications, 2001), 59–85.

"Setting the Stage: Government-Industry Creation of the Japanese Fax Industry," Japan Industry and Management of Technology (JIMT) Series Working Paper 97-03-1, University of Texas, March 1997.

"Technological Innovation and Failure: The Case of the Fax Machine," *Proceedings of the Conference on Business History* (Rotterdam: Centre of Business History, Erasmus University, 1995), 61–77.

"The Failure of Fax: When a Vision Is Not Enough," *Business and Economic History* 23, 1 (Fall 1994), 272–82.

"Facsimile's False Starts," *IEEE Spectrum*, Feb. 1993, 46–49.

Like the proverbial kid in a candy store overwhelmed by the feast of tempting opportunities before him, I have had more fun doing this research than a historian should. I've also taken longer than a productive historian should, as my wife and children can attest. This book is dedicated, with more appreciation than words can express, to Lisa, Alex, and Caroline.

Abbreviations

BFA   Broadcasters' Faximile Analysis
BFT   binary file transfer
CCITT   International Consultative Committee on Telegraphy and
            Telephony of the ITU
FOD   fax on demand
FOIP   Fax over the Internet Protocol
G_1 1968 analog, 6-minute per page CCITT fax standard
G_2 1976 analog, 3-minute per page CCITT fax standard
G_3 1980 digital, subminute CCITT fax standard
G_4 1984 fully digital, 20-page per minute CCITT fax standard
IETF   Internet Engineering Task Force
ITU   International Telecommunications Union
MPT   Japanese Ministry of Posts and Telecommunications
OA   office automation
PT   Picture Telegraphy
PTT   Postal, Telegraph, and Telephone administration
SG XIV   Study Group XIV of the CCITT
TR-29   Facsimile committee of the Electronics Industries Association (later,
            the Telecommunications Industry Association)
VOIP   Voice over the Internet Protocol

Introduction

I find—and no doubt you have found it too—that in this subject
there is a great difference between theory and practice.

—Alexander Graham Bell, 1875

In an age of instantaneous information and images, it is hard to appreciate the magic that millions in the 1930s experienced upon seeing photographs of distant disasters appear the next day in their newspapers, or the excitement in the 1980s of watching an exact copy of a letter emerge line by line from a machine connected to the telephone network. By accomplishing these contemporary miracles, the fax machine helped create the accelerated communications, information flow, and vibrant visual culture that characterize our contemporary world. Most people assume that the fax machine originated in the computer and electronics revolution of the 1980s, but it has actually evolved, albeit unsteadily, since the 1840s. This book tells the multigenerational, multinational story of that device from its origins and describes how it changed the world, even through its decline in the twenty-first century.

The basic concept of a facsimile, or fax machine—a machine that electrically transmits an image—has not changed since 1843. The three main components were, and remain, the scanner-transmitter, the transmitting medium, and the receiver-recorder. What changed greatly were the enabling and supporting technologies, the competing technologies and services, the social and economic environment, and the expectations and assumptions of promoters, patrons, observers, and users about markets and applications.

In 1863, Jules Verne imagined the Paris of 1960, a capitalist anti-utopia where fax machines were common.[1] Verne's fiction raises the question: If facsimile was such a known and desirable concept, why did it take so long to succeed commercially? It was not for lack of trying or resolve — over the decades scores of inventors and companies developed and marketed fax machines.

What happened was a global tale involving the diffusion of people, ideas, research, manufacturing, and applications from Europe to America to Japan and back. These shifts reflected larger movements of technology, manufacturing, capital, and consumption. A history limited to one country would be dreadfully incomplete. Where facsimile was, so was the center of telecommunications "high technology," an impressive phrase signifying a transformative technology that was on the cutting edge of the new and — or so proponents promised — the profitable.

This book tells the story of fax chronologically, from the earliest devices through what is now assumed to be the technology's final days, with four unifying themes over its history of a century and a half: the inadequacy of technological accomplishment to assure market success; the role of visionaries and the persistence of innovators, patrons, promoters and investors in the face of repeated failure; the impact of seemingly unrelated aspects of mass culture and commercial context; and the ultimate unexpected breakout appearing at almost the same time as technological obsolescence became inevitable.

Histories of technologies usually focus on success, on the triumphant progress of a technology from a dream into a world-reshaping reality. These histories tend to minimize, if not exclude, failure. As faxing's history illustrates, a technology does not just emerge, like Athena full-blown from the head of Zeus, but it has to be pulled, pushed, promoted, mashed, prodded, poked, shifted, cajoled, and otherwise shaped. Its path to the market is usually far more twisted and unexpected than its proponents expect — assuming the technology actually works. A constant theme running through fax's history is failure — of ideas, technologies, projects, and companies that never succeeded. Inventors, innovators, investors, and advocates did not intend to fail, but often they did.

Faxing is often viewed as a failed technology in two ways. First, since 1843, inventors and entrepreneurs repeatedly tried to promote fax in multiple marketplaces and usually failed, sometimes even before offering products. Not until the mid-1980s did faxing finally exceed its promoters' promises. Second, in the late 1990s and early 2000s, fax lost its primacy and independent existence to digital communications in the form of the Internet, the world wide web, PDF, cellphones, and other technologies. In both ways, fax was a "normal" technology, experiencing many unsuccessful attempts at commercialization and ultimately succeeded by technologies that provided the same or better services more easily, more cheaply, and with higher quality.

For most of its long history, facsimile was a theoretical solution searching for a practical problem, a classic case of promoters developing a technology for envisioned but elusive markets, of supply outpacing demand. This dominance of technology push over market pull stemmed from the underlying reality that for generations faxing remained too expensive, unreliable, slow, or otherwise inadequate as compared to its competitors. Not until transmitting photographs for newspapers in the early twentieth century and the explosion of business faxing in the 1980s did market demand—actual users—really pull and shape fax technology.

It is difficult to overstate the importance of economics for technology. The best technology is worthless if it is too costly to be used. For most of fax's history, competing services and technologies cost much less and performed better. As Picture Telegraphy operators discovered in the 1920s–30s, their only customers were those whose need for speed and 100%-accurate reproduction in sending an image outweighed the very high expense—up to a hundred times more costly than competing technologies. As important as improvements in technology, the post-1960s liberalization of telephone networks encouraged innovative services and reduced operating costs and barriers to entry, greatly improving the economics and competitive advantage of faxing.

Fax engineers and entrepreneurs cannot be accused of business naiveté. Many were successful telegraph, telephone, and computer engineers and managers who recognized that commercial success depended on economic as well as technical feasibility. As a rule they failed, outperformed by competing technologies: the letter, telegraph, telephone, telautograph, telex, airmail, express delivery, and faxing's ultimate digital vanquishers, which offered faster, cheaper, and easier communicating. Wisely not putting all their eggs in one basket, many developers, manufacturers, and users of fax machines also developed, manufactured, and used competing technologies. Telegraphy, television, photography, photocopying, and computing closely intertwined with facsimile as actors decreased their uncertainty (and hedged their bets) by pursuing competing options, often based on complementary technologies.

Competition within facsimile also became the norm. Starting with Alexander Bain and Frederick Blackwell in the 1840s, there was no area or niche too small (or perceived to have potential) for competitors to emerge and fight for market share. Unfortunately, while competition accelerated the rate of innovation and offered consumers choices, faxing suffered from over-competition for most of its existence. When too many overly enthusiastic entrepreneurs entered the imperfect, uncertain fax market, they often took business from existing firms instead of expanding demand. Incompatibility among competing machines worsened the problem of over-competition by confusing, not exciting, potential customers and fragmenting demand. Faxing was unique because this startup phase, so common with new technologies, literally lasted for generations before fully compatible machines emerged.

The history of facsimile embodies Michael Brian Schiffer's concept of the cultural imperative, "a product fervently believed by a group—its constituency—to be desirable and inevitable, merely awaiting technical means for its realization."[2] Optimism about its potential resided in an evolving and expanding group of enthusiasts and advocates whose visions outran the technically and commercially possible until the 1980s. Such proponents and observers hailed fax as a "coming wonder" in the 1930s, a "Cinderella" in the 1940s, a "commercial infant" in the 1950s, a "sleeping giant" in the 1960s, and an "aged infant" in the 1970s.

Despite decades of technological and commercial failures, proponents of fax persisted. The image that comes to mind is a World War I battlefield covered with the dead and wounded from the previous charge while the next wave of soldiers prepare to go over the top into no-man's land. What motivated individuals, firms, and governments to continue to commit resources to try to create new technology and markets, given fax's daunting history? Why was there always another line of firms ready to jump into the harsh reality of a skeptical, uncertain market?

Of answers stated and unstated, the most important was "This time it will be different." This time, promoters claimed, they had improved the technology, improved the economics, targeted a more receptive audience, were taking advantage of a new, more hospitable environment, and so forth. Those claims were usually true but not significant enough to enable success. But if those faults were recognizable after the fact, why were they not visible earlier?

The theoretical attractions and advantages of facsimile did not really change over time. What did change was the practical realization of those advantages as embodied in actual machines and experienced not in isolation but in the context of evolving competitors, comparative economic calculations, and societal expectations about communications as various meanings, standards, markets, technological capabilities, and societal demands changed. The goals of easy operations, good quality copy, high speed, and low cost changed little over time, but the actual expectations grew significantly. Like other technologies, the rising capabilities of fax machines raised assumptions of what the technology should do. As the exact definitions of those goals increased (e.g., acceptable picture quality grew from 25 to 65, then 96, then 200 lines per inch, and acceptable speed went from 20 minutes for a small picture to 20 seconds for a page), expectations continually outpaced capabilities for decades. What was once technologically novel soon seemed normal, marginal, or obsolete.

The gap between inflated expectations and technological and commercial feasibility also contributed to faxing's continuing failures. Overly optimistic predictions about their transformative and commercial promise normally precede and accompany the development of new technologies and products. Such selling was necessary to attract interest and investment to convince people and organizations to switch from their current technology to this promising technology. Overcoming that inertia, whether personal, institutional, or technological, demanded optimism and hyperbole in order to persevere. Even here, visions of a fax-filled future had to compete with visions of other technologies like television and computer-based office automation that drew attention and resources away from faxing.

Fax promoters, like technology promoters elsewhere, motivated potential customers with fear and hope. Firms pursued faxing for fear of falling behind their competition, of missing a new opportunity, of being blindsided by a new technology, of not keeping current with a potentially important field. Effective promoters played the fear card well, using competitors' actions to legitimize their own efforts and justify the allocation of resources within their organizations.

The hope of hitting the Next Big Thing induced governments, organizations, firms, and individuals to invest. I experienced the most visible demonstration of these great expectations while doing research at the International Telecommunications Union library in Geneva. Helpful staff wheeled to my table carts stacked with documents, the physical manifestation of decades of efforts of hundreds of engineers and managers to create standards for Study Group XIV of the International Consultative Committee on Telegraphy and Telephony. Only a small percentage of this material actually concerned facsimile; most focused on videotex and teletex. Those uneven stacks of paper offered visible proof of the different priority manufacturers and telecommunication administrations had given facsimile and its competitors. If the quantity of documents measured success, videotex and teletex should have been far more popular than facsimile.

Faxing's projected and actual markets changed over the decades, reflecting shifts in technology, potential users, perceived need, and economics. Time after time, proponents envisioned applications years or decades ahead of their successful realization. The original goal, a goal that persisted through the 1950s, was replacing the telegram with an exact copy of the original message. In the late 1800s, a new market emerged, transmitting photographs for newspapers as fast as telegraphed words. The bigger twentieth-century dream, slowly emerging at RCA and AT&T in the 1930s but not widespread until decades later, was, as Xerox told its stockholders in 1971, that "facsimile transmission may well become as indispensable to the office as the telephone itself."[3]

The changing capabilities and economics of fax technology laid the foundations for changing concepts of its applications and users. The many names facsimile has had since its inception (shown in table I.1) expressed not only its evolving uses but also promoters' envisioned markets.

Competing visions and resources did not always win. Fax proved to be a stealth technology in the 1980s in the United States. Its sudden, explosive growth surprised advocates of digital communications, who had predicted its decline and invested in competing technologies. The unexpected success of analog faxing represented—for digitalization advocates and industry analysts—an embarrassing demonstration of how demand-driven markets can bypass a theoretically superior technology in favor of less sophisticated but more practical products.

Because competitors' fax machines had to be compatible in order to communicate, creating standards was essential to faxing's diffusion. Forging—and forcing—such agreements was one of the most contentious challenges in fax's evolution, a wonderful example of "nature red in tooth and claw" as firms and countries used standards as a strategic competitive tool. Achieving that compatibility proved to be extremely difficult politically as well as technically and hindered the diffusion of faxing. Only after the adoption and successful implementation of the G3 standard in 1980 could faxing truly become a commodity.

Once actual instead of anticipated demand began to pull faxing, its development and diffusion accelerated. In a textbook case of demand-driven innovation, fax machines became commodities mass produced by the tens of millions as new manufacturers entered the market and more businesses, organizations, and individuals realized they needed those machines. The ensuing market specialization, shrinking product life cycles, and relentless reduction of costs benefitted users and pushed the diffusion of faxing. The innovation continued as faxing became increasingly integrated with computers, creating new markets like fax-on-demand and internet faxing.

In a seeming paradox, increasing the complexity of fax machines increased the base of users. However, the paradox proved more apparent than real: well-designed and well-packaged complex technologies shifted specialized skills and knowledge from the operator to the machine. This blackboxing of the technology put more capability and opportunity in the hands of individual users while reducing the expertise needed to fax, just like automatic transmission reduced the skill and effort needed to drive a car compared with a manual transmission.

TABLE I.1
Facsimile names

fac-simile ("make similar")	wirephoto, telephoto
copying telegraph	radiophotogram
pantelegraph ("all-telegraph")	message facsimile
autographic or automatic telegraph	document facsimile
picture telegraph	fax

Blackboxing vastly expanded not only the audience of consumers, creating growing network externalities, but also enabled entrepreneurs and experimenters to develop their own faxing applications. As faxing became a disruptive technology—"cheaper, simpler, smaller and, frequently more convenient to use,"[4] a virtuous circle of better technology attracted more users, which improved the value and economics of faxing, producing more technological innovation.

The fax machine may not automatically be associated with the study of visual culture, but faxing's ability to replicate and reproduce images, repeatedly if needed, over distance and time made the technology an important actor in creating the modern and postmodern worlds. Telegraphy only transmitted printed letters in a rigid format. The visual component of faxing—the ability to reproduce not just documents but images too—played an important role in fax's triumph over telegraphy as people took advantage of its flexibility and openness. Faxing vastly expanded rapid communication beyond the telegraphed word, becoming part of the visual culture of not only the office but, especially in Japan, the home for tens of millions of users.

Facsimile's users did not exist; they had to be created. Successfully constructing those users—figuring out who would want to fax, what they would fax, and convincing them to fax—proved to be as challenging as developing the actual technology. Like many other technologies, faxing first found success not as a general-purpose tool for many customers but in filling smaller, more specific niches. These users included newspapers, militaries, and large corporations, all entities with geographically widespread operations and willing to pay a premium for rapid and accurate transmission of information and images.

In the 1920s–30s, faxing revolutionized newspapers by publishing photographs of distant events at the same time as the telegraphed story. Four decades later in the United States, modern faxing built on the rapid expansion of the copying culture made possible by the Xerox 914 plain-paper photocopier. Aided by its ease of use and dropping prices in the 1980s, faxing quickly became much more than a basic business tool as people took advantage of their ability to edit, write, scrawl, change, annotate, and illustrate anything they could put on paper. Anyone could and did fax. Faxing revolutionized office and, increasingly, personal life by enabling the immediate dissemination of information and images.

Another form of innovation, driven from below by users, both benefitted from the diffusion of faxing and accelerated that diffusion as users created new applications unforeseen and unimagined by manufacturers and vendors. This innovation in use, a democratization of technological opportunity, proved even more intriguing as innovation of the actual technology.

Faxing's widespread success ironically contributed to its demise by inculcating offices and individuals with the expectation not only of rapid transmission of documents but also easy access to electronic databases and images. When the reality of digitalization and office automation finally began to match decades of promises, faxing faded quickly but unevenly, increasingly replaced by a range of easier, more capable, and less expensive technologies.

Yet even as the sea of digitalization rose, islands of faxing remained. Nowhere was this more evident than in Japan, where faxing diffused and evolved more deeply and lasted longer than in the West. Japan's three ideographic and phonetic languages favored the use of facsimile, but here too widespread acceptance had to follow the development of favorable economics as well as feasible technology. Once sparked, however, fax's flame burnt brighter and longer in Japan than anywhere else, reflecting cultural and age-dependent factors as well as economics and access to technologies.

Facsimile's story begins in 1843 with a patent and ends after 2000 when a standards committee for facsimile merged into a committee on data transmission standards. The first year represents legal recognition of facsimile as a concept; the latter, its bureaucratic absorption into computer-based communications. If I had completed this book when first intended, I would have missed one of the most exciting aspects of facsimile's history, the long-predicted, chaotic, and delayed triumph of digitalization over faxing. My procrastination has thus enabled me to report the obsolescence and decline of faxing, a story as fascinating as its long, uncertain, and unpredictable rise.

First Patent to First World War, 1843–1918

> I trust the day is not far distant when a merchant in London can
> take a penholder in his hand, [and] write a letter of advice or a com-
> mercial order, which will be instantly committed to paper in Paris,
> Berlin, or Vienna.
>
> —*Alexander Bain, 1850*

The world's oldest fax message sits in the quiet archives of the Institution of Engineering and Technology at 2 Savoy Place in London, carefully preserved in a
non-acidic envelope. The white lines on a blue background are still sharp since
more than 160 years ago when Frederick Bakewell dipped the paper in prussiate
of potash and a stylus traced distinct letters on the 1×12-inch slip.[1] This message, a
stunning technological accomplishment, represented the promise of facsimile —
a promise that enticed many inventors but rewarded few.

From the first patent in 1843 through the first years of the twentieth century,
facsimile changed from a possible early telegraph trajectory to a perennially
promising but underperforming technology facing an established giant. Facsimile's heralded virtues of accurate, authentic, and errorless transmission remained
constant. What changed in a long, legally contested, and grueling path from
concept to functioning system were its technical components, proposed markets,
and worsening comparisons with the competing technologies of the telegraph
and letter. Despite numerous efforts by inventors and some state support, pushes
to develop facsimile technology never created corresponding significant pulls by
market demand.

Beginning with Alexander Bain's British patent and his contentious battle with
Frederick Bakewell over priority, faxing images attracted over a score of inventors
in Europe and America but never achieved commercial success. In an effort to in-

crease the speed and accuracy of messages, the French telegraph administration pioneered regular fax telegram services in the 1860s–70s with the pantelegraphs of Caselli and Meyer, but those efforts proved short-lived. Customers valued easy and inexpensive transmission of a message above its authoritative and authentic visual reproduction.

This episode in telecommunications history illustrates how promoters, patrons, and potential users normally backed competing technologies to reduce their risk and maximize their options. In an uncertain, changing environment and with multiple technological options, a firm or government was foolish to back only one horse in a race. Not until the capability to transmit photographs evolved after 1880 with a wave of innovation caused by the selenium photoelectric cell did the promise of a new market, newspapers, emerge in Europe and the United States. Two decades more passed before the first systems competent enough to warrant actual newspaper use emerged, but commercial success remained elusive for Picture Telegraphy (PT) until after World War I.

The technical challenges proved to be persistently formidable. To send a fax, the message had to be prepared, scanned, converted to electrical impulses, transmitted to the receiving machine, and then reconverted to impulses and reproduced. Before the photocell, a stylus haptically scanned a message that contrasted electrically or physically with its medium. A telegraph circuit transmitted the message to a receiver, which copied the message onto chemically treated paper. This simple explanation hides the myriads of technical, economic, and legal challenges that inventors, craftsmen, and promoters had to recognize, understand, and solve in translating a promising idea into a commercially viable reality. The technical challenges proved so daunting that most projects failed even before trying to demonstrate their economic feasibility against telegraphy.

Compared with a conventional telegram, preparing a fax message took more time, effort, and materials such as metallic paper and special ink. The sender did not simply dash off a note and hand it to the telegraph clerk but wrote with insulating ink on a metallic paper or created a three-dimensional message by etching a photographic negative or sprinkling a finely powdered shellac over an adhesive ink, which was then heated to melt and dry.

Transmitting faced major challenges of synchronization and circuit quality. Inventors devoted much effort to synchronizing the transmitter and receiver for accurate reproduction. Weights or pendulums powered the clockwork that moved the message or stylus, but achieving uniform motion proved very difficult. Not until the 1870s did electric motors become sufficiently reliable and affordable to replace clockwork. The poor quality of telegraphic wires and inductance caused by lines too close together often garbled transmissions, causing "mistakes, repetitions, general confusion, and consequent delay."[2]

Recording was the final technical challenge. Despite problems of preparation and a short life of a few days or weeks, paper soaked in an electrochemical solution best turned the weak received signal into an image comparable to those produced with pens, pencils, and other means of recording. Different solutions produced different colors (nitrate of manganese produced a brown mark; prussiate of potash, a blue).

The actual machines themselves proved frustratingly demanding to construct, operate, and maintain. As R. S. Culley, chief engineer of the British Electric and International Telegraph Company, warned in 1864, "The value of a particular system must be estimated, not by the beauty or the singularity of its effects, but by pounds, shillings, and pence; that is, its cost, the expense of working and maintenance, its speed, correctness, and freedom from derangement."[3] This remained equally true nearly a century later when *Electronic Engineering* warned in 1947, "Perhaps the most important factor in a really first class Picture Telegraphy system is the standard of mechanical design and workmanship which goes into the scanning and optical systems."[4]

BAIN AND BAKEWELL

Geographically, faxing began when Scotsman Alexander Bain moved to London and the Florentine Giovanni Caselli, to Paris. These isolated inventors moved to centers of electrotechnical innovation to find markets, mechanics, and money, though they also sought markets internationally. Despite several American efforts, Europe reigned as the center of facsimile development.

Alexander Bain's life exemplified the best and worst of rapidly industrializing Britain in the early nineteenth century. Poor and unschooled, he educated himself, learning about electricity from public lectures. Caught in the murky nexus of patent priority between research and commercial exploitation, Bain received large amounts of money but lost it in unsuccessful litigation and died a pauper.

He harnessed electricity to control and monitor railroad track safety, clocks, shipboard sounding devices, and musical instruments. In telegraphy, Bain employed chemically treated paper as the receiver, first for his fax machine and later for the far more successful chemical telegraph. As a later fax pioneer, RCA's Richard H. Ranger noted in 1926, Bain was "so basically correct that . . . generally, we are all following in his footsteps."[5] Like a good entrepreneurial inventor, he filed patents and sought markets for his inventions in Britain, the United States, and continental Europe.

Unfortunately for Bain but typical of a newly developing technology, other inventors also filed patents and sought markets. Bain clashed with Charles Wheatstone over who invented the electric clock and telegraph. He clashed with Frederick Bakewell over who invented a practical fax machine. And, most disastrous financially, he clashed with Samuel Morse over the telegraph in America. At one time, a third of the telegraph lines in the United States used Bain's system, but the legal triumph of Morse's system banished Bain into semi-obscurity. Bain's life was perhaps best expressed by the *Telegraphic Journal and Electrical Review*, which eulogized him as "a marvellous inventive genius, a skilful watch and clockmaker, but unfortunately in his later years, an improvident and intemperate mechanic."[6]

Alexander Bain grew up in Watten in northern Scotland, one of thirteen children and became an indentured apprentice to a watchmaker. In January 1830, he walked 19 kilometers to attend a lecture on "the electric fluid." His interest sparked, he began experimenting with electricity and, disgracing his family, broke his apprenticeship to go south to Edinburgh. In 1837 he went to London and worked as a journeyman clockmaker. He continued his interest in electricity, attending public lectures at the Adelaide Gallery and the Polytechnic Institution, which demonstrated the latest advances in electrotechnology.[7] In 1838, combining his trade with his curiosity, Bain started experimenting on electric clocks and telegraphs. There was no better location. London was the center of research and applications for the rapidly expanding field of telegraphy.

Bain had built a model of an electric printing telegraph by late 1838. Seeking funding for further development, the clockmaker approached the editor of *Mechanics Magazine*, William Baddeley, who introduced Bain to Charles Wheatstone in August 1840. Wheatstone, after witnessing a demonstration, agreed to finance two working models. Along with navy Lieutenant Thomas Wright, Bain received patents in 1840 and 1841 for applying electricity to clocks, signals, printing and railroads.[8] The agreement and patents embroiled him in the first of the many arguments and lawsuits that consumed most of his time and resources.

Bain conceived of his "Electro-Chemical Copying Telegraph" in early 1842 and received patent 9745 on May 27, 1843, for "Certain Improvements in Producing and Regulating Electric Currents and Improvements in Electric Time-pieces, and in Electric Printing and Signal Telegraphs," which covered several different inventions, including the first "fac-simile" system and the printing telegraph that embroiled him with Wheatstone.[9]

Returning to Edinburgh to make and sell his electric clocks, Bain in 1846 erected the first telegraph line in Scotland from Glasgow to Edinburgh, which also marked the first time a telegraph operated an electric clock.[10] That year he opposed a bill to establish the Electric Telegraph Company, claiming that Wheatstone had violated his confidence and usurped several of his inventions. After investigation, the House of Lords suggested that the company offer Bain compensation. The company gave him £7500 for his patents, £2500 for options on future patents, and a directorship, which Bain soon resigned. By mid-1851, the Electric Telegraph Company had paid Bain £20,486 (approximately $3 million in current dollars).[11]

Bain's major successful invention was the 1846 automatic chemical telegraph, which used chemically treated paper to receive messages.[12] So widespread was this paper that an electrical journal in 1885 could refer to "ordinary Bain paper" without defining it.[13] While not the first to conceive of electrochemical telegraphic recording, Bain was the first to devise a practical system. Compared to other systems, his chemical telegraph did not need electromagnets to make and break contact or needles to indicate letters, offering potentially faster and simpler machines. More importantly, Bain provided Henry O'Rielly and his New York and New England Telegraph Company with the technology and the 1848 American patent to compete with Morse in connecting New York and Boston in 1848–49. In 1852, Bain's system operated over approximately 8,000 of the 27,177 miles of telegraphic lines in the United States.[14]

Bain's success was short-lived. Hostile American courts declared the Scotsman had infringed on Morse's patent. It is doubtful if any non-American could have prevailed against Morse.[15] In a decision that had less to do with justice than business, O'Rielly agreed to use only Morse's system in 1852, freezing Bain out of the American market. Bain returned to London and his prior occupation as a clockmaker. Eight years later, he traveled again to the United States for further patent battles with Morse that bankrupted him.[16] In the early 1870s he returned to Scotland, where he died a poor man in 1877.

Bain's failures were threefold: ideas in advance of the enabling technologies of the time, being "not a commercial man,"[17] and personal shortcomings. Although it was only delicately hinted at, Bain suffered from intemperance and a ready willingness to sue, unfortunately against better funded opponents.[18] This readiness to take offense may have stemmed partly from a sense that elite members of society were taking advantage of this self-made Scotsman. Particularly rankling was the Royal Society's refusal to let Bain report on his 1842 discovery of using the Earth as a conductor because he had applied for a patent, thus moving from the world of science into the world of commerce. The Society had, however, allowed Wheatstone to speak in November 1840 about his telegraph, which was also patented.[19]

Was Bain unusually litigious or did he invent in areas so competitive that rivals existed in abundance and lawsuits offered another way of gaining a tactical or strategic business advantage? When inventors thought about the same issues, problems, and challenges at the same time, they often independently developed similar solutions.[20] Nonetheless, Bain had more than his share of disputes with prominent and minor figures in telegraphy in Great Britain and the United States.[21] This pattern of prickly, litigious relations diverted the inventor's attention and resources from improving and commercializing his inventions. In his obsession with receiving appropriate recognition, Bain followed the path of too many independent inventors to the detriment of his commercial success.

Bain actually developed two fax machines, his unbuilt 1843 model and a much improved version a few years later. In Bain's 1843 "improvement for taking copies of surfaces," the transmitting and receiving plates consisted of a metal frame filled with short, insulated wires bound together with sealing wax and ground smooth. The message on the transmitting plate was made of printers' type. A pendulum moved a stylus back and forth across each plate. The end of each swing activated clockwork to drop the frame so that the next swing scanned a new line. The patent stated, "[I]t is evident that a copy of any other surface composed of conducting and non-conducting materials, can be taken by these means."[22]

Bain did not successfully construct his initial fax machine because of the challenges of synchronization, lack of precision in operation, and, despite his efforts to eliminate them, its need for multiple wires.[23] In 1846–48, he built a new fax machine with a cylinder replacing the flat plate. Instead of raised type, the sender wrote a message with a non-conducting ink on tin foil or paper coated with Dutch metal (a thin leaf of brass). Weights unwinding by clockwork synchronized the pendulums and moved the stylus gradually across the rotating cylinder. Bain's copying telegraph amazed observers by sending images through a wire.[24]

But so did another machine. In December 1848, Frederick Collier Bakewell, listing himself as a gentlemen, applied for a patent "For improvements in making communications from one place to another by electricity," which he received in June 1849.[25] Bakewell also employed a rotating cylinder instead of a flat plate and non-conducting ink on a thin sheet of tin foil. The sender wrote his message on a varnish-coated piece of tin foil with an ink containing caustic soda, then wiped the foil with a wet sponge. The foil was wrapped around a six-inch-diameter cylinder, and a stylus scanned the message as the cylinder rotated. Where the soda had removed the varnish, the stylus touched the actual foil, making a circuit.

Synchronization was simple: a thin paper strip running the length of the foil produced a white gap at the receiver. By adding or removing weight to keep the gap in a straight line for each new line, the receiving operator visually synchronized his machine. For more exact synchronization, an electromagnetic regulator linked by a second telegraphic circuit to its counterpart at the receiver allowed Bakewell to send images the 190 kilometers from London to Bath "with perfect accuracy."[26]

An American observer quickly noticed a similarity to Bain's machine.[27] So did Bain. In 1848–50, Bain and Bakewell exchanged heated public letters in technical journals and the London *Times*, accusing each other of theft of intellectual property, dishonesty, incompetence, and betrayals of trust. Anyone who thinks engineering is boring work done by boring people should read their correspondence.

Bain claimed that his patents of 1843 and 1846 covered Bakewell's invention because they stated "most distinctly that *any surface* composed of *conducting* and *non-conducting* materials might be used for this purpose." Furthermore, Bakewell had frequently visited his workshops in his capacity as a writer about technology, and in 1847 Bain had asked him to write an overview of his inventions, including contemplated improvements. Once Bain had left for the United States, however, "'Mr. Bakewell's Copying Telegraph' was pompously announced!"[28]

Bakewell responded that not only had he not stolen Bain's ideas, but Bain had offered him £500 for his ideas. The only similarity between his invention and Bain's 1843 patent was a stylus moving over a message. Indeed, Bain had modified his 1843 concept so significantly by 1848 that "scarcely anything of the original invention remains." Bain used raised metal type on a flat plate, while Bakewell employed non-conducting ink on tin foil wrapped around a cylinder. Obviously Bain had not thought of writing on foil because otherwise he would not have used a system that required setting type.[29]

The public claims, counterclaims, and invective continued until late 1850. The combatants may have decided their priorities were elsewhere. More likely, the *Times*' refusal to accept further letters except as advertisements cooled their ardor for public dispute.[30] Bakewell referred to himself as the "inventor of the copying electric telegraph" in his 1853 history of electrotechnology, and his 1860 history of modern inventions completely ignored Bain's 1843 patent, although crediting the Scotsman for the electric clock, electrochemical printing, and printing telegraph.[31]

Despite years of demonstrations, including at the 1851 Crystal Palace exhibition, the *Electrician* declared in 1862 that the Bakewell Cylinder Printing Instrument, "although beautiful and ingenious, has not been found a commercial success, and has fallen into disuse."[32] Ironically, that year Bakewell applied and received a second patent for improvements to his 1848 copying telegraph patent. Bakewell received twenty-one patents from 1832 to 1869, but only two involved electrical communications, lending credence to Bain's accusations.[33]

For Bain, the priority was his electrochemical telegraph. Unlike facsimile, it proved technically practical and even financially profitable, but its promotion and application demanded Bain's presence abroad. With his demonstration before the president of France in April 1850 resulting in the French government purchasing his patent, France proved as lucrative as the United States.[34] Bain did not return to fax.

Bain patented his facsimile telegraph in 1843, when electric telegraphy experienced the typical early years of a new, rapidly evolving technology with minimal barriers to entry. An astounding range of experimentation and development of new products occurred in an exciting and confused environment of promising leads, dead ends, and cross-fertilization of ideas, partnerships, patents, and promotion. More than sixty versions of telegraphs appeared by 1837, and the application of imagination, ingenuity, and imitation continued throughout the century. The telegraph evolved from a needle that pointed to a letter and demanded a circuit for each letter to single-circuit lines that rapidly transmitted a message as perforations on paper, marks on chemically treated paper, or sound.[35]

By 1850, both the Bain and Bakewell machines, while considered technological marvels, were doomed. The reason was simple. Instead of one form of a new technology among many, they now competed against an established, effective, and rapidly expanding telegraph technology. The fluidity of the early years had vanished as telegraph industry coalesced around the technology of Morse and Wheatstone and Cooke. The technical, economic, and legal environment had changed, and facsimile telegraphs, even if they had operated flawlessly, could not have overcome the huge investment and diffusion of the standard telegraph.

The Morse system benefitted not only from economies of scale in manufacturing thousands of its inexpensive sounders but also from the training of operators and users. These lower costs as well as institutional inertia benefitted telegraphy as careers and companies increasingly evolved around dots and dashes instead of images. Nor did telegraphy stagnate technologically. New inventions and applications greatly increased the efficiency and spread of telegraphy, continually raising the barrier that facsimile had to jump to succeed.

Yet faxing continued to attract inventors because standard telegraphy had significant shortcomings, which persisted into the twentieth century. By transmitting an exact reproduction of the sender's message, faxing offered 100-percent accuracy and the authority of the original—a major attraction for telegraphers struggling with illegible handwriting, unfamiliar languages, codes, ciphers, and simple human error. Bakewell claimed his system offered "authentication of telegraphic correspondence by the signatures of the writers, freedom from the errors of transmission, and the maintenance of secrecy." This lure of exact reproduction maintained its attraction over time: Eight decades later another English fax promoter promised, "The sender need not fear that 'Mother-in-law dead hurry' will be altered to 'Mother-in-law dead Hurray,' or something like that."[36]

Fax inventors were not naive idealists attracted by the theoretical simplicity of faxing and unaware of the technological and economic dominance of telegraphy. What is striking is how many fax inventors had successful telegraph inventions to their credit. Inventors, knowledge, and ideas flowed easily between faxing and telegraphy. Consequently, most fax developers set specific competitive benchmarks based on telegraphy. William E. Sawyer stated in 1876 that a successful copying telegraph should transmit as quickly as a Morse system (25 words per minute), use ordinary paper, and not demand expensive and complex equipment.[37] Thomas Edison promised his system would equal Morse's in speed; transmit "outline Photographs" and any language; retransmit the chemical strip of a received telegram; demand no special preparation to send messages; and prove simple, reliable, fast, and practical.[38]

Achieving those criteria proved quite another matter. What science popularizer Dionysius Lardner wrote in 1867 still held true in 1887 and 1907: "They are at present more matters of curiosity than of practical utility. . . . This form of telegraph has proved too slow in operation, and too uncertain and difficult in management, to allow of practical application; besides which, it requires the apparatus to move onward at exactly the same speed at each end, a result almost impossible to attain."[39]

The challenge was not to send a facsimile telegram; the challenge was to send a facsimile telegram comparatively more easily, less expensively, or in some other way that was superior to the established methods of telegraphy.

FRANCE: FAILED COMMERCIALIZATION

Those difficulties did not stop inventors from trying, telegraph companies from testing, and writers from speculating. In his 1863 novel about the twentieth century, Jules Verne credited "Professor Giovanni Caselli of Florence" with developing the fax machine.[40] Verne, always attuned to current technological developments, was well informed: Abbe Giovanni Caselli had not only faxed images from Paris to Lyon but had created the first fax machine to enter regular service, transmitting thousands of images before the service stopped in 1867.

Caselli's pantelegraph ("all-telegraph," a universal telegraph that could transmit any image or message) was neither the first facsimile machine nor the first on the continent. Gaetano Bonelli and Lucy-Fossarieu preceded him. What Caselli did was to combine a good design and excellent engineering with a state need and high-level patronage. While the pantelegraph was an impressive technological accomplishment in itself, Caselli's commercial service demonstrated a state promotion of facsimile as part of ongoing efforts by the French government to improve the speed, accuracy, and reliability of telegram transmission.

Caselli was not an aberration with his interest in facsimile, however. In 1858, at the office of the French newspaper *Moniteur*, P. de Lucy-Fossarieu demonstrated his system. A stylus scanned the message, which was written with insulating ink on a copper plate. To avoid the inconvenience of preparing and marking moist electrochemical paper, a pen marked the receiver's plain paper. An electromagnet kept the pen off the paper until triggered. Even compared with Bain's 1843 system, it was comparatively crude, with a crank-and-cog system providing power and synchronization. The uneven movement of a crank was one reason that the machine quickly faded from sight.[41]

Far more impressive (but also impractical) was Gaetano Bonelli's typo-telegraph. To speed transmission and minimize the synchronization problem, the inventor, the Director-General of Sardinian Telegraphs in the mid-1850s, used a comb with several teeth instead of a single stylus. Each tooth needed its own circuit, which added to the complexity and cost of his system. His ultimate version had five teeth (compared with eleven for an earlier model), requiring five separate circuits, although a short message of twenty to twenty-five words took only 15–20 seconds to transmit. The five separate circuits had the advantages of speed and little distortion by poor synchronization (which only expanded or compressed the letters, enabling messages to be read). Bonelli also employed type for the message, which was put on a little car and rolled on rails on a two-meter oak table past the comb, creating a more three-dimensional contrast for the styli but increasing the time and effort needed to prepare the message. The five circuits, not to mention the time needed to prepare the type, meant Bonelli's system was simply uneconomical, as tests in England and Italy proved.[42]

If not for an unsuccessful political struggle, religion in Italy would have been the stronger and the international advance of electrotechnology the weaker. Giovanni Caselli, born in Siena in 1815, had entered a religious order in 1836 and moved to Parma in 1841 to tutor the sons of Count Sanvitale. His participation in the failed 1848 revolution for Italian unification caused him to flee that city to Florence in Tuscany. He switched from politics to scientific research and

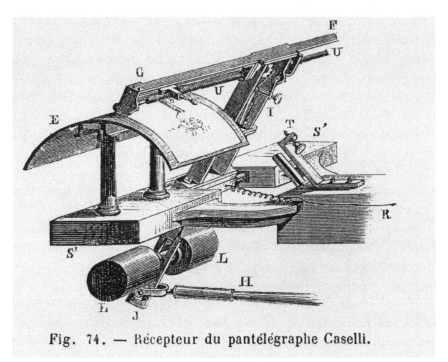

Fig. 74. — Récepteur du pantélégraphe Caselli.

The French telegraph administration operated the world's first fax service in 1865 with Caselli's pantelegraph. Louis Figuier, *Les Merveilles de la science* (Paris: Furne, Jouvet, 1867)

taught physics at the university. Like many others, he experimented with telegraphy and, in 1856, built his pantelegraph.[43] His inspiration was Bakewell's copying telegraph, a fact that English observers often noted.[44]

Although Caselli received some financial support from friends and attention from the local newspaper, his need for technical and financial support—a problem even more demanding in the nineteenth than in the twentieth century due to the lack of dedicated venture capital institutions— prompted a move to Paris in 1856. It was a wise move, for Caselli found Paris, which vied with London as the world's most hospitable city for electrotechnology, a welcoming and supportive environment. By 1858 knowledge of Caselli's work had penetrated Parisian scientific and engineering circles.[45]

To transform his ideas into practical, functioning equipment demanded craftsmanship of a high level. Fortunately for Caselli, physicist Léon Foucault arranged for him to meet Paul-Gustave Froment, whose workshop produced some of that era's most precise and impressive electrical equipment. Starting with a working model in 1858, Froment built Caselli's machines. Indeed, Caselli's successful demonstration of transmitting images 80 kilometers from Florence to Livorno in 1860 "appears to have been due, in a great measure, to the perfection of the instruments constructed by M. FROMENT."[46] This painstaking mechanical design and construction, not radically new concepts, played a significant role in the success of Caselli.

Having a technology that works is not enough to succeed commercially. Inventors often need patrons whose approval can provide resources and legitimacy. One way of gaining patrons was successfully demonstrating the technology. Caselli's most important demonstration occurred on January 10, 1860, when Emperor Napoleon III visited Froment's workshop.

For the visit, Caselli transmitted a drawing of the emperor on a local circuit. Meant to attract royal attention, the demonstration succeeded admirably. The emperor authorized a 140-kilometer transmission from Paris to Amiens, which included a musical score and accompanying note by Rossini. Caselli's politically adroit transmission of a likeness of the Empress Eugénie did not hurt his cause either, even though "the picture was considerably interrupted by messages traveling the same course, and had dots and dashes all over it, but was nevertheless recognizable."[47] This second successful demonstration prompted Napoleon III to state that the pantelegraph "did great honour to Italy, and was a discovery of which France herself might be proud."[48]

Illustrating the international community of electrotechnology, knowledge and praise of Caselli's system quickly spread in the technical as well as popular press and "excited the interest of many philosophers in Europe."[49] Nor was non-French interest confined to reporting. Like Bain before him and others like Korn and Belin after him, Caselli marketed his technology internationally, trying to create as many markets as possible. The Electric Telegraph Company experimented with the Caselli system in England in the mid-1860s but rejected it as too slow and difficult to operate. In Russia, the tsarist government was sufficiently intrigued to purchase two units, but, just as the first Russian railroad only moved between the tsar's summer palace and St. Petersburg, so too did the first fax system in Russia serve only these two imperial residences. Commercial application remained far from the Russian official mindset.[50] By allowing inventors to test their systems on company lines, the telegraph firms provided otherwise unobtainable facilities and real-world experience while obtaining firsthand knowledge of a potentially useful technology. Caselli also received an American patent but never tried to commercialize it.[51]

The interest of the emperor aligned with the Administration des Télégraphes,

If there were no problems with synchronism and interference, the quality of a Caselli transmission could be quite good. Deutsches Museum, Bildstelle

which wanted to learn whether pantelegraphy could handle the rapidly growing demand for telegrams better than the Morse system could. In the 1850s, France had two tiers of equipment, dial telegraphs on low-volume lines and Morse machines on high-volume lines. To handle rapidly growing demand in the 1860s, the Administration needed more capable equipment.[52] Slow and deliberate action established faxing's legal and administrative parameters. Tests between Paris and Lille and Paris and Marseilles contributed to its operating knowledge. Not until April 16, 1865, did service officially begin between Paris and Lyon, with service between Paris and Havre added two weeks later.[53]

Strikingly beautiful in appearance, the complex Caselli apparatus was dominated by a two-meter pendulum with a seven-kilogram iron bob that provided power and synchronization.[54] Two electromagnets on opposite sides of the frame regulated the bob's motion. A second half-meter pendulum controlled the local battery that activated the electromagnets. Instead of Bakewell's cylinder or Bain's initial flat plate, Caselli used two curved metal tablets to hold the original message and facsimile. The large pendulum moved a lever that moved a tablet under a stationary stylus. Each revolution also turned a screw so the tablet moved in two dimensions under the stylus. While the stylus touched the surface of the metallic paper, the main battery was shunted; when the insulated writing broke the contact, current passed to the receiver. The clips, which kept the tin-coated "silver paper" flat on the transmitting tablet, maintained the circuit. The receiver used an iron stylus on chemically treated paper. A regular Morse set enabled the operators to communicate.

Caselli continued to experiment and modify his system, learning from his experiences.[55] He never quite overcame the technical challenges of synchronization, inductance, and reverse polarization. For synchronization, as in Bakewell's system, the receiving operator adjusted the pendulum swing by keeping a line at the edge of the paper vertical. More demanding was inductance, the electromotive force created by changes in the electric current. When transmission lines were long or a circuit opened and closed quickly, inductance could make a faxed message blurred and jagged at the edges. Caselli solved this problem by setting weak auxiliary batteries in the main circuit of the main, more powerful battery, but in reverse. He also experimented, with less success, with a rheostat to combat problems caused by rapid charge and discharge.

Bakewell's system transmitted current when the sending stylus touched the conducting foil, creating a white message on a colored background. In "one of the most original and ingenious parts" of his system, Caselli reversed the polarity of the current so the main circuit remained closed until the stylus touched the ink. Then the opposing current of the auxiliary batteries immediately stopped the line transmission.[56] The benefits were twofold: sharper lines and a more aesthetically pleasing arrangement of dark lines on a white background.

When the pantelegraph system operated smoothly, American engineer Franklin L. Pope noted, "Fine close handwriting, such as you would put on a postal card, was transmitted with reasonable rapidity and with very great perfection."[57] The standard message sheet of 111 by 27 millimeters (4.4 by 1.1 inches) held 25–30 words. Since the stylus moved at one revolution per second and the spacing was one-quarter millimeter, a message required 1 minute and 48 seconds under optimum conditions, which rarely occurred. Using smaller messages with finer writing, operators achieved a peak performance of sixty messages an hour, but twenty to twenty-five messages proved more realistic, although traffic rarely reached even that level.[58]

Practical problems doomed the world's first commercial fax service. Contemporary and later observers like Thomas Edison and William Sawyer considered its synchronization the system's "great defect." Poor synchronization often blurred messages, sometimes to the point of illegibility.[59] The resulting retransmission reduced the message's timeliness and increased the cost to the telegraph agency. Preparing a message intimidated senders because it demanded more time and effort than sending a regular telegram. They had to write carefully on the form to avoid spotting or crinkling the foil paper. Removing an error meant carefully scraping the ink off and dusting with a feather.[60]

The shortcomings of the pantelegraph extended beyond the technology. The telegraph administration charged, quite reasonably, by the square centimeter. At 20 centimes per square centimeter, a message cost at least 6 francs plus another 6 to 24 francs for the metallic sheet. In contrast, a regular telegram cost 2 francs for 20 words.[61] The high cost, compared with a regular telegram, reduced the financial attraction, while the pantelegraph's proclaimed advantages of security and authentication through signatures did not convince businesses to switch from telegrams.

The telegraph administration had selected the Paris–Lyon route, thinking that handwritten messages and signatures to ensure authenticity would benefit the financial transactions between the two cities. The premise proved both correct and erroneous. In 1866, 4,853 of 4,860 Caselli transmissions between Lyon and Paris involved finance or business. Given the opportunity to send exact reproductions of their messages, however, most users continued sending less-expensive regular telegrams with codes and ciphers providing security and authenticity.[62] Considering the 2.8 million telegrams sent in France in 1866, the 4860 Caselli transmissions appeared minuscule.[63]

Furthermore, only three cities employed the expensive fax equipment versus thousands of standard sets used in the bureaux across France. Physically insignificant compared with the two-meter pendulum and other components of the Caselli pantelegraph, the simple Morse sounder had enormous advantages of easier use, much lower cost, less interference in transmission, and an already-developed infrastructure as well as users who had by now incorporated the standard telegram into their business routines.

Disappointed, the French telegraph authority discontinued Caselli's system in 1867. Caselli returned to Italy, refusing an appointment in the French telegraphic service. He settled in Siena, where he became the city school director before his death in 1871.

Caselli did, however, spark the first Japanese interest in facsimile. In 1862 and again in 1864 a Caselli system impressed visiting Japanese delegations by transmitting Japanese. The envoys sent information back to Japan, but no action occurred.[64] The 1867 Paris international exhibition produced the first Japanese order for fax machines, albeit more from political embarrassment than design. Humiliated by the appearance of a delegation from Satsuma, the major political rival to the Bakufu throne, the official Japanese delegation demonstrated its importance by ordering a pair of fax machines from Mathias Hipp, a clockmaker and electrotechnical inventor living in Switzerland. The machines arrived in 1868 to a new government, which fired the officials who brought the Hipp machines, thereby ending interest in faxing. Neither the demonstration of a d'Arlincourt system at the Imperial College of Engineering in 1878 nor the 1884 visit by a Japanese minister to Thomas Edison sparked government or private interest.[65]

In France, interest continued. The potential of a government market inspired a new wave of post-Caselli French fax systems, which experimented with conical pendulums, flywheels, tuning forks, and other new concepts and components. Creators ranged from telegraph engineers like Chevalier Guyot d'Arlincourt to Jean J. E. Lenoir, the inventor of one of the first practical internal combustion engines. Lenoir's 1867 fax machine had a simplicity of design that markedly contrasted with Caselli's and a very innovative receiver. A rubber cylinder was coated with indigo ink, and a thin piece of tracing paper was wrapped around it. Instead of a stylus pressing down when current passed through, an electromagnet held a tracing point above the cylinder. When the current from the transmitter turned the electromagnet off, the point fell to mark the paper. Turning the electromagnet on lifted the point from the paper. Synchronization depended on two conical pendulums, a flywheel at the receiver, and a relay that needed the combined current of the main batteries at both ends. This left synchronization overly dependent on the variable quality of the telegraph line.[66]

These machines did not go unnoticed by the engineering community: At the 1867 Paris Exhibition, the fax machines of Caselli and d'Arlincourt received gold medals, and Lenoir's a silver medal; while Cyrus Field and David Hughes received the grand prizes for their telegraphic accomplishments. D'Arlincourt won medals for his fax inventions at the Paris expositions of 1867, 1878, and 1884, and the 1873 Vienna exhibition.[67]

In 1869, the Commission for the Perfection of Telegraphy (Commission de Perfectionnement du materiel telegraphique) studied the facsimile systems of

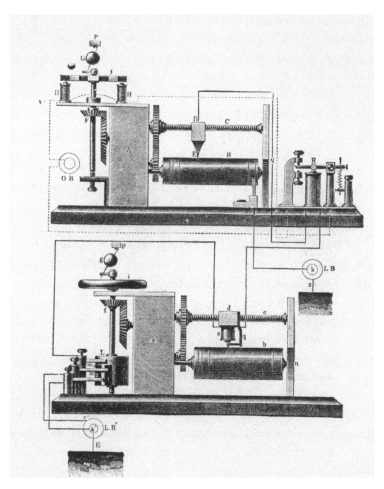

Like every pre-photocell machine, the Lenoir automatic telegraph scanner and receiver physically scanned a message by measuring differences between electrically conducting and non-conducting areas and reproduced the image at the receiver. Notice the elaborate gearing. George B. Prescott, *Electricity and the Electric Telegraph*, 7th ed. (New York: D. Appleton, 1888)

Lenoir, Dutertre, Cook, and Meyer, and rated Meyer's the best in quality, speed, and synchronization.[68] French telegraph clerk (*controleur*), Bernard Meyer (1830–1884) created "one of the most ingenious and effective" solutions. Instead of Caselli's stylus tracing over a curved plate, a spiral rib rolled over the message on a rotating cylinder. Only one point of the paper contacted the rib at a time. At the receiver, an electromagnet, when activated by the transmitting current, pressed the paper up against the inked rib. A conical pendulum provided synchronization.[69] Meyer pioneered the first use of this "helix and spiral" system which played such an important role in the twentieth century with the photoelectric cell.

The French telegraph authority experimented with Meyer's system in actual operations between Paris and Lyon for a "considerable time" in 1868–69. In Britain, the Electric Telegraph Company decided against adopting the Meyer equipment but thought it worth watching because "[it] is susceptible of great improvements and may eventually be of important service" in transmitting ordinary telegrams.[70] Indeed, telegraph administrations and companies in Europe often tested fax systems to determine if they merited investment. While that determination was invariably negative, data, not conviction in the innate superiority of telegraphy over facsimile, guided those judgments.

Although it performed better than Morse's system, Meyer's fax system failed commercially because other approaches performed even better. The real threat to the Morse operator came not from the fax machines of Caselli and Meyer but from improved methods of sending Morse code. Meyer's system handled 75 dispatches per hour, but the automatic printing systems of Wheatstone and Baudot in the 1870s handled 100 and 200 messages respectively at a third of the cost.[71]

The pantelegraph itself did not have a future, but its principles and experience did. Meyer transferred his skills and knowledge into multiplexing, sending several messages simultaneously, and multiple telegraphy, using multiple instruments on the same line. In 1872, he demonstrated his first multiple telegraph, transmitting between Paris and Lyon. Perhaps the greatest contribution of Meyer and pantelegraphy to telecommunications was "to the contemporary technological milieu as another utilization of synchronicity in telegraphy and as a stepping stone for the creation of a new form of fast telegraphy."[72]

AMERICAN EFFORTS

Across the Atlantic Ocean, faxing also attracted many inventors, including Alexander Graham Bell and Thomas Edison. Both developed systems but found themselves distracted by their more successful inventions.[73] Whether their fulltime dedication would have altered fax's trajectory is doubtful: given the state of technology, the obstacles were simply too great. By any criteria, regular telegraphy clearly trumped fax.

Nor did inventive success elsewhere guarantee success with fax, as Edison demonstrated with his autographic telegraph. Edison optimistically predicted to possible patrons in 1868 that $500 to $800 and one year would enable him to develop and produce a $200 set.[74] A well-publicized August 1868 visit by a Chinese delegation to Boston may have stirred Edison's interest, but he claimed that he had already worked on fax for nearly three years to transmit Chinese characters. Most likely the visit reignited his interest. Neither that low cost nor fear of possible competitors attracted a sponsor until 1870, when the Gold and Stock Telegraph Company funded Edison to develop a fax system to compete with Morse's system and also extend its business to Constantinople. The $3000 enabled him to open a telegraph manufacturing shop in Newark.[75]

Progress proved slow. With Patrick Kenny, Edison built two machines and applied for a patent in 1881.[76] Kenny displayed their Fac-simile Telegraph at the 1881 Paris exhibition but, fearing public failure, declined to demonstrate the system on a telegraph circuit between Paris and Brussels.[77] After that only outside inquiry, including an 1896 attempt by William Randolph Hearst to transmit images between New York and Chicago, revived Edison's interest. Despite several inquiries and negotiations between 1891 and 1896, Edison never sold any of his fax machines or patents.[78]

For his fax system, Patrick Delany transferred his concept for multiplex telegraphy, which he had demonstrated at the 1884 Philadelphia electrical exposition. A 12-pointed stylus traced an insulated message on tin foil. A wire connected each point to a disc with 84 insulated contacts, divided into 6 sets of 14 (two contacts synchronized the transmission of each set). Inside the disc, another wire rotated and touched the contacts of each set. This method divided the message into 12 sequential segments which a single wire transmitted to a receiver with a similar stylus. The major advantage of this system was speed, an advantage pursued at the cost of complexity. Like Bonelli, Delany proposed an idea far too complex and expensive for practical use.[79]

More conceptually, Franklin Pope argued that if facsimile could combine the speed of conventional telegraphy with the low cost of mail, "we should have a practical solution of the much discussed problem of 'cheap telegraphy.'"[80] Such a solution envisioned the mechanization and automation of telegraphy by removing the human operators. Six decades later, Western Union's Desk-Fax system began to realize Pope's vision.

SELENIUM AND NEWSPAPERS

In the 1890s–1900s, the solution of facsimile technology and the problem of market demand edged closer together as new generations of machines proved increasingly practical and a real consumer appeared—newspapers seeking faster delivery of photographs.[81] As was painfully common with promising technologies, expectations and hopes outran technical and economic reality. Despite lack of competition, transmitting photographs over telephone lines did not become truly viable commercially until the 1920s–30s.

Interest in faxing photographs came from competitive newspapers. Since the *Illustrated London News* appeared in 1842, drawings and, increasingly, photographs had helped sell papers. Getting the pictures to the paper quickly provided an opportunity for faxing. Words traveled at the speed of a telegraph message, enabling newspapers to print news just minutes after it occurred. Images, however, still traveled at the speed of their physical carrier, lagging behind a telegraphed story by days or weeks.[82] Sending photographs at the same speed as articles promised to radically alter the role of photography in newspapers—and sell more newspapers.

Sending a photograph was much more challenging than sending a message because it demanded accurate rendition of the photograph's grey shades (or tones) and not just the black-and-white (or on-and-off) of text and line drawings. Researchers followed two distinct approaches, both involving the application of new technologies. Scanning by light using a photoelectric cell ultimately triumphed. More practical throughout the 1920s was haptic scanning of a three-dimensional photographic negative. The two approaches were not exclusive since inventors used elements of both, depending on the capabilities of the latest innovations.

While the new technologies received the most public attention, the medium also changed from telegraph lines to telephone lines, from sending discrete pulses to analog waveforms. The result was greater capacity to transmit data, a capacity needed because of the greater amounts of data scanned in photographs.

In 1873, English electrical engineer Willoughby Smith and his assistant, Joseph May, serendipitously observed that selenium's electrical resistance depended on how much light illuminated it.[83] The most impressive aspect of this discovery was how quickly experimenters in Europe and the United States began trying to transmit still and moving images electrically via selenium.[84] This competition intensified with Frenchman Constantine Senlecq's 1877 telelectroscope and rumors that Alexander Graham Bell's photophone transmitted images by telegraph. Unveiled in August 1880, the selenium-based photophone transmitted sound by light; however, the idea never succeeded commercially. Senlecq used a camera obscura to focus light on a small piece of selenium held by two springs in an electric circuit. A thick iron plate vibrated at the receiver to push a stylus on paper.[85]

The ability to scan by light instead of touch inaugurated the modern fax age in 1881, when English physicist Shelford Bidwell transmitted an image via a selenium cell.[86] Bidwell quickly advanced from transmitting geometrical patterns cut from tin foil to simply shaped black-and-white photographs painted on a two-inch-square glass slide. A lantern projected the image onto a small brass box which enclosed the selenium cell. A platinum-covered brass spindle moved the box so that its pinhole passed over the entire projected image.[87]

More than two decades elapsed before the first selenium system transmitted a picture for a newspaper. The lag resulted not from lack of effort but from lack of feasibility. Selenium was difficult to handle, proving sensitive to aging, temperature changes, and other environmental shifts. Most serious was the inertial lag between exposure to light and the drop in resistance. Theoretical understanding necessarily followed experimental expertise.[88] In 1894, AT&T's Thomas D. Lockwood, reflecting over a decade of efforts in Europe and America, concluded that this "Visionary Telegraphy" represented "a branch of non-practical electricity" in which "no real progress has been made, and in my opinion it is very doubtful whether any practical progress ever will be made."[89]

Creating and physically scanning three-dimensional negatives proved more practical. By the 1890s, advances in the photography and printing industries made transforming a photographic negative into three dimensions feasible. A light-sensitive gelatin was coated on a copper sheet and then exposed to a photographic negative. The light turned the gelatin insoluble. Hot water washed the unexposed and partially exposed parts of the gelatin away, leaving a three-dimensional film with its contours reflecting the tones of the original. Preparing the three-dimensional film was both an art and a science, requiring a skilled operator.[90]

Cleveland's Noah S. Amstutz was the first to scan and transmit a three-dimensional negative. A draftsman and printer increasingly involved in the growing field of electrical engineering, Amstutz demonstrated a working system in 1891 and a more advanced system in 1895.[91] For him as for others, the recorder proved the weakest component.[92] Chemically treated papers did not produce the photoengraved blocks necessary for printing. The faint signals reaching the recorder proved too weak to actuate a stylus cutting through a gelatin or wax film on a copper plate to create an engraved image. The development of amplifying vacuum tubes solved that problem—two decades later. Frustrated, Amstutz returned to printing.[93]

Despite such setbacks, facsimile contributed to attract inventors. While some revisited old solutions, including scanning with multiple styli, others were more novel, such as Belgian engineer H. Carbonelle's 1905 patent for a machine that provided three scanning options: a non-conductive ink on metal foil, the differences in a gelatin's thickness, or the differences in electrical resistance of the metallic salts on a photographic plate or film.[94] An ocean away in Cleveland, the Electrograph etched a message in a zinc plate and then filled those hollows with a hard insulating material so the sheet was smooth. A stylus moved "very much as the reproducing stylus of the phonograph is caused to travel along the sound record."[95] The flat surface meant the stylus did not jump, a serious practical problem with the three-dimensional gelatin.

The machines that moved beyond the laboratory left potential users unimpressed. In 1899, five American newspapers experimented with Minnesota watchmaker Ernest A. Hummel's telediagraph but concluded that the low resolution and slow speed did not warrant the expense.[96] The Cleveland Facsimile Telegraph Company and the International Electrograph Company in New York actually offered the Electrograph for sale in 1901–2. Failing to attract customers, both machines and firms vanished into obscurity.

The first actual applications occurred in Europe, but not in Paris or London but in Munich. Dr. Arthur Korn not only created the first practical selenium system but proved to be an excellent promoter, building Picture Telegraphy networks between newspapers over telephone wires. A physics lecturer (*privatdozent*) at the University of Munich, Korn began working on facsimile in 1901 and transmitted his first long-distance photograph in 1904. His interest derived from childhood reading about predictions of television and his earlier research on photographing sound vibrations.[97]

To transmit photographs, Korn reached into his memory of Edison's phonograph from a book of inventions he received for his eighth birthday. He bought a second-hand phonograph and replaced the wax cylinder with one of glass and the phonograph needle with a beam of light. Instead of the phonograph needle revolving around the wax cylinder to create sound, a light beam passed through a film negative wrapped around the revolving transparent cylinder onto a selenium cell.

In 1906, Korn made two significant advances. The one less heralded was his receiver. By careful design and experimentation, he turned a string galvanometer, which twisted in proportion to the fluctuations in the current flowing through it, into a shutter. This partially solved the challenge of weak signals that had defeated Amstutz. Most praised was his ingenious solution to selenium's inertia. Installing a second smaller selenium cell in the transmitter circuit compensated for the main cell's inertia. Korn's mathematical training proved invaluable in determining the optimum balance between the two selenium cells, which reduced the transmission time for a small photo from 42 minutes in 1904 to 12 minutes in 1906.[98]

Korn made the transition from researcher to entrepreneur in 1907 at the French Academy of Sciences by transmitting the image of President Clement

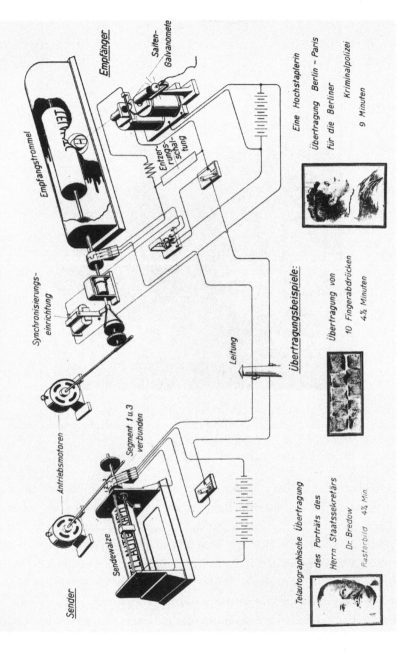

This schematic of Korn's teleautograph depicts the basics of facsimile: scanning, transmitting, and reproducing an image. Deutsches Museum, Bildstelle

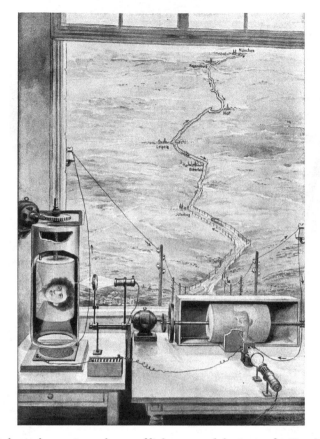

Instead of physical scanning, a beam of light scanned the image for Korn's selenium-based system, which used electric motors for smooth movement of the cylinders. Deutsches Museum, Bildstelle

Armand Fallieres live, before Fallieres and other guests, from Paris to Lyon and back to Paris. This key demonstration, replete with telephone poles inside the auditorium to show that the transmitter and receiver were actually connected, was a resounding success. In October, Korn transmitted Fallieres' photograph from the *Lokal Anziger* in Berlin to *L'Illustration* in Paris, which purchased the French rights to his system. Korn also astutely sent a transmitted photograph to Bidwell, who publicly praised the German, thus providing legitimacy and publicity. The low resolution (25 lines per inch) meant portraits were the preferred subject, reinforcing the traditional newspaper use of photographs.[99]

In November, Korn's system crossed the English Channel. The *Daily Mirror*, which in 1904 became the world's first newspaper illustrated only by photography,

Korn transmitted this portrait, which has a resolution of 25 lines per inch, in 1903. Deutsches Museum, Bildstelle

received a photograph of King Edward VII from Paris, exciting the paper's editors and readers.[100] Faxing saved the *Daily Mirror* a day—a lifetime in the newspaper business—publishing photographs from Manchester or Paris. The next year, the Copenhagen *Politiken* and Stockholm *Dagnes Nyheter* received photographs from Berlin.[101]

Acclaim for Korn's system did not hide its slow speed and serious susceptibility to interference. Transmissions proved acutely sensitive to circuit quality, a recurrent problem for facsimile. The signal was so weak—as low as one to two milliamperes—that interference from nearby poorly insulated telephone and telegraph lines marred transmissions with Baudot codes and other marks, often requiring resending. Trans-Channel transmissions faced another operational problem: obtaining a quality long-distance telephone line. The average conversation only lasted a few minutes, but transmitting a photograph demanded at least thirty minutes, and the Post Office hesitated to tie up a circuit for so long. Often the *Daily Mirror* obtained a circuit only after midnight, which reduced the timeliness of the transmissions.[102]

Responding to the transmission challenge, Korn reverted to a scanner based on Bakewell's concept of a stylus tracing insulating ink on a metal plate. The current of 10–20 milliamperes, an order of magnitude more than his selenium system, enabled reliable long-distance transmission. Korn's switch to physical contact reflected a wider consensus that selenium was too demanding for practical use. In late 1908, Korn installed his new system in Paris, while his colleague, Dr. Bruno Glatzel, installed its twin at the *Lokal Anzeiger*. The Prince of Monaco and *L'Illustration* also installed a Korn system to link Monte Carlo to Paris to receive daily crime reports—and to generate publicity.[103]

Despite these successes and his hope that "electrographs will soon become a common phase [*sic*] of the illustrated Press of many countries," Korn never left the university to commercialize his system, shifting his attention instead to a grander project, transatlantic transmission.[104] Consequently, he never acquired the industrial base of resources and supporters so important to competitor Edouard Belin in France. The inability of a protégé, English chemist Thomas Thorne Baker, to operate an economically viable service confirmed the wisdom of Korn's decision to stay in academe. Thorne Baker used his "telectrograph" in 1909 for the *Daily Mirror* in London, Manchester, and Paris, sending scores of photographs until the high cost of long-distance telephone circuits caused the paper to end the transmissions in 1911. His efforts to convince American newspapers to adopt his system proved equally unsuccessful. *L'Illustration* also abandoned picture telegraphy due to the low quality and high cost.[105]

More successful ultimately was French photographer Edouard Belin. Starting with public demonstrations in 1907 at the Societe Francaise de la Photographie and continuing for more than three decades, Belin linked the pre- and postwar worlds of Picture Telegraphy and message facsimile. His different machines and applications reflected the evolving technologies and the relentless push by developers and users to improve the equipment. His attention to facsimile, however, wisely did not deter him from other more profitable areas: Belin's firm, the Establissements Edouard Belin, produced precision electrical and optical equipment and dominated French efforts in developing television.[106]

Belin's success demonstrated how the best commercial technology was not always the best technology. Like Korn, Belin first experimented with a selenium-based scanner but switched to scanning three-dimensional negatives. In Belin's "teleosterography," a stylus pushed a small platinum roller over a rheostat that had five, and later twenty, copper plates, separated by layers of mica. Each plate corresponded to one tone of gray.[107] To obtain the right resistance, Belin replaced the roller and rheostat with a microphone to convert tones into electrical impulses. His most crucial innovations were altering the thickness of the relief for easier scanning and using a large microphone to capture those variations. These modifications were as much art as science, the results of repeated experiments done with painstaking attention to detail.[108]

Like Caselli and Korn, Belin recognized the importance of publicity. In 1907, two cabinet ministers watched him transmit a portrait of President Fallieres from Paris to Lyon and back. A January 22, 1908, demonstration resulted in a gold medal from the Society for the Encouragement of National Industry (Societe d'encouragement pour l'Industrie Nationale). In April 1914, Le Journal published the first news photograph with his equipment.[109]

Although already hailed in 1910 by Thorne Baker as "one of the most indefatigable workers" in facsimile, Belin did not succeed commercially in creating any market for faxing until the mid-1920s, when his simplified telestereograph, devised in 1909, entered actual service sending messages.[110] The long time from his early efforts to sales reflected the need both for large improvements in the actual technologies and for the French telegraph administration and newspapers to convince themselves to invest in the expensive equipment.

Even less successful than these attempts to harness telephone and telegraph were the numerous prewar efforts to send pictures via radio. The attraction of circumventing distance without wires was as obvious as the growing ability to transmit code and voice. The technical challenges for faxing in this early stage of radio's development proved far too great to even contemplate a commercial service. Early equipment remained simply too crude and indiscriminate to send and receive images, the equivalent of sculpting ivory with a jackhammer. Even demonstrating the concept stretched the limits of radio equipment.[111]

TELAUTOGRAPH

Not all efforts at transmitting writing failed. In the 1890s–1900s, a new competitor and close technological relative took potential markets from faxing. In the writing telegraph, or telautograph, a set of relays and rheostats transmitted attached to a pen transmitted a message as it was written. At the receiver, a similarly configured pen duplicated those movements.[112] Advocates of the telautograph contrasted its "simple and versatile" design and construction with the "beautiful but complex instrument" for faxing, while still providing the same advantages of signatures, secrecy, accurate reproduction, authenticity, and lower labor costs.[113]

Despite the telautograph's proclaimed simplicity compared with faxing, significant government and commercial use began only after 1900, more than a decade after Elisha Gray, thwarted developer of the telephone, founded the Telautograph Company in 1888. To overcome its very limited range, promoters defined markets as within a building or a similarly short distance. The military proved a small but crucial early market. The American army employed telautographs in the 1890s for fire control communications in coastal forts, and the navy used them for internal ship communications. By 1922, more than 10,000 telautographs provided internal communications for banks, businesses, and government.[114] Fax machines did not reach that number until the 1950s.

The slow evolution of the telautograph had a fourfold significance. First, its long time from demonstration to commercialization showed the generic challenges of transmitting and reproducing images. Second, the telautograph's success demonstrated that a market existed for an affordable method of transmitting written messages. Third, the telautograph and not fax machines filled that market. And finally, facsimile's failure highlighted the technological and economic challenges of sending images reliably over long distances.

෴

Looking backward in 1914 revealed seven decades of failed efforts to develop and commercialize facsimile. As much as that technology had advanced, conventional telegraphy had advanced even further, solidifying its grip on rapid printed communications. Users and telegraph companies had developed ways to send telegrams securely despite the lack of an exact reproduction of the original message. Korn's photocell-based machine opened the market of photographs for newspapers and magazines, but the high cost, slow speed, and low resolution deterred successful use. Fax seemed a failed technology, a tempting idea that attracted many inventors but found few users.

Looking forward, however, a new generation of inventors equipped with new technologies saw renewed opportunities in newspapers and messages. A few tentative successes had occurred, but major technological and economic obstacles still faced developers. World War I would change that situation dramatically by forcibly accelerating the development of electronics, setting the stage for the first large-scale and sustained applications of facsimile.

CHAPTER 2

First Markets, 1918–1939

Scientifically, the problems were all solved years ago; economically,
it is just beginning to appear that a solution may be possible.

—*J. W. Horton, 1929*

Viewing an experimental facsimile newspaper broadcast in 1939, an unidentified Californian rhapsodized, "To be granted the permission to see it in operation makes me feel like I had been permitted to ride in the first automobile constructed or in the first airplane that had successfully taken the air."[1] She was not alone in her wonderment about the magic of facsimile. In a fusion of high technology and high culture inconceivable even two decades earlier, the Radio Corporation of America (RCA) saved the Boston Symphony Orchestra in December 1937 when the full score of Jean Sibelius' "Uko, the Fire Maker" failed to arrive by mail from Finland. RCA arranged for a European Picture Telegraphy service to fax the score from Helsinki to Berlin, where RCA Communications radiophotoed it to New York for printing. A courier then sped the reconstituted sections to Boston. The *New York Times* proclaimed that Finnish newspapers "declare that America again has shown herself a country of limitless possibilities and resources."[2] RCA President David Sarnoff, who had brought Toscanini to New York City and the city's Metropolitan Opera to millions of living rooms by radio, must have been ecstatic.[3]

Such achievements made the interwar decades the most exciting period in the history of facsimile until the 1980s. Facsimile efforts blossomed across the globe. In 1928 developers pursued at least nineteen systems in Europe, the United States, and Japan, with four reaching commercialization.[4] Two major forms of facsimile appeared, Picture Telegraphy (PT, also called telephotography and wirephoto) and direct telegram replacement, and a third, message facsimile, was conceptualized. PT, although more technologically demanding, appeared first because it found a protected niche market, rapid transmission of photographs for newspapers willing to pay premium prices. By 1930, newspaper photographs and the occasional document crossed large parts of the globe in minutes via telegraph, telephone, and radio circuits, finally matching the speed of the telegraphed word.

The number of fax machines actually was quite small—less than a hundred in 1933[5]—but their significance extended far beyond their numbers because their customers were newspapers. By providing images at nearly the same speed as stories and privileging action photos over more formal photographs, faxing helped transform the visual culture of newspapers and their readers.

Replacing telegrams, the dream of early fax pioneers, became partially feasible via systems pioneered by Edouard Belin in France and Western Union in the United States. More novel was message fax, transmitting documents between offices. Although promising a wider audience and technically simpler than PT, faxing for businesses faced such formidable technical and economic challenges and strong competition from telegraphy and mail that its realization remained on the horizon.

The technological accomplishments would have astounded anyone familiar with facsimile in 1880 or 1900. Yet visions, expectations, and plans continually outpaced the technology, leaving an impressive series of failures, including facsimile broadcasting of newspapers and transmitting from airplanes. Numerous efforts to transform facsimile technology into more capable, competitive, and robust machines failed. The technology push did improve fax machines, but compared with competing technologies, faxing still underperformed despite the predictions and investments of its growing number of promoters. Market pull remained far weaker than technology push with the exception of portable transmitters for newspapers.

Failures stemmed from misconceptions about the market and technological prematurity. Overly optimistic estimates of demand, and especially the cost and time needed to develop and build machines, bordered on naiveté if not self-deception. Technological prematurity occurred when a technology functioned poorly in its desired application. While the underlying concepts might be sound, the implementing technologies were not. Often the deficiencies were not discovered until the product left the controlled environment of the laboratory and entered the messy, unpredictable real world. These failures were increasingly economic as well as technological, indicating the growing maturation of fax technology, albeit still not at the levels of reliability, capability, practicality, and affordability needed to successfully compete against increasingly capable rivals.

One way to increase the chances of success was to redefine the audience from competitive to niche markets where faxing received greater resources (including users willing to pay the high costs) and support, giving it an opportunity to mature and develop. These protected environments allowed a fragile and expensive technology to survive. For facsimile, that protected niche was both institutional and technological. The monopolies of national Postal, Telegraph, and Telephone administrations (PTTs) and AT&T removed the immediate need to make profits while ensuring that their skilled workers controlled all facets of faxing. European and Japanese PTTs integrated fax into their existing telegraph and telephone networks and cooperated through the International Telegraph Union. Faxing even became a visible component in Japanese techno-imperialism in China. In the United States, the Federal Communications Commission (FCC) provided the regulatory framework within which radio, including fax broadcasting, and television developed.

Because commercial operations—and profits—seemed more realistic, corporations now became major players, establishing research groups and devoting significant resources. Individual inventors like Austin Cooley and John L. V. Hogan played important roles only by linking up with larger firms or earning enough income to keep their own companies afloat. Similarly, xerography inventor Chester Carlson, whose second patent, granted in 1942, described a fax machine, succeeded only after finally finding corporate sponsorship.[6] Not all the usual suspects were present. Unlike radio but like television, amateur experimenters, deterred by the high entry barriers of expertise and equipment, did not play a major role in facsimile.[7]

Actual applications occurred first in Europe due to easier access to craftsmen, greater scientific expertise, and very competitive newspapers. In 1928, German, British and French firms and inventors claimed eleven of the eighteen Picture Telegraphy systems known to AT&T, with American firms developing the others.[8] By the 1930s, American research and applications outpaced their European counterparts because of the country's more competitive telecommunications market and geography. Whereas European PTTs promoting facsimile had to worry about the new technology hurting their other services, telegraphy and mail existed institutionally separate from American firms promoting facsimile (with the significant exception of Western Union, which turned facsimile into an extension of its telegraph network). The greater distances between major American cities gave PT the advantage of speed compared with airmail letters and trains. Despite interest in other countries, faxing's high cost and other uses for that necessary investment meant that the technology remained limited in geographic scope.[9]

Competition, perceived and real, accelerated the development and employment of facsimile technologies. The fear that competitors might succeed induced PTTs and firms to invest resources in order to understand faxing, guide its path, preempt or battle competitors, and possibly even profit. Newspapers' efforts to scoop their rivals and be the first with the latest photographs were the most important factor pulling the development of PT. Prestige and financial rewards—selling thousands of extra copies with the latest pictures—provided the incentive for papers. While exerting their considerable political influence to encourage PTTs to offer PT service, newspapers also hired engineers, developed their own equipment, and funded research. Fearing technical dependence on a rival or the reputation of a technological laggard, media groups funded competing lines of equipment and developed their own PT services.

During these interwar decades, visions of a fax-filled future spread from the realm of extrapolating engineers to an expanding range of potential users. Like other emerging technologies, such visions greatly contributed to facsimile's development by creating constituencies, generating interest, and attracting resources. Their existence and persistence provided guideposts that shaped planning, perceptions, and expectations of future markets. Most prominently, RCA envisioned and invested in the rapid transatlantic transmission of photographs, weather maps for ships and aircraft, broadcasting newspapers by radio, and replacing telegraphy. RCA's extensive promotion of facsimile was so enthusiastic that its 1934 annual report warned that "complex financial, commercial and operating problems" meant future, not immediate, applications.[10]

RCA was not alone. An impressively wide range of proposals, experiments, and demonstrations promoted the potential of instantly transmitting images and original documents by facsimile. Fingerprints and photographs for police, X-rays from hospitals to distant doctors, verification of signatures for banks, maps for the military, and reducing labor costs all found champions if not markets.[11] Not all predictions concerned business. In 1921, *Vanity Fair* humorously informed readers that an accurate picture by telephotography might prevent embarrassing situations, such as a dangerously sexy French maid for a son's house or a misrepresented fiancé.[12]

Before faxing could extend the realms of visual culture, however, the technology had to work. This chapter begins with the major changes in the actual technologies and then explores their attempted applications in Europe, the United States, and Japan. As the first market pull began from newspapers, fax promoters pushed the technology into other markets. While photographs for newspapers succeeded, other proposed applications, including faxing weather maps, newspapers, and documents, found competing technologies, particularly the telegraph, more attractive. Perhaps the most convincing indication of the growing technical feasibility of faxing, however, was serious military interest. Even here the lack of active peacetime deployment demonstrated the challenges that faxing had to overcome.

TECHNOLOGY

Every area of facsimile advanced technically, responding to developments in supporting technologies like radio and to specific challenges in scanning, transmission, synchronization, recording, and even data reduction. Frustratingly, operators lacked full control over their environment. Most obvious was the transmission circuit, but other problems came from surges and other unwanted variations of the electric currents powering equipment and carrying data. Evolving technologies both helped and hurt, since what benefitted facsimile also benefitted telegraphy and television. Firms developing fax also developed its competitors, wisely not placing all their bets in one technological basket.

Facsimile benefitted greatly from advances in electronics during World War I, especially photoelectric cells and vacuum tubes. The first practical photoelectric cell (or "optical microphone") appeared in 1904, and postwar development provided facsimile engineers with many options for scanning images and reproducing grey scales. Arthur Korn extended his expertise to put photocells in weaving machines to create color images and to warn operators against overly tight threads.[13] The vacuum tube amplified the minute current from a photocell for transmission and detection at a recorder, solving the problem of an inadequate signal at the recorder. The photocell and vacuum tube rejuvenated scanning photographs by light. Of nineteen systems in 1928, eleven scanned by light, and the rest by stylus.[14]

Two types of optical scanning emerged, direct and reflected. In direct scanning, the operator wrapped a photographic negative of the image around a glass cylinder. A beam of light passed through the negative onto the scanning photocell mounted inside the cylinder. In reflected scanning, a beam of light hit the document or image wrapped around a cylinder and was reflected into a photocell mounted above the cylinder. This demanded a more efficient optical system but avoided the time and cost of turning a document or image into a photographic negative for transmission. By 1935, reflected light dominated.

The photocell was only one component of a complex and interdependent scanning system. Switching from a caesium to a rubidium photocell for RCA's electrolytic recorder necessitated redesigning its optics to handle the lower sensitivity of the new photocell.[15] The differences among systems centered on these components. When minute fluctuations distinguished gradations of grey, minor variations in the light source, electric current, or sensitivity of the transmitting and receiving instruments created ruinous errors. Converting the transmitted signals into a recorded image employed oscillographs, galvanometers, Karolus cells, and other types of shutters to control the light hitting the recording medium.[16] Of the multiple methods of recording, newspapers preferred photographic paper or film. Where resolution was not as important, less costly carbon and electrochemical paper prevailed.

If the essence of engineering was controlling the environment, then faxing suffered from bad engineering. Facsimile faithfully recorded the imperfections of a telephone circuit that normally people mentally filtered in a conversation. Flawed transmissions necessitated costly and time-consuming retransmissions. Thus, just as important as improving the fax machines was improving the telephone, telegraph, or radio circuit. The demands of facsimile sparked research into understanding, measuring, and modifying these environments as well as theoretical studies on information. Applications thus pushed research.[17]

Facsimile telegraphy used existing cables but suffered from low resolution and slow speed. Telephone lines provided greater capacity, but faxing's sensitivity to crosstalk, delay distortions, and other circuit shortcomings often ruined the image. Ensuring good transmission pushed the development of electrical filters to dampen or eliminate unwanted interference and circuit variation. Even so, flawless transmissions demanded high quality, conditioned lines, "nursed and coddled like a precious pet."[18] The problems were not all technical: operators, trained to break into conversations to warn about elapsed time, had to be instructed not to interrupt and thus ruin a wirephoto transmission.[19]

Reliable, coherent transmission over radio circuits through a poorly understood and highly variable atmosphere proved an even more daunting technical and scientific challenge.[20] The sensitivity of faxing to radio circuits (including atmospheric noise and interference) often meant resending flawed images. One-fifth to one-half of the radiophototelegrams transmitted between Amsterdam and Bandung in September 1933 required retransmission.[21]

As speeds and distances increased, so did the challenge of synchronization. Vacuum tubes enabled tuning forks to synchronize transmissions, but tuning forks demanded a closely monitored and controlled physical environment to maintain accurate calibration, often demanding hours to attain (for the Marconi-Wright system, the temperature had to vary by less than 0.05 degree Fahrenheit).[22]

Several researchers recognized that much scanning was wasted on empty white space and duplicating much of the previous line, but the field of conceptualizing information was so young that this was not even a problem without a solution.[23] Not until Claude Shannon's 1948 theory of information and the development of vastly faster electronics in the 1970s did data reduction become feasible. Like many predicted applications, the solutions to sending unnecessary data remained far in the future.

Engineers, technicians, and craftsmen struggled to make affordable, precise, and robust equipment. Save for Western Union's simpler machines, production never reached the numbers needed for mass production or even large lots. The inability to machine parts to sufficiently high precision dogged designers. The 1925 AT&T scanner moved over a lead screw that had 100 threads per inch, the same number as the scanning line resolution. This fine thread wore out rapidly, however, increasing maintenance costs. To eliminate this problem, AT&T's second-generation system initially employed a screw with 25 threads per inch. This coarser screw produced "drunkenness," narrow irritating lines, on the received picture, so AT&T returned to 100 threads per inch.[24]

COMPETITION

Fax evolved in a rapidly changing environment of actual and predicted competition. Engineers, managers, investors, and consumers judged facsimile not just by itself but in comparison with regular and air mail, telegraphic and telephonic services, the established standard (AM) radio, and the promise of FM radio and television. Outside of news photos, fax fared poorly regardless of country. Very few people or businesses needed rapid but expensive image transmission when alternatives offered adequately fast service at much less cost and effort. Not all competition actually existed. Television, for example, proved more attractive than facsimile in attracting resources and attention.

The mail remained the main method of communicating the written word and images. In the United States airmail shipments grew from 7.7 to 19.5 million pounds between 1930 and 1937 and another order of magnitude by 1945. Those years saw the handling of 28, 26, and 38 billion pieces of regular mail respectively. Long-distance telephone calls increased from 13 million in 1919 to 49 million in 1929 and 59 million in 1939, showing both the need for immediate long-distance communications and a way to meet that demand orally.[25]

Facsimile's most serious competition remained telegraphy, which had advantages of decades of familiarity by customers, tens of thousands of skilled telegraph engineers and operators, reliable equipment, and codes and other social structures that integrated telegraphy into office operations. From 1919 to 1929, the number of telegrams grew from 139 to 234 million before the depression years reduced that to 189 million in 1939.[26]

Nor had telegraph technology stood still. In 1932, Western Union introduced telex ("*tele*printer *ex*change"), which enabled offices to send and receive telegrams directly. To prepare a telex, a skilled operator typed the message on a teletypewriter, which punched a paper tape. The operator usually printed the tape before transmission to detect any errors, which were fixed by splicing the tape. Average speed was 60–100 words per minute, excluding preparation, verification, and correcting errors. A key reason for telex's acceptance by businesses was its "automatic answerback" function, which verified a message's receipt, thus giving a telex legal contractual status. European PTTs and AT&T offered similar services.[27] Unsurprisingly, telex grew into an essential domestic and international tool for businesses and governments.

A competing hybrid approach, transmitting pictures as telegraphic code, peaked in the early 1920s. The advantage was obvious—an enormous infrastructure already existed to handle telegrams. The problem was equally obvious: How could a telegraph convey tones and not just black and white? The answer was a code that conveyed the tone and its location on a grid. An image was scanned for black, white, and greys, then encoded. The coded data was transmitted as a telegram and reconverted into a photographic negative.

The concept was not new—Nevel Maskelyne patented a six-tone system in 1909 in Britain, but an artist not a machine reconstituted the image.[28] This approach enabled Korn to transmit across the Atlantic Ocean in 1921 for the American navy, discussed below. Such a system enabled the *Los Angeles Times* to print a more impressive image of the July 2, 1921 Jack Dempsey-Georges Carpentier boxing match in New Jersey than Belin's faxed transmission to Paris.[29] Employing a photocell system to scan and reconstitute an image, the Leishman Telegraph Picture Service Company sent hundreds of its "photograms" nationwide in the early 1920s. After AT&T opened its PT service in 1925, Leishman suffered.[30]

Dating back to a shared selenium ancestry, television and facsimile shared many commonalities, including researchers, patrons, and bureaucratic classification. The Federal Radio Commission, precursor to the FCC, lumped facsimile and television in its visual experimental broadcasting category as did Bell Telephone Laboratories (BTL), AT&T's research and development arm, through the late 1930s.[31] Developers often had to explain, not always with success, that facsimile was not television.[32]

Many fax enthusiasts, such as Korn and Sarnoff, envisioned facsimile as a logical stepping stone along the great electromagnetic chain of being to television, technologically, financially, and geographically. Just as pre-electronic generations of fax inventors worked on telegraphy, so too did several inventors, including Ernst F. W. Alexanderson and Vladimir Zworykin, work on both facsimile and television.[33] The Establissements Edouard Belin led the French efforts in television as well as facsimile. RCA, the major American promoter of television, was the major developer of fax technology while BTL's television research benefitted its wirephoto equipment.[34]

"Facsimilists" perceived television not as a competitor but as an expensive "personalized delivery of motion pictures" that offered only "fleeting, moving images on a screen, whereas fax provides a recorded duplicate of each transmitted page."[35] Television advocates like Orrin J. Dunlap, Jr., the New York Times Radio Editor from 1922 until 1940, viewed fax broadcasting as a supplement for television, printing programs and synopses.[36]

Where fax broadcasting and television parted ways was attention. The development of television took resources, frequency allocations, and advertising from both fax broadcasting and the emerging FM radio field. Television consumed over $10 million of RCA's resources before World War II, an order of magnitude greater than its fax research. Although Sarnoff promoted facsimile, it clearly never excited his imagination like television, whose electronic transmission without any moving parts clearly captivated him. The fax machine, its commercialization frustrated by complicated electromechanical mechanisms, was not in the same league.[37]

PICTURE TELEGRAPHY

Whereas prewar PT was experimental and newspaper-to-newspaper, by the late 1920s facsimile technology had developed sufficiently that, pushed by newspapers, European PTTs and AT&T established public PT services. In June 1928, the German Ministry of Posts convened a meeting of European PTTs to create an international Picture Telegraphy network. The delegates agreed to work with the International Telegraph Union (ITU, which became the International Telecommunications Union in 1932) to establish common rates, standards, and equipment compatibility. By 1936, more than a dozen PTTs participated, some via radio circuits.

For the first time, pictures flowed among European capitals and major American cities as quickly as words. Financially, these services lost money, reflecting high costs and limited use. Economics, however, proved less important than

The Siemens-Karolus Picture Telegraphy system in the Berlin post office literally needed a room of supporting equipment. Deutsches Museum, Bildstelle

prestige and other concerns. Indeed, the Austrian government heralded the first international service, opened between Berlin and Vienna on December 1, 1927, as a large technological step to Austro-German union.[38]

Europeans used German Siemens-Karolus or French Belinograph equipment, with the exception of the *Daily Express*, which paid AT&T $55,000 to trump the *Daily Mail* and its Siemens equipment by two weeks in transmitting pictures from London to Manchester.[39] Leipzig physics professor August Karolus had developed his system in the 1920s with the Telefunken wireless firm and then the German electrotechnical giant Siemens, thus ensuring the corporate support to turn his idea into a practical reality.[40] Absent from these efforts was Korn, who concentrated on radio facsimile with the Italian and Prussian police.[41]

In Britain a wide range of fax advocates inside and outside the Post Office combined to overcome Treasury reluctance about creating a service that would lose money. The forces favoring PT included enthusiastic postal employees, companies promoting their equipment, and continental PTTs trying to create a wider and thus more valuable network. Administrators viewed PT as maintaining its monopoly on international communications and enhancing Post Office standing in the ITU.[42] The most important external advocates were the politically influential newspapers, which, ironically, were also the first to desert the public system for their own private operations. The more than three thousand people who signed the guestbook in 1929 in the "Picture Palace" in the most vibration-free part of the London Central Telegraph Office indicate that this interest was not limited to just a few. British service officially began on January 7, 1930 to Berlin.[43]

AT&T's January 1928 survey did not mention Japan. Eleven months later, a Nippon Electric system transmitted photos of Emperor Hirohito's accession. The introduction of faxing exemplified the Japanese approach to modernization: gathering information about foreign activities; arranging demonstrations of foreign equipment; cooperating with foreign firms and engineers; and developing the expertise to create their own system. Military attachés in particular served as active conduits of information. Indeed, the American navy's fax expert, Stanford Hooper, refused a Japanese navy request to visit because, he said, "My experience is that the Japanese merely learn from us, maybe purchasing an initial order for the purpose of copying the design at home, and later make all their own equipment."[44]

As in Europe and America, the major interest came from newspapers. Japanese newspapers approached AT&T in 1925 and 1927 but found the firm uninterested, reflecting AT&T's misgivings about its system's poor quality.[45] Instead, the Ministry of Communications and the *Mainichi* newspaper tested Korn and Belin systems in 1925–26 with Belin himself. Belin visited Japan en route to China to lecture and promote his fax technology in China in 1926.[46] Although the first transmission between Tokyo and Osaka occurred in May 1925, the Ministry of Communications did not grant licenses for newspaper services until mid-1928.[47] The spur was the Imperial Accession Ceremony of the new emperor on November 10, 1928, in Kyoto. Over the three weeks of ceremony, PT systems transmitted 253 photos to Tokyo and other cities, an impressive display of the robustness of the equipment and its operators.[48]

At Nippon Electric starting in 1926, Yasujiro Niwa and Masatsugu Kobayashi developed the first indigenous Japanese fax system. Niwa was one of the major figures in Japanese communications from the 1920s through the 1960s. His 1924–25 stint at BTL introduced him to PT.[49] *Mainichi* intended to use its Belin system, but its poor synchronization made the paper switch to Niwa's system, a decision eased by Niwa's prior testing with the paper. After the coronation, *Mainichi* employed the Nippon Electric system to transmit photos between Tokyo and Osaka.[50]

By 1933, newspapers and the Ministry of Communications' public service between Tokyo and Osaka, opened in August 1930, operated ten Siemens-Karolus and six Nippon Electric sets.[51] The Ministry continued to add cities, and in 1943 it reached Manchuria, reflecting a "follow the flag" policy. In 1932, the army opened a radiophoto circuit for documents between Tokyo and Mukden in Manchuria. This expanded capacity fit the army's need for more communications and propaganda as it expanded into China.[52]

For most countries, joining the international network was adequate. More ambitious and supporting a domestic manufacturer, the French government established a national PT service in 1935 with eleven fixed and two portable sets—all Belin. French newspapers also procured their own equipment, owning ten fixed and portable sets in 1936.[53] Private systems emerged too, to the detriment of public services. British newspapers continued to embrace PT but did so by purchasing their own equipment and either leasing domestic lines or reserving international trunk lines just like a telephone call. In 1936, British newspapers owned twenty-seven Siemens-Karolus and ten Belin sets (eight portable) compared with the one Post Office set.[54] These sets allowed regional editions to print the same photographs as London papers.

PT technology proved more impressive than its economics. Transmitting photos, whether by telegraph cable, telephone circuit, or radio, proved to be far more expensive than conventional telegraphy and mail, as table 2.1 shows. Even when the Japanese Ministry of Communications offered a small "1-yen telegram" in 1934, only 20 percent of the cost of a regular PT transmission in 1929, that price was still two orders of magnitude more than a letter.[55]

Operating deficits were normal but minor compared with PTT profits. In Britain, the Post Office usually recouped only a sixth of expenditures, leaving the Post Office to swallow most of the £3000–3500 (approximately $15,000–17,500) annual costs as well as the original equipment cost.[56] The exception was AT&T which lost $500,000 in its first 28 months because of its long-distance conditioned circuits.[57]

Improved airmail service from the continent to Britain hurt PT. While PT saved a few hours of time—assuming good transmission—"cheaper and sufficiently rapid Air Mail Services" carried original negatives and physical articles that could weigh three pounds before equaling the cost of one phototelegram. Picture agencies after 1930 noticeably increased their telegraphic traffic from Paris to London to notify clients when their pictures would arrive at the Croyden airfield.[58]

High cost was not the only problem. In Britain, transmitting a 3×5-inch photograph took only twenty minutes but total time averaged two hours. The difference included preparation, waiting for a high-quality line, ensuring a good connection,

TABLE 2.1

Comparative costs of Picture Telegraphy

System	Cost	Alternative	Cost
London to Paris phototelegram	1–4 pounds	airmail	3–4 pence/ounce
Tokyo to Osaka photogram	5–8 yen	letter	3 sen
New York to Berlin photogram	$57	letter	6 cents/ 2 ounces
London to New York photoradiogram	£10	telegram	0.9 pence/word
New York to Chicago	$15–50	airmail	5–10 cents/ounce

Sources: Accountant General to the General Secretary, Dec. 30, 1927, File 4, Minute 7970/26, M10060/1928, Post 33 2371B, and "Picture Telegraphy Service to Germany. Instructions to Office of Acceptance," Jan. 1, 1930, File 1, Minute 2198/1939, PO 33 5441. British Telecom Archives; Keijiro Kubota et al., *Fakushimiri bakujoho chosa* [Facsimile back information survey] (Tokyo: NTT, Nov. 2004, 3rd ed.), 10.1.3.3; Long Lines submission to FCC, "Telephotograph," 1936 85 08 02 01, AT&T Archives; "Radio Facsimile Service of R.C.A. Communications, ca. Aug. 1931, Clark Radioana Collection, Smithsonian Institution Archives.

and retransmitting if the received image proved flawed. Like PTTs elsewhere, the Post Office limited hours of phototelegram transmission to avoid disrupting voice traffic. In 1930, delays averaged 38 minutes, with some transmission waiting more than 100 minutes. Interference and other circuit problems necessitated retransmissions of up to half of all images.[59] These unpredictable delays frustrated newspapers, which operated under tight deadlines.

Although newspapers were the initial market, fax advocates envisioned wider applications for PT. Repeating arguments from the nineteenth century, faxing telegrams promised exact images, reduced labor costs, and eliminating both transcription errors and the tedious chore of counting words. The Norwegian PTT estimated that shifting to PT could reduce its telegraphers from 117 to 16.[60] RCA conceived its transoceanic radiophoto market as businesses willing to pay for the rapid and error-free transmission of illustrated material and time-sensitive documents.[61]

Extensive advertising campaigns encouraged banks, police, doctors, and other businesses to fax "drawings, plans, diagrams, prescriptions, signatures, greetings, advertisements, prospectuses and long messages in small type."[62] AT&T transmitted pictures as gifts to important politicians, military officers, scientists, and engineers. Its 1928 campaign in January sent 2,400 real estate brokers a letter describing a sale made possible by a picture transmitted from Cleveland to Boston. The next month 2,000 department stores and their buyers in Chicago and New York received the pamphlet, "Telephoto and Fashions," which proclaimed the advantages of rapid transmission of the latest styles. In March, 256 steel manufacturers learned how PT could transmit drawings, specifications, and other graphics. Regardless of the audience, the results universally proved "almost negligible."[63]

This misplaced optimism of expanding PT from newspapers to other applications ran aground on the shoals of unfamiliarity and high cost. Very few businesses could justify paying ten to one hundred times the cost of a letter or telegram to save time or send an image. Instead, AT&T reported that domestic and international PT transmissions consisted primarily of "press pictures of groups of people, damaged buildings, aeroplane and train wrecks, sporting events, and other types containing considerable detail." Non-picture traffic consisted of financial data and advertisements transmitted "with barely sufficient quality to assure legibility at the receiving end."[64]

A market of news photographs meant long periods of quiet alternated with sudden spikes of saturation. In Britain, faxed photographs, both private and public, grew from 387 in 1931–32 to 796 in 1938–39, and averaged nearly 600 photographs annually.[65] The abdication of Edward VII generated one-quarter of 1936 British traffic in three weeks. Like Europe, Japanese PT traffic was "relatively small, except in the case of some notable event."[66] For special occasions, such as a royal wedding, PTTs arranged extra equipment and close international coordination to minimize delays.[67] For the 1936 Olympics, Japanese and German engineers cooperated to radio wirephotos between Berlin and Tokyo, an impressive technical and propaganda feat commemorated by an illustrated photoletter from Adolf Hitler to Nippon Electric.[68] Reflecting the Nazi emphasis on visual propaganda, German circuits carried 12,000 wirephotos in 1938, nearly 1,100 abroad.[69]

Some PTTs justifiably feared losing telegram traffic, the essence of their monopolies, if businesses sent photographs of coded messages instead. Although British tests showed that a message had to be at least seventy words before PT became more economical, the fear was occasionally justified. Coded pictures reduced telegraphic traffic from Java to Holland and comprised a significant part of pre-Depression New York–London radiophoto traffic.[70]

Even before land-based PT networks, photographs had bridged the Atlantic Ocean via underwater telegraph cable and radio. The attraction was obvious: compared with a ship, a faxed transmission arrived days earlier. Pride of place for the first transmission went to the Bartlane system of coding pictures developed by the *Daily Mirror*'s Harry G. Bartholomew and Maynard D. McFarlane. In 1920, Bartholomew, who became the paper's chairman in 1944, transmitted two pictures to the *Daily Mirror* from the America's Cup yacht race. A selenium cell classified every spot on the negative by tone and turned it into a ten-letter group. These groups were telegraphed to London and converted back into a photograph.

With each line cranked forward by hand, preparing one photo took thirteen hours to produce what the paper admitted were "imperfect and not wholly accurate" pictures. Nonetheless, this represented a stunning technological achievement.[71]

The evolution of this system exemplified the long distance between the demonstration of a concept and actual commercialization. Not until 1925 did Bartlane images regularly cross the Atlantic. The problem was not resources (researchers at AT&T informally learned development cost over $250,000), but reliably transmitting a photograph with enough resolution to satisfy readers.[72] Failed negotiations to use the Marconi's Company's radio circuits meant that Bartlane photos traveled under, not over, the Atlantic Ocean via Western Union. Through April 1939, 438 photographs had crossed the Atlantic, approximately one every two weeks.[73]

Western Union replaced the Bartlane system in April 1939 with its own higher resolution cablephoto service, which cost only $91,000 to develop. Reflecting wartime coverage, clients paid approximately $70 per image to transmit more than 800 photographs—one a day before halted by American entry into the war.[74]

The first successful radiophoto experiments occurred in Europe with Danish watchmaker Thorvald Andersen transmitting photos from Copenhagen to the London *Daily Express* in August 1920. Reflecting the interests of newspapers, the first pictures were of a monarch, a political leader, and an actress: King George, Lloyd George, and Irene Vanbrugh. European and American firms soon developed commercial services.[75]

Research on transatlantic radio facsimile began at General Electric, a parent of RCA, in 1923. According to company folklore, the president, Owen D. Young, wondered why a newspaper could not be transmitted across the Atlantic Ocean as easily as a conversation in one "zip," but he then demurred, "Not being an engineer, I am not interested in details; that is your job." David Sarnoff offered a different genesis tale of Young touring RCA's Radio Central and telling Sarnoff that he wanted to see a full message or newspaper "flashed" instead of dots and dashes for transcribing. A technical expert then said, "It is splendid to have an imagination utterly unrestrained by any limitations of technical knowledge." Regardless of which—or both—tale proved true, company officials promptly took the hint.[76]

Under the guidance of General Electric's Ernst F. W. Alexanderson, Richard H. Ranger sent the first photograph over the RCA transatlantic radio circuit on November 30, 1924. Ranger moved to RCA, and RCA Photogram Service commenced from New York to London on May 1, 1926.[77] Even for the 1920s, the quality of the images was poor. Nonetheless, RCA expanded service (including Berlin in 1932 and Moscow in 1941) while improving speed and transmission quality.[78]

Compared with the Bartlane system, RCA's system was a huge success, averaging three transmissions daily in 1934, mostly for American newspapers or where speed saved money, like a drawing for a ship's broken engine. The 14.5 million international telegrams and 27,000 international telephone calls that year, however, dwarfed the number of photoradiograms.[79]

AT&T AND WIREPHOTO

Although interested in facsimile since the 1890s, AT&T began serious laboratory research only in 1920.[80] Four factors triggered its decision: appreciation of technological advances, pressure from newspapers, the emergence of competing services, and a sense that a market was finally developing. AT&T constantly feared that if it did not pursue facsimile, the corporation would find itself frozen out of a rapidly expanding and profitable field of photographs for newspapers. Future information theory pioneer Harry Nyquist also predicted lesser markets of technical drawings, criminal identification, and legal and financial papers.[81]

Although telegraph-based systems attracted newspapers, AT&T initially focused on Belin, who tested his system with the *New York World* and *St. Louis Post-Dispatch* in 1920.[82] In 1923, the *World* asked AT&T about renting long-distance circuits to receive photographs from other newspapers equipped with Belin transmitters.[83] AT&T, which judged the Belin equipment the most advanced, considered forming a partnership with the *World* to promote this service but instead decided to introduce its own equipment and service.[84] Other newspapers inquired about PT services, partly due to the well-publicized Belin tests but also to reduce the expense of conveying photographs from isolated events.[85] So intense were newspaper interest and inquiries that BTL president Frank B. Jewett justly feared introducing a technologically immature system into an unprepared telephone network in order to satisfy newspapers.[86]

On July 25, 1923, AT&T vice president Walter Gifford authorized spending $55,000 for BTL to develop and Western Electric to manufacture equipment.[87] Judging by the rapid escalation of costs, AT&T greatly underestimated the technical challenges of both its PT equipment and sending images over its circuits. As engineers refined specifications, modified equipment, and tried to control capricious telephone circuits, costs grew. In mid-1924, BTL estimated development costs at $230,000, but they exceeded $700,000 by 1928. Through 1935, AT&T, including Western Electric, spent $1,371,382 on PT development.[88]

AT&T opened its new service among New York, Chicago, and San Francisco with Calvin Coolidge's presidential inaugural in March 1925, the first inauguration broadcast live by AM radio. By February 1927, Type 70-A Telephoto sets served Boston, Cleveland, Los Angeles, St. Louis, and Atlanta.[89] Traffic did not meet expectations because PT proved slow, costly, and unreliable. To avoid interfering with voice service, AT&T initially restricted transmissions to 4–10 a.m., greatly reducing its timeliness to newspapers. To mark the completion of the eight-station chain, AT&T expanded the available hours to sixteen.[90]

High costs discouraged use. Initially, sending a picture from New York to Chicago cost $50 and $100 to San Francisco, including twenty-five words for description. These rates dropped significantly in mid-June (e.g., New York–San Francisco fell to $60). When all eight stations became operational, AT&T reduced its rates to $15–50 depending on the destination or $105 for all stations, a major drop but still three orders of magnitude more than a six-cent letter and about the same cost as sending a person by train.[91]

One major mistake stemmed from misjudging the market. AT&T's system directed light through a photographic film on a glass cylinder onto a photocell mounted inside the cylinder. AT&T expected customers to bring negatives. Some newspapers did, but many clients submitted prints, which had to be photographed first, adding time and increasing AT&T's cost. Not all problems were AT&T's fault. Poor originals, often action shots suffering from rushed processing, produced poor-quality transmissions.[92]

Even as it expanded, AT&T contemplated whether to improve its existing equipment or develop a new generation. Pressure came from large financial losses and the realization that European advances had outstripped AT&T's technology.[93] Tellingly, AT&T refused to sell its equipment to outsiders, considering it was not "a reasonably stable commercial system which, when operated by another company, will be satisfactory enough to them to reflect credit on the designers and producers."[94]

The promise of reduced operating costs helped swing the decision to develop a second generation of equipment. Using ordinary instead of high-quality circuits crucially promised to improve the economics and expand geographic reach — high-quality circuits did not link every city. Achieving this goal again proved more challenging and costly than predicted. Depression-mandated work week reductions slowed development. More important were design changes and unanticipated legal and technical problems such as possible patent infringements and ensuring a uniform flow of electricity.[95]

Scanning by reflected light brought AT&T into the mainstream of fax scanning, but the new machine did not meet other AT&T's goals. The desired fivefold improvement in speed to two minutes for a 5×7 photo over regular circuits became four to five minutes over dedicated circuits. Nonetheless, the Type 70-B equipment proved faster, more flexible, easier to operate, and useable on ordinary circuits, although dedicated lines produced better results.[96]

But AT&T decided against operating its new equipment itself due to low commercial demand and growing interest by the Associated Press in a private service. Unlike the European PTTs, prestige and participation in a prominent technology did not constitute sufficient grounds for AT&T to continue its service. The corporation quietly discontinued its PT service in mid-1933 and instead began to supply equipment and lease long-distance lines.[97]

Utilizing this second generation of AT&T telephoto equipment, wirephoto emerged from a favorable conjunction between "the picture people" of journalism and fax promoters searching for a market. The decision by newspapers to adopt the new technology of PT was neither automatic nor easy. Photography, like any technology, was not without its drawbacks, especially cost. Improvements in technology—smaller cameras, faster films, and flashbulbs—meant less static pictures, thus fueling demand for pictures of events and action. Improvements in distribution—photo services such as International News Photos (1910) and picture magazines such as *Life* (1936)—made pictures increasingly familiar to newspaper readers. Yet only 300 of the 2,100-plus American newspapers in 1935 had their own photography departments.[98]

Wirephoto raised questions of newspaper culture as well as costs. The increasing incursion of photographs sparked arguments about the "emergence of a visual culture that challenged the predominance of the printed word in public discourse," with some journalists criticizing the "picturization of news stories" for detracting from good writing and lowering the tone and value of newspapers.[99] Nonetheless, newspapers kept increasing the presence of photography because of readers who preferred the ease of " 'look pictures' and 'listen radio' " to the effort of reading an article.[100]

As pictures became more important in the competition for circulation and scoops, so did rapid delivery. Newspapers chartered airplanes, express trains, motorcycles, and even pigeons in the race to publish the latest photographs. Wirephoto consequently seemed very alluring, promising transmission from San Francisco to New York in minutes instead of 85 hours by train or 24 hours by plane.[101]

Photojournalism enthusiasts created the newspaper demand for wirephoto. The most important figure was Kent Cooper, General Manager of the Associated Press (AP) since 1927, who had long promoted new technologies to accelerate news gathering.[102] He maneuvered his member newspapers into accepting AP-distributed illustrations, photographs, and, ultimately, wirephoto. That he accomplished this despite the directly competing financial interests of some owners was a tribute to Cooper's vision, diplomatic skills, and determination.

When AT&T introduced its PT service in 1925, Cooper suggested AP offer a photo service. Rejected by a conservative, cost-conscious board of directors, he proposed what became the AP Feature Service in 1927. Subscribers received a story and appropriate pictures by mail. In 1928, Cooper established the similar AP News Photo Service.[103]

In October 1933, AT&T offered to lease specially conditioned long-distance lines and sell its new PT transmitter-receivers to news agencies. To ensure an AP monopoly, Cooper suggested that AT&T send an identical letter to the four newsphoto services. Cooper drafted the letter, which set specifications that only AP could meet.[104]

The 1934 decision of AP's board of directors to adopt wirephoto sparked a battle. The opposition coalesced around John Francis Neylan, the aggressive general counsel for Hearst newspapers, fifteen of which belonged to AP.[105] Hearst's possessions included International News Photo, a newsphoto agency. Another foe was Roy W. Howard, head of Scripps-Howard, whose NEA Service operated Acme News Pictures. Neylan and Howard did not hide that their newsphoto agencies competed with wirephoto; indeed, they proclaimed that their newsphoto experience informed their opposition.

The other major photo agencies, the New York Times' Wide World Photo and Pacific and Atlantic Photos, operated by Colonel Robert R. McCormick's Chicago Tribune and Captain Joseph R. Patterson's New York Daily News, remained neutral. Cooper deliberately withheld information from Adolph S. Ochs, the owner of the New York Times and an AP board member, fearing that informing Ochs would prompt Wide World Photo to develop its own service. That did happen—but it was because of the embarrassment and slight Ochs felt.[106]

Opponents attacked not the concept but "the extravagant and commercially impossible telephoto business."[107] The direct cost was $3.5 million for five years, which included an annual $56 charge per mile for thousands of miles of leased line and twenty-five machines at $16,000 each.[108] The indirect costs would be incurred by rival news photos agencies developing their own wirephoto systems. Newspapers would have to either subscribe to a wirephoto service or lose circulation to competing papers that did. In the early years of Franklin Roosevelt's presidency, one goal of government and business was to eliminate "wasteful" competition. From a Depression-based perspective, wirephoto certainly seemed to fit that definition.

Like a classic arms race, opponents simultaneously pleaded with AP to avoid a costly technological innovation and warned that they would, if necessary, follow suit. The result would be journalistic stalemate but at much higher cost to everyone. Scripps-Howard General Manager W. W. Hawkins thought wirephoto was an extravagance until employees' pay cuts were restored. Was the AP management implying, "Let 'em eat pictures?" he asked. And wirephoto was expensive: Cooper privately estimated it cost three to fifty times more than mail service. Another fear was that there would be too few newsworthy photos to warrant a dedicated system. Frank Knox of the *Chicago Daily News* claimed that at most two pictures a day were worth the cost of wirephoto.[109] He did not realize that wirephoto would greatly expand the definition of a news picture and redefine timeliness for photographs.

Newspapers' fear of the proven technology of radio and the potential technology of television also played a role. AM radio threatened the newspapers' monopoly on quick-breaking "flash" news. Television, which entered limited commercial service only in 1939 but whose promise had been touted for years, also appeared menacing, albeit in a more shrouded, unknown form. Both wirephoto advocates and opponents used these threats to justify their positions.[110]

But publishers were newspapermen as well as businessmen, and the desire to be the first with the latest was very powerful. Just as important, newspaper readers increasingly wanted pictures. Pictures sold papers, and that sold advertisers.[111] At their 1934 annual meeting, AP members voted to proceed with wirephoto.

The opposition failed, partly due to the proselytizing efforts of Norris A. Huse, the AP News Photo Service editor who suggested the name "wirephoto." He spent much of 1934 visiting editors across the country and overcoming their understandable skepticism. Cooper had the vision, AT&T the equipment, but Huse clinched the deal.[112]

Although remaining outside the AP debates, AT&T had problems too. Development and manufacturing of the twenty-five sets cost significantly more and demanded more time than anticipated. As telephone engineers quickly realized, the new system involved not just the actual wirephoto equipment but also developing and installing delay equalizers on the leased circuits, switching facilities, and other equipment. AT&T lost over $35,000 instead of earning $28,000 on selling the machines. In addition, Long Lines spent nearly $400,000 conditioning long-distance lines.[113]

Four months later than planned due to slow delivery of equipment from Western Electric, the first wirephoto transmission on January 1, 1935, showed the survivors of a plane crash.[114] Supported by a management emphasis on proper and periodic maintenance, the system worked well. Close AT&T–AP cooperation led to a stream of modifications that increased reliability and efficiency.[115] AP editors and publishers praised wirephoto for promptly providing pictures. The major complaints came from smaller papers scooped by larger papers in nearby cities and losing sales, exactly as Neylan warned.[116]

By early 1936, the three major newsphoto agencies had introduced their own competing systems that operated over regular long-distance lines. Although their picture quality was lower, so were capital and operating costs: Wide World Photos claimed its transmitter leased for less than "a good office boy" and the receiver "at the price of the least efficient reporter."[117] Inductive and acoustic coupling avoided the AT&T prohibition on direct contact with its circuits, a ban issued to safeguard its circuits and not to prevent competition.[118]

Most audacious was the *New York Times*, which had semi-secretly experimented with Austin Cooley's system. When AT&T refused the paper's request to use regular long-distance lines for wirephoto transmission, the paper took another tack.[119] Publisher Arthur H. Sulzberger asked AT&T president Walter Gifford a deliberately vague question about what he considered interference and took Gifford's response as permission to proceed. After the fact, Sulzberger asked for AT&T's "good will and cordial cooperation" with the *Times*.[120]

The fact was the first public demonstration of Cooley's system. By chance, the newspaper was secretly testing the system between San Francisco and New York when the naval airship *Macon* crashed in the Pacific Ocean on February 12, 1935. The San Francisco office transmitted crash photographs to New York, where they appeared on February 14, dumbfounding AT&T and AP.[121]

In June 1935, 57 of AP's 1,300 members subscribed to wirephoto. A year later, AP had 150 wirephoto subscribers compared with 160 for its competitors. In 1940, wirephoto served 726 newspapers in the Americas, 120 directly and 606 by a combination of faxing and mail.[122] As subscribers increased, however, so did AT&T's revenues from leasing—wirephoto accounted for a quarter of AP's 1940 $1.7 million telephone bill.[123] Facsimile had finally earned a profit.

As important as the number of subscribers was the number of photographs sent. Wirephoto transmitted 60–70 photographs daily by March 1936.[124] AP considered it a "constant, continuing service," while its competitors viewed wirephoto as an occasional "flash and bulletin service."[125] Indeed, the first Hearst transmission was a "flash news" photo of murderer Edith Maxwell.[126] These perspectives were not *ex post facto* justifications but reflected how their promoters viewed the role of photography in journalism. Clearly, AP wanted to justify its expensive system. But wirephoto also fulfilled Cooper's expectation of pictures normally transmitted along with a story. And it was Cooper's vision that ultimately prevailed. As *Fortune* stated in 1937, "What really caused the commotion was the dawning realization among publishers that pictures had become quite as important as news itself, and that journalistic necessity was rapidly requiring that pictures be fresh as the news, riding right behind it on the wires."[127]

BROADCASTING FAXPAPERS

Wirephoto was the exceptional success for facsimile. Other markets proved more demanding, setting technical and financial criteria that faxing could not meet but its competitors could. That did not stop proponents from trying. Starting in the late 1920s, American newspapers, radio stations, radio manufacturers, and fax promoters cooperated to create the "newspaper of the future." The concept seemed appealingly simple: a radio station scanned and broadcast a specially prepared newspaper. In a subscriber's home, a recorder attached to an AM radio printed the newspaper. To use the contemporary analogy, the facsimile scanner was the microphone, and the receiver, the loudspeaker.

An earlier attempt had failed in Britain. In 1926, Thomas Thorne Baker and Captain Otho Fulton proposed "wireless phototelegraphy as a public service" for the newly formed British Broadcasting Corporation (BBC). When experimental broadcasts began in 1928, excited investors oversubscribed Wireless Pictures' stock offering and twenty-seven prominent locations displayed Fultographs and the received photos for public viewing.[128] However, the public bought only 700 of the 3,000 receivers manufactured, and transmissions stopped in October 1929. In the classic chicken-and-egg situation facing a novel technology, the BBC refused to offer a regular service until more people brought receivers, and people refused to buy the £22 machines (compared to a penny for a newspaper, which provided much more information and sharper photographs) until the BBC guaranteed its use.[129] Fulton suffered a similar lack of success in Europe and America.[130]

More optimistic American proponents envisioned revolutionary possibilities such as a national newspaper published by a network of transmitters.[131] The National Resources Committee's 1937 analysis of technological trends considered fax broadcasting one of the thirteen most socially significant new technologies.[132] M. H. Aylesworth, president of the RCA-owned NBC radio network, proclaimed in 1934, "The day will come when one will turn on the facsimile receiver when retiring, and in the morning the paper tape will tell the story of what flashed through the sky while you slumbered. It will contain road maps, fashion designs, comic sketches for the children, and no end of things, for whatever a pen can portray facsimile radio will handle."[133]

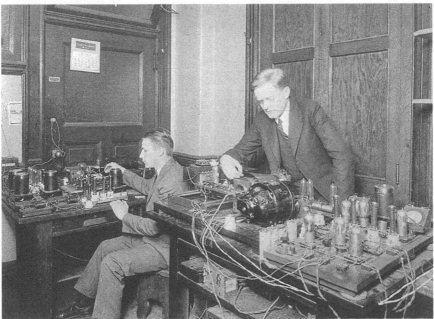

Black boxes hid the complexity of the electronic and mechanical components of a fax system as shown by these two 1924 photos of a Fultograph demonstration and RCA engineers working on their equipment. George H. Clark Radioana Collection, Archives Center, National Museum of American History

Fax broadcasting's direct customers consisted of radio stations and newspapers. In the early 1930s, when some newspapers considered banning listings of radio programs, radio stations saw fax broadcasting as a potential weapon, providing not only a listing of programs but a tabloid newspaper to visually supplement regular programs.[134] As their ownership increased to 211 of the 743 radio stations in 1938, newspapers perceived fax broadcasting as combining the speed of radio with the permanence and imagery of the newspaper.[135] Newspapers owned most of the two score radio stations that experimented with fax, approximately 5 percent of all stations.

This newspaper interest grew from familiarity, opportunity, and fear. Wire-photo had familiarized managers and engineers with the concept and value of fax. The opportunity was expanding circulation and advertising. The fear was of being surprised by a new medium, as had happened with radio in the 1920s. If fax broadcasting lived up to its promises, newspapers wanted to control this potential threat. As journalist and future historian Bruce Catton wrote to a colleague in 1938, "This thing is now coming along at a fast enough rate to make newspaper men like you and me experience a mild shiver or two when they look to the future."[136]

The technological impetus came from a small cadre of fax engineers who were substantially accomplished in radio electronics and, for the participating newspapers and stations, from a sense of technological pace-setting and a willingness to experiment with new technologies. The McClatchy Broadcasting Corporation considered that local goodwill and nationwide publicity more than recouped its $75,000 outlay.[137]

Several fax engineers were independents of limited resources who, if one application did not bear fruit, tried another. Pride of place for the first fax broadcasting went to C. Francis Jenkins and Austin Cooley. In 1924, Jenkins faxed newspaper headlines and radio schedules as part of his *"radio service to the eye"* but focused on obtaining navy contracts.[138] Cooley's "Ray-foto" aimed more at "the far-sighted radio enthusiast" than the 1920s couch potato, but it nonetheless had over a score of radio stations experimenting in 1926–28. Abandoning this work because of static and slow speed, he switched to developing wirephoto for the *New York Times*.[139]

Also active were John Vincent Lawless Hogan and William G. H. Finch. Hogan, a former lab assistant for Lee DeForest in 1906 and developer of high fidelity broadcasting, founded Radio Inventions, Inc. in 1928 to incorporate his laboratory and consulting practice. Initially researching both facsimile and television, he dropped the latter when the expense proved too great for an independent.[140]

In 1934, Hogan demonstrated his "radio pen," which printed on a 3-inch-wide paper strip, with the *Milwaukee Journal* and its WTMJ.[141] After failing to interest AT&T and RCA in his invention, he switched to other areas.[142] Before forming Finch Telecommunications in 1935, Finch worked as a radio engineer for the Hearst newspapers and the FCC. While most successful with fax broadcasting, Finch also promoted a 4-inch-wide receiver for communicating with airplanes and police cars.[143]

The most prominent corporate engineer was RCA's Charles J. Young, the son of General Electric president Owen D. Young. After graduating from Harvard, he joined GE's Radio Department in 1922 because of his longstanding interest in radio. Before becoming involved in facsimile, he worked in studio acoustics and developed a carbon microphone for WGY, GE's Schenectady station.[144] By September 1928, Young's engineers at General Electric had established the parameters of a broadcast facsimile service and highlighted the technical challenges that hindered commercialization over the next two decades: slow speed, low recorder resolution, lack of synchronization, and, most importantly, high recorder costs.[145]

RCA researchers always considered economics. In 1932, Young's Facsimile Section estimated the cost of a 9-inch-wide carbon paper receiver at $93.00 in a production run of 1000 and $58.07 in a run of 50,000, with $19.70 for an additional AM radio and loudspeaker. The estimated price for consumers was $250. For comparison, a teletype printer cost $350.[146] But few if any consumers bought teleprinters. They did, however buy newspapers, which cost a few cents, and basic AM radios, which cost several dollars—communications technologies the RCA study ignored.

After considerable testing and exploration of potential markets, RCA almost introduced a national fax broadcasting network similar to its Red and Blue radio networks in 1935–36, but deployment was halted due to unfavorable economic projections and a poor carbon paper recorder.[147] Consequently, RCA was surprised in October 1937, when the McClatchy Broadcasting Company applied to the FCC to experiment with Finch equipment at its KFBK (Sacramento) and KMJ (Fresno) stations. By late December, eleven stations nationwide had FCC experimental licenses.[148] Stung, RCA reentered the field with equipment essentially unchanged from 1936, and by December 1938, three stations broadcast with RCA equipment.[149]

Experimenting was expensive. The FCC required 50 receivers for these broadcasts, which RCA sold for $250 each. Together with the $5,000 for the scanner and transmitter, a radio station spent $18,000 before its first transmission, and operating expenses easily doubled that amount.[150]

THE MOVING FINGER PRINTS

This is a Finch radio facsimile set with the cover removed. The paper contains a dry chemical compound which responds to electric impulses by darkening. The motor-driven stylus sweeps back and forth across, carrying a varying current, thereby gradually building up the printed text and illustrations being transmitted.

William Finch's radio receiver used 4-inch-wide chemically treated paper to receive faxed images. Science Service Historical Images Collection, National Museum of American History, courtesy of RCA

The actual product was a condensed 4- to 6-page newspaper, taking 15 to 20 minutes to transmit an 8-inch-wide page, or 3 to 5 square inches per minute. At that rate, one observer recalled, "it seemed to us that it took ages" to produce a page. To generate publicity, stations provided the receivers to their executives and prominent community members. In eleven months of operations, the Sacramento *Radio Bee* broadcast to 580 homes, chosen from over a thousand volunteers.[151]

RCA and Finch envisioned that the direct customer—newspapers and radio stations—would purchase the scanner and receivers and then rent the receivers to the ultimate customers, their readers. RCA sold its elegantly packaged recorders with an AM radio. The Crosley Radio Corporation, a pioneer in low-cost AM receivers, manufactured and marketed Finch's recorders as the Crosley Reado Printer for $79.50 at R. H. Macy's or $49.50 as a mail-order kit.[152]

The recorders proved to be technically troublesome as well as expensive. Mechanical flaws forced WHO (Des Moines) to return forty-eight of its fifty Finch receivers. More frustrating, especially for FCC approval of commercial operations, the two systems were incompatible, despite efforts by the Radio Manufacturers Association to establish common standards.[153] To avoid interfering with regular AM broadcasts, the FCC limited fax broadcasting to midnight to 6 a.m., which restricted faxpapers to a morning edition.[154] These post-midnight broadcasts did not provide a unique service—the morning newspaper delivered far more information and advertising at far less cost. FM broadcasting theoretically allowed continuous service, providing the latest "flash" news, but it did not become technically feasible until the 1940s.

At the 1939 New York World's Fair, Finch claimed his WXBF broadcast to 2,000 receivers, the largest faxpaper audience ever. RCA broadcast four fax edi-

Charles J. Young led RCA efforts to commercialize faxing in the 1920s–30s, including this newspaper receiver. Science Service Historical Images Collection, National Museum of American History, courtesy of RCA

RCA's envisioned market of people willing to pay hundreds of dollars to receive an abbreviated newspaper by radio never materialized. George H. Clark Radioana Collection, Archives Center, National Museum of American History

tions daily, but its black-and-white television received more attention.[155] RCA had sold 12 scanners and 300–350 receivers by December 1941; 22 stations had used Finch equipment. The numbers were deceptively positive. Once the novelty and promotional value had faded, so did enthusiasm. McClatchy estimated the recorder had to drop below $75 for significant sales because fax broadcasting was "much less spectacular and less entertaining" than television.[156] Only sixteen of the more than forty stations the FCC licensed in December 1939 actually transmitted in 1940, and only two in 1941.[157]

What technical problems did not do, World War II did. On April 16, 1942, the FCC and the War Production Board restricted the operation of radio stations. These restrictions did not hurt AM fax broadcasting—the FCC already viewed it in the past tense—but did delay the development of FM. Only W8XUM in Columbus continued broadcasting, and that but two three-page papers weekly to two machines.[158]

WEATHER MAPS

Responding to "perhaps the most urgent technical problem concerning meteorology," the creation of an Atlantic Ocean weather service, the 1920s and 1930s saw a wave of government and corporate experiments faxing weather maps to ships, land-based receivers, and airplanes in the United States, Europe, and Japan.[159] The technical challenges proved great: receiving equipment had to shrink in size and workload while handling vibration, electrical interference, restrictions on antenna size and location, and radio signals that varied with the season and weather.[160]

Technically, faxed weather maps worked. But the International Meteorological Organization and national weather bureaus also developed less costly, less technically demanding, and more reliable approaches to collect weather data from ships and turn that data into useful maritime information. Standard nomenclature and procedures enabled the easy and accurate creation of weather maps, thus reducing the advantage of fax's imagery. By 1934, this international system of codes enabled ship's officers to report data and prepare their own maps.[161]

On land, the result was similar. The United States Weather Bureau had broadcast daily forecasts since 1918 for shipping and military aviation, and since 1926 for civil aviation. Starting in 1928, the Bureau rented thousands of miles of teletype networks for transmitting observations and receiving forecasts.[162] Its goal, reported a 1934 National Research Council study, was transmitting weather data "by any method that is economical, rapid and efficient—by telegraph, telephone, teletype, radio, or all of these agencies combined."[163] Notably absent was facsimile.

The Weather Bureau might have experimented more with facsimile if the Great Depression had not slashed its budget. When firing hundreds of employees and closing scores of stations, an expensive new technology received low priority.[164] The Bureau was not opposed to fax. When it was revamping its procedures in 1942, the Bureau hoped to fax its maps, but could not obtain the necessary wirephoto circuits and equipment.[165] Not until 1943 did planning a national radio fax network begin; it was completed in 1946, thanks to military support.[166]

FAX TELEGRAPHY

Not all attempts to create new markets for faxing failed. While PT did not replace telegrams, less expensive forms of facsimile directly replaced some telegrams, the original goal of Bain and Bakewell. With Belinograms and Telefax, customers themselves wrote and faxed a telegram. Less visibly, trunk offices faxed telegrams to each other out of the public eye.

Six decades after inaugurating Caselli's service, the French PTT in 1924 again made telecommunications history with its Belinogram service between Paris and the cities of Strasbourg, Lyons, Bordeaux, Nice, and Marseilles. Unlike Caselli, Belin offered easy message preparation by customers and did much better operationally and financially, impressing the British Post Office with its "very considerable" use.[167] The service emphasized its advantages of "unchallengeable authenticity"—the assurance of a signature, exactly transmitting drawings and other black-and-white images, and retaining the original message.[168]

Customers wrote their messages at a telegraph office with an adhesive ink. Powdered shellac was dusted over the ink, and the message placed over a small electric heater for 20–30 seconds to melt the shellac. When the shellac cooled, a stylus scanned the three-dimensional message for sending. Belin later introduced a photoelectric scanner, which substituted a regular pencil for the special ink and shellac, further simplifying and speeding the procedure.[169]

Even more attractive than the preparation was the price. A 4×5-centimeter message cost 15 francs but only 5 francs for one-third of that space and 10 francs for two-thirds. In contrast, a regular telegram cost 3.5 francs for 10 words, 6 francs for 20 words, 13.5 francs for 50 words, and 26 francs for 100 words. Since 100 words could be squeezed into a full Belinogram, its economics proved compelling for all but the shortest messages, in marked contrast to the expensive Caselli service.[170]

Across the Atlantic, Western Union introduced facsimile gradually as an extension of its existing telegraph system instead of offering radically new services. Starting in the 1930s and continuing through the 1950s, Western Union became the world's largest manufacturer and user of facsimile equipment. While employing the Bartlane and its own system to compete with RCA's transatlantic radiophoto service, Western Union's major domestic interest in faxing was exactly reproducing a message to avoid transmission errors, cause of the company's expensive "Accuracy First" program. Faxing also promised lower labor costs by eliminating operators and message boys, faster message handling, and expanded telegraphy service. Crucially, Western Union considered facsimile "automatic telegraphy," a version of the telegram and not a separate form of communication.[171]

Although a few company engineers began investigating facsimile technology after 1925, formal research commenced only in 1934 as part of a larger modernization to compete with the growing threats of telephony and airmailed letters.[172] In 1935–36, the company tested facsimile circuits between Buffalo and New York, and New York and Chicago. On January 1, 1937, its Public Message Facsimile Service opened on these circuits to handle graphic material such as manuscripts, charging by the square inch. By 1939, these circuits handled nearly 300,000 faxes annually (compared to over 139 million regular telegrams) and 1.5 million in 1945, though very few customers designated their messages as faxes. Instead, Western

Union faxed regular telegrams at standard rates.[173] A similar service in the Soviet Union faxed more than 100,000 telegrams between Moscow and 17 cities in 1938.[174]

More visible was Telefax, an effort to eliminate the expensive "last mile" to a telegram's ultimate recipients. Since 1915, a Morse repeater (replaced in the 1930s by teleprinters) directly connected large customers to a branch or central office, eliminating the need for messengers. Western Union viewed Telefax as serving customers—"patrons" in its language—too small to justify their own teleprinter while reducing the number of messengers and branch offices.[175] The New York office of Brown Harriman & Co. received the first commercial Telefax on August 30, 1938. Western Union installed the equipment free and charged standard telegram fees. Two hundred Telefax circuits in New York, Chicago, Atlanta, and San Francisco handled more than one million telegrams in 1942.[176]

Incompatibility with normal office procedures initially stymied Telefax. The difficulty of making notes on the red recording paper and the impossibility of writing on its black back proved serious shortcomings for firms that commonly added notations as a telegram was transferred internally. An extensive research effort solved this problem with Teledeltos, an electrically sensitive, silver-colored paper that needed no processing and kept permanently.[177]

Western Union did not pursue domestic facsimile telegraphy alone. In 1934, RCA built a New York–Philadelphia radio circuit at the experimental frequencies of 85–105 megacycles (then called ultra-short wave radio) to carry two fax and three telegraph channels. RCA intended to compete with Western Union and AT&T by offering high-speed transmission for any kind of documents. Vice President William A. Winterbottom predicted that by 1940 RCA would operate a nationwide facsimile system among major American cities with teletype circuits linking smaller towns.[178] The circuit opened in 1936, but RCA discontinued faxing because of the slowness and low resolution of its carbon paper recorder. Experimental faxing resumed in 1939 with a new electrolytic recorder.[179]

FAXING DOCUMENTS

While PT and faxing telegrams attracted the most resources, a third, more revolutionary market began to develop, if primarily in the ideas of engineers and promoters: faxing documents directly from office to office. As reporter Martin Codel enthused in 1934, "The future day can be envisioned when a businessman scribbles a note, or his secretary types a letter, inserts it in an automatic radio-facsimile transmitter machine and knows it will be delivered in a matter of seconds in a distant city as an identical reproduction of the original."[180]

His optimism was definitely unwarranted. Apart from the formidable technical and economic challenges, Western Union quickly discovered with Telefax that message fax faced a great problem in gaining business acceptance: creating legible copies from faxed images, essential for normal information flow in offices. That problem remained unsolved until the Xerox 914 photocopier in the 1960s. In the 1930s, proponents of message faxing had more immediate challenges.

Possibly the most glaring indictment of facsimile's technological immaturity was the inability of RCA and Western Union to create a commercially viable duplicator. The wide range of copying technologies testified to the growing demand for reproductions of documents in offices. The technological logic also seemed sound: If a facsimile system created a duplicate at a distance, why not set the scanner and recorder side by side to meet the growing market for reproducing documents? Such a system should be easier to develop and operate because it avoided the problems and costs of synchronization and transmission. In reality, faxing's slow speeds and high cost compared with competing stencils, hectographs, and other copying technologies meant this market also remained a mirage.

In the mid-1930s RCA built three different fax duplicators, based on its carbon paper recorder, to perforate stencils. Conventional duplicating methods easily outperformed the 15–20 minutes RCA needed to create one master. The development of its electrolytic recorder and the need to copy regular radiotelegrams revived interest in 1939.[181] Spurred by market surveys and discussions with potential customers like Prudential Insurance, RCA built two Duplifax machines in 1941, one for the Navy and the other for its Camden Sales Department. After 1945, RCA resumed Duplifax research and reached a speed of one page per minute, but instead the company decided to focus on electrostatic powder printing, which became its profitable Electrofax paper.[182]

Western Union introduced its Multifax to make carbons, stencils, or a hectograph master sheet in 1941. Its percussion recording needed 12 minutes to hammer tiny dots into letters and images on one page at 100 lpi and 15 minutes for 150 lpi. Except for noise, Multifax worked well in an office setting. Nonetheless, its slow speed and high cost ($2,000 to buy and $45 a month to rent) compared with alternatives soon sank Multifax.[183]

Message facsimile also attracted AT&T's interest, although its early studies concluded telegraphy remained much less expensive. In 1938, AT&T renewed its fax research, responding to a BTL report noting the firm's lead in most communications technologies except facsimile. Although the initial step, testing rivals' equipment, was modest, the long-term goal was developing equipment for business applications.[184] "The *big unknown*, of course, is what this business purpose may be," BTL admitted, but it argued that tests would determine if facsimile could, as some had claimed, render "the printer as dead as the dodo."[185]

In 1939–40, Western Electric and the New York Telephone Company tested RCA machines. Facsimile proved slow and unreliable, was unable to make carbon copies, and demanded too much skilled labor to operate and maintain. As important, users had to change their work patterns to benefit: the telephone company's Cost Accounting Bureau refused to duplicate its best copies by a fax machine because that upset its routine flow of paperwork. The obvious technical and economic problems of faxing often obscured such organizational challenges. As fax technology improved, however, shaping work routines to the new machine became increasingly important. These results did not deter BTL, which proposed a major development program to produce an economically viable machine that would be as fast as a teletype, simple to operate, and able to make duplicates.[186] Although the war halted this proposal, the concept of facsimile for business had been planted.

MILITARY FAXING

The military has historically played a major role in the development, promotion, deployment, and diffusion of new technologies ranging from the light bulb in Russia to space travel. Military interest in a technology can provide legitimacy as well as resources and a market. Faxing seemed an ideal military tool, able to transmit images and messages over long distances and even across oceans and skies. Like newspapers, the military often regarded cost as less important than capability and reliability. Unfortunately for faxing, two decades of military experiments proved almost uniformly disappointing. The equipment, although promising, proved too costly, unreliable, and fragile, especially compared with competing technologies like the teletypewriter.

Military faxing had been insignificant before World War I. The French army considered d'Arlincourt equipment for fortress communications in the late 1870s, a role fulfilled by telautograph equipment in the United States Army.[187] In 1910, Thorne Baker suggested that a portable PT system could provide photographic reconnaissance.[188] Germany pioneered wartime facsimile when Kaiser Wilhelm followed the Eastern Front via maps and photographs transmitted from army headquarters in Brest-Litovsk by a Korn system. Its army also experimented with faxing artillery targets from airplanes by radio with sets by Siemens & Halske and Max Dieckmann.[189] In the United States, Leishman's work sparked the Navy to consider radio faxing to transmit secret information while Belin developed cryptographic equipment for the French army.[190]

Military interest in radio facsimile grew sharply after the war, reflecting greater desire for better communications and technical developments in both faxing and electronics. In 1921, the U.S. navy tested Korn equipment lent by the Italian navy. These transatlantic transmissions employed a selenium cell to transform a photograph into a lettergram of hundreds of coded five-letter groups, each representing a tone and location. At the receiving end, an operator typed the groups into a machine "which resembles very closely a typewriter" to reconstruct the image, but unfamiliarity with the code often garbled transmissions.[191]

The same year, Belin faxed sketches and handwriting of Premier Aristide Briand and General John Pershing from the Lafayette radio station outside Paris to Bar Harbor, Maine, and from the navy radio station at Annapolis, Maryland, to Lafayette. The navy concluded that the equipment was "quite complicated and in laboratory form" but warranted funding to turn it into a practical system.[192] The British War Office, like its French counterpart, also bought and tested Belin equipment and followed the technology.[193]

The leading American proponent of naval photoradio was Lt. Commander Stanford C. Hooper, head of the Navy's Radio Division.[194] The Naval Research Laboratory funded independent inventor C. Francis Jenkins, who transmitted photographs from a navy radio station to his Washington, D.C., laboratory in May 1922, and research by Westinghouse and other corporations.[195] Although peacetime budgets pinched funding, Hooper's office hoped "a keen interest on our part helps to keep such developments progressing more rapidly."[196]

American naval engineers envisioned facsimile providing rapid and accurate fleet communications while and reducing the labor- and time-intensive demands of telegraphy. Automatically faxed messages would eliminate hundreds of the skilled but all-too-human radio operators and save time spent encoding, decoding, and, in the case of weather maps, reconstruction. Hooper hoped to reduce the time for map transmission and reconstruction from ninety to seven minutes.[197]

Of twenty-one naval radio research problems in 1926, "radio television and photograph communication" ranked sixteenth. Highest priority was transmitting orders and information quickly and reliably among formations and within ships, a function not considered for facsimile.[198] In 1926–28, tests with Jenkins and RCA equipment on ships and the airship *Los Angeles* proved, according to Lt. Commander Joseph R. Redman of the battleship *Texas*, "not worth the time and effort."[199] Experiments with cruisers in 1931 confirmed that appraisal.[200]

The more the Navy learned, the less attractive fax appeared. The complex equipment's demanding synchronization, susceptibility to vibration (including gunfire, an occupational hazard on a warship), slow speed, and high bandwidth all argued against it. As a 1938 report concluded, fax equipment proved so "susceptible to trouble from so many different sources [that it] can hardly be considered thoroughly reliable, the primary requisite of any basic military communication system."[201]

The army independently reached similar pessimistic conclusions. Although in early 1920s the Signal Corps envisioned a bright future for faxing images by radio, the wealthier Air Service first experimented, attracted by faxing images from an airplane to a ground station for rapid reconnaissance.[202] Aerial faxing proceeded in stages. In October 1925, an Army plane photographed a target at Ft. Leavenworth, Kansas, developed the film, and dropped it to soldiers on the ground, who relayed the photograph by an AT&T PT transmitter to New York thirty minutes after the image was snapped, stirring much excitement in the Air Service. As tests in 1930–32 demonstrated, faxing from aircraft proved much more challenging due to weight and size limits and problematic radio transmission.[203]

Telegraphy deterred more prosaic but technically easier applications of faxing.[204] Like PTTs, the military prudently pursued multiple paths to reach its goal of faster automated and error-free messages. Navy engineers pioneered teleprinter radio transmission in 1922 and encouraged the development of teletypewriters, as did the Army Airways Communications System. The military's eager embrace of teletypes, integrated into existing operating procedures, reduced the attraction of facsimile.[205]

⌁

Judged strictly by numbers, facsimile was a minor technology. Less than a thousand transmitters and receivers existed in 1940 (excluding the failed fax broadcasting receivers). Their impact was greatly out of proportion to their numbers, however, because they enabled newspapers to print the latest photographs with the latest stories, visually transforming the news and strengthening the role of photographs in newspapers.

This truly was a golden age of enormous accomplishments, of facsimile entering daily operations, some even profitable, and becoming integrated into corporate, state, and international structures. By 1937, Western Union engineering vice president Ferdinand d'Humy stated in internal correspondence, "Without reservation I foresee *FACSIMILE* as a prime mover for transmitting the 'written word' by telegraph. This transition is inevitable. It is only by the aid of this means that Western Union will be able to regenerate an ascending place in the field of communication." D'Humy envisioned Western Union subscribers directly dialing and faxing one another instead of going through human operators.[206] Unfortunately for Western Union, World War II delayed the realization of those goals until the 1950s, too late for faxing to change the firm's decline.

Yet these years were also an age of under-fulfillment, of numerous unsuccessful attempts to commercialize facsimile in a dazzling range of potential applications. Even as PT became an integral part of newspapers, promoters pushed the facsimile into new markets. Often, like fax broadcasting and sending documents, potential customers supported these efforts, providing a friendly site for experiments. Yet as a rule, these new applications failed, victims of unfavorable technology and economics.

The emergence of new problems—providing products that fit office procedures and needs—was a sign of progress and promise. Prewar equipment would not have fit in a newspaper office, let alone a business office. That potential users as well as fax promoters considered placing 1930s equipment into their offices was a gigantic expansion in the conceptual, if not actual, diffusion of fax technology.

Why did so many efforts fail? Imagination was not the problem. Nor can a common shortcoming with new technologies, the failure to think in economic terms, be charged. Technological prematurity, however, played a large role. The machines worked, but they performed below technical expectations or user competency, especially in terms of operations and maintenance.

While fax technology improved greatly, competing technologies still proved faster, more reliable, easier to operate, and less costly, and they provided higher-quality output. Whether 1919, 1929, or 1939, the telegram retained overwhelming advantages of speed, installed infrastructure, and lower costs. Until facsimile matched those advantages or found niches where telegraphy did not compete, faxing remained a solution looking for a problem.

CHAPTER 3

Facsimile, 1939–1965

The atomic bomb and facsimile, being demonstrated by *The Miami Herald*, are symbols of the new age.

—*E. W. McAlpine, Australian Consolidated Press editor-in-chief, 1947*

When hailing the *Miami Herald*'s fax newspaper broadcasting in 1948—"It will educate, entertain and enlighten, it will increase thinking and reflection, we hope; it will take man from the slums of living to the heights of luxury-living"— the rival *Miami Beach Sun* envisioned peace, not war.[1] The technology that made the *Herald*'s faxing possible, however, rested on wartime advances. World War II proved to be a turning point for fax, shifting the technology to a faster evolutionary path. Like the citizens who donned uniforms, facsimile entered World War II as an amateur, filled with potential, and emerged as a veteran, confident of its place. By war's end, the American military had purchased hundreds of fax machines and become the largest patron and user of faxing, a role it retained through the 1960s.

Market demand began to pull fax applications, though still significantly less than technology push. Experimentation and multiple efforts at commercialization marked these decades as manufacturers and potential users again tried to turn fax from an expensive, specialized technology into an economically affordable, widespread service. As this chapter shows, these rising expectations and experiments with message faxing by manufacturers, telephone companies, large users, and the International Telecommunications Union (ITU) constituted an unprecedented interest in expanding faxing beyond its immediate postwar markets. While technological leadership remained in the United States, a vibrant Japanese fax industry also emerged in these decades.

Most impressive was what did not happen, as technology push failed to translate into market pull, best illustrated by faxing's inability to take advantage of wartime opportunities, the unsuccessful 1944–48 attempt to again commercialize fax newspapers, RCA's 1949 Ultrafax, AT&T's abortive 1957 service, and the Post Office's Speedmail in 1959–61. Less obvious were the many experiments and attempts at commercialization, ranging from weather reports to cotton farmers in Tennessee to transmitting X-rays from rural doctors to a Philadelphia hospital.[2] The many failures represented strides forward because they reflected attempts by potential users to pull the technology forward.

The technological challenges—high equipment and operation costs, poor reproducibility, and slow speed—remained daunting. Yet the technology did change. For the United Nations charter meeting in 1945, the *New York Times* faxed a 4-page special edition to San Francisco. Eleven years later, the newspaper faxed a 10-page edition to the Republican National Convention there. One page without photographs demanded 34 minutes in 1945 but only 2 minutes with photographs in 1956.[3]

The major markets remained transmission of images—photographs and maps—but increasingly a new market developed for the transmission of documents. A growing number of customers appeared whose need for quick delivery of information with 100 percent accuracy—orders to bomb the Soviet Union or handle a power plant—outweighed the higher cost of acquiring and operating fax machines compared with telex. Other early applications, usually by large organizations, employed fax to provide more efficient use of limited resources like library books and military weather maps by centralized distribution.

Additional indicators of progress appeared, such as commercially successful fax-based duplicators like the Winchester Industries' Faxcoa and Gestetner Gestetfax, and changing technical indicators like measuring speed in minutes per page instead of square inches per minute.[4] The 40,000 Western Union Desk-Fax machines gracing office desks in the United States, Britain, and Cuba provided the most radical demonstration of faxing's potential. Some organizations even used faxing to rethink how they communicated. When longtime fax developer Austin Cooley proclaimed in 1954, "Facsimile now is where the teletype was in 1918," he meant fax's best days were yet to come.[5]

WORLD WAR II

The war represented a tremendous opportunity for fax, providing it with a market: the military, whose rapidly expanding needs overwhelmed normal economic concerns. Faxing offered nearly real-time transmission of maps, drawings, and photographs, three vital components of the American military's information flow. This opportunity was only partly realized, however, because the fax industry proved unable to manufacture reliable machines on schedule. Delayed delivery of capricious equipment increased opportunities for competitors, especially teletype, to satisfy military needs instead of facsimile.

Nonetheless, World War II accelerated and reshaped faxing, halting civilian development while redirecting resources toward military needs and ensuring American dominance. Two immediate casualties were European public picture telegraphy and Western Union's Telefax expansion.[6] In America, the Army's Signal Corps Laboratory received responsibility for facsimile development on July 10, 1940, becoming the technology's main patron and customer (the Navy independently developed and deployed machines), a role that continued after the war.[7]

Facsimile encountered the normal problems of mobilizing a technology for war. The Signal Corps had to define its needs, choose equipment, ensure its compatibility with other systems, contract its manufacture, provide adequate maintenance and spare parts, train operators, educate users, and fend off uninvited solicitations and inventions.[8] So uncertain was the Signal Corps of military demand for faxing that it asked suppliers in 1942 and 1943 to quote costs for production runs ranging from ten to a thousand units.[9] Three machines appeared: the RC-120 to transmit 7 × 7 3/8 inch images in 7 minutes, the AN/TXC-1(a modified RC-120) to send weather maps up to 12 × 17.5 inches in 20 minutes, and the RCA RC-58B, which transmitted a message on a narrow paper tape at the equivalent of 42 words per minute.[10]

In March 1942, the Signal Corps contracted with Times Telephoto Equipment (TTE) to supply 300 RC-120 units. Production proved distressingly slow. The last unit arrived only in July 1944, more than a year and a half behind schedule. Alden Products Company, a Massachusetts electrical equipment manufacturer, which received a second-source production contract because TTE was so late, and RCA also lagged behind schedule.[11] One cause, common to many wartime contracts, was shortages of components as sophisticated as photocells and as basic as wire.[12] More problematic for faxing was moving from customized to large-scale production of the "extremely intricate mechanical assemblies" due to a lack of manufacturing experience, equipment not designed for easy production, and an inadequate work force. TTE never produced more than seven machines a week and averaged three, each unit "hand fabricated, individually adjusted and nursed into operation."[13]

Facsimile's inadequacies did not end with production. Equipment proved fragile and unreliable, demanding significant maintenance—assuming it reached its users. Losing equipment during shipping proved to be a surprisingly serious problem, a problem that was minimized by sending complete sets as one shipment.[14] Operators had to be educated in the equipment's proper (and demanding) operation, maintenance, and repair, so the Signal Corps established a training school.[15]

Most importantly, the military had to figure how to best incorporate faxing into its operations. Prewar experiments such as the 1938 Second Army exercises at Fort Knox, Kentucky, and larger 1941 Louisiana maneuvers provided some appreciation of faxing's capabilities, but the main applications appeared elsewhere.[16] Sending weather maps and reconnaissance photographs proved to be facsimile's most important direct military contribution. Starting in 1942, different commands tested facsimile, with the Air Weather Service and Anti-Submarine Command the most enthusiastic. In June 1942, the Signal Corps began faxing submarine situation maps daily for the Anti-Submarine Command.[17] In late 1942, the Weather Service began faxing weather maps to the 8th Weather Squadron at Presque Isle, Maine, the departure base for bombers ferrying to England. This marked the start of an extensive effort to fax current weather maps to airfields. By 1945, a ten-station network connected military meteorologists in Britain, France, and Germany.[18]

Not all expected applications proved feasible, especially air-to-ground faxing. Testing of different equipment from 1939–42 subdued military interest because of the slow speed, low resolution, and poor synchronism of transmitted photographs.[19] Instead, a reconnaissance aircraft had to return to base to have its photos developed and then transmitted by fax or messenger.

Military concern about confidentiality generated research into fax encryption, principally in prewar France but also in wartime America and Japan, until secrecy, technical challenges, and higher priorities slowed this area.[20] Time, not technological troubles, halted another application. The war ended before the 200 portable fax transmitters Westinghouse developed for the Office of Strategic Services for covert operations saw service.[21]

Faxing's most public contribution was accelerating the delivery of photographs from the Office of War Information and the military to newspapers, providing the home front with pictures to accompany the latest news from overseas.[22] One unexpected consequence, ending the debate about the role of photography in newspapers, was that the war confirmed news photography as a legitimate and increasingly essential part of journalism.[23]

Those photographs flowed over Army and Navy wirephoto systems. In June 1942, the Army Communications Service established a domestic network, and in March 1943 it started radiophotoing pictures from Africa and Europe to the

United States for newspaper distribution.[24] In July 1944, the Navy Public Information System introduced its more technically challenging radiophoto service in the Pacific.[25] By war's end, the Army Communications Service had faxed 11,533 images. Faxing, however, remained a minor component of military communications. In August–September 1944, 736 telephotographs flowed between the European theater and the War Department, but so did 357,000 teletyped messages.[26]

American fax machines served the Canadian, British, and Russian militaries.[27] For China, Western Union supplied the Army with $90,000 of Telefax equipment, enough to equip eight cities, to avoid the problems of converting Chinese into telegraphic code and back.[28] Like other militaries, the Luftwaffe relied on teletype and telephone for internal communications, though some faxing of weather information between weather centers and to the air staff occurred. A peacetime lack of interest in tactical reconnaissance meant little promotion of airborne facsimile, which did not change during the war.[29] Japanese navy interest did not extend beyond research. In contrast, the army had used phototelegraphy for propaganda in Manchuko in the 1930s and, in 1943, introduced a document service on its Tokyo–Manchuko circuit, the world's first dedicated service.[30] Japanese defeat and postwar occupation stopped that application.

The military never viewed facsimile by itself, but only as part of a larger system of communications that included teletypewriters, telephones, switchboards, and radios. Teletype networks actually deepened their role as the backbone of rapid, written communications, to the detriment of facsimile. In the British army, teleprinters expanded from 181 in 1939 to more than 11,000 in 1945.[31] Even the telautograph, rejected by the prewar Air Corps, found non-combat applications that further reduced opportunities for faxing.[32] Teletype's dominance continued after the war. In 1960, the Air Force's thirteen domestic teletype networks made it Western Union's single largest customer, with 250,000 wire links; the largest private user, General Electric, had 87,000 links.[33]

POSTWAR TECHNOLOGICAL DEVELOPMENT

American military involvement in fax expanded after the war, becoming the world's largest market and research patron through the 1960s. The Signal Corps and Navy developed and deployed hundreds of units ranging from lightweight field transmitters to high-speed equipment that faxed a 32×40-inch weather map in 20 minutes.[34] The military did not develop facsimile in isolation from the wider fax community but gave contracts to every major fax firm and kept abreast of their activities and interests. With few exceptions, these contracts were unclassified.[35] Although most contracts went to component and system development, theoreti-

The United States Army sponsored the development and construction of fax
machines ranging from this portable model to models capable of handling large
weather maps. U.S. Army photograph

cal research also advanced.[36] These contracts kept firms funded if not flush. Without this military involvement, the fax world would have been much smaller and
would have evolved more slowly.

From war's end into the 1960s, fax equipment did not change significantly, but
researchers did establish the groundwork for the revolutionary changes realized
in later decades. The major challenges in 1945 remained much the same as in
1965: cost, speed, reliability, and recording quality. The great precision demanded
in building electromechanical systems meant high acquisition costs. Dependabil-
ity demanded careful, costly, and often extensive maintenance. At each munici-
pal airport its pilots used, the Air Force had to station several trained technicians
to maintain its fax weather recorders, an expensive use of personnel.[37]

Equipment and operations therefore remained expensive, restricting interest and application. Transmitters and receivers sold for $1000–$8000 each, with
document systems at the low end and higher quality weather map systems at the
high end.[38] Predicted cost reductions proved as illusory as anticipated markets. In

1952, Hogan Laboratories expected its redesign to reduce prices to only $1000–$1500 for a transmitter and $300–$400 for a recorder. In 1954 its non-synchronized transmitter and recorder actually cost $3000, with another $1000 for synchronizing equipment.[39] The firm's ever overly optimist Elliott Crooks predicted in 1954 that faxing would become economically competitive in 1955 with office copying because tens of thousands of fax machines would provide rapid, accurate, written communications from desk to desk, office to office, and plant to plant.[40]

Reliance on the vacuum tube, "the wheel horse of progress in the telegraph industry," meant the equipment remained large enough to require separate transmitters and receivers.[41] Western Union's 1951 Telecar needed a station wagon to carry the radio, facsimile receiver, and auxiliary equipment. The transistor promised lower power consumption, smaller size, lighter weight, and greater reliability: Transistors reduced the weight of NEC's 1963 FT-68T phototelegraph transmitter to 31 pounds, a savings of 9 pounds, as well as increasing its operating time and ease.[42] As early as 1954, the Signal Corps promoted transistors for faxing, so transistors appeared in military equipment years before civilian equipment.[43]

In its 1945 postwar research goals, the Signal Corps reflected the frustrations of its wartime experience and the long-term hope for the facsimile community when it stated, "The ultimate in facsimile scanning of an original copy is the elimination of mechanical moving parts."[44] This meant electronic digital, not electromechanical analog scanning. Analog machines scanned area, not data. The varying amplitude of the continuously transmitted signal contained the information about the white, black or grey of the scanned line. Digital scanners generated discrete data, dividing each line into hundreds of cells. The more cells scanned, the finer the resolution with the tradeoff of transmitting more data. Because this data was digital, it was far more amenable to manipulation and compression than analog data. Compression shortened transmission time by reducing waste (white space) and redundant data. The latter was the segment of an individual letter or image that repeated what was above and below it.

The "considerable application" anticipated for digital faxing prompted military funding of both its theoretical and empirical foundations.[45] BTL and RCA researchers used Claude Shannon's seminal 1948 theory of information in the mid-50s to maximize information flow. Other researchers statistically represented scanned data—especially David A. Huffman, whose seminal code in 1952 formed one leg of the foundation of the game-changing 1980 CCITT G3 standard.[46]

Seeking faster sending, military funding paid for research into increasing transmission speeds over telephone and radio channels by increasing bandwidth and channel capacity. More significantly for the future of fax, the Signal Corps also funded investigations into reducing the amount of data that had to be transmitted for a coherent message. This research included white-line skipping, variable velocity scanning, and run-length coding, which statistically analyzed one scanning line to predict the content of the next.[47] The commercial payoff for this research did not occur until the 1970s–80s.

Despite their theoretical attraction, electronic scanning and data compression were, in the words of BTL, "challenges to our imagination"—challenges 1950s electronics technology proved incapable of meeting.[48] By the mid-1960s, a robust and growing base of research had evolved, albeit still generations of coding and computer technology too early for practical and affordable application. The supporting software and technologies—especially signal processing and efficient coding algorithms—did not exist.[49]

Telephone circuit quality continued to bedevil facsimile, limiting its attractiveness and raising its costs. The postwar decades were a period of major expansion of telephone networks and advances such as direct dialing of long-distance calls instead of using operators, but there was also growing frustration about quality. Compared with the 1920s–30s when picture telegraphy developed, the telephone system was considerably larger and electrically more complex. As new generations of users discovered, facsimile was more sensitive than voice and less forgiving than a person. Bell engineers had "conditioned" high-quality circuits to reduce blurring, streaking, and smearing during wirephoto transmissions. Even so, measured by the number of complaints, wirephoto was one of the most troublesome services AT&T offered in the 1950s.[50] Ominously for advocates of business faxing, BTL estimated in 1956 that 75 percent of regular lines were inadequate even for local faxing.[51]

Even the military could not afford to condition every circuit, so the Signal Corps funded research to measure and correct signal distortions. The result was greater theoretical and practical understanding of circuits by the facsimile and telecommunications communities.[52] Research on facsimile over radio circuits included bouncing fax signals off the ionized trails of meteors entering the earth's atmosphere.[53]

Expanding bandwidth offered an alternative to the technically challenging, economically daunting approach of compressing data. Building from military and civilian research in the 1930s and the communications demands of World War II, microwave technology promised transmissions of scores of faxes—or hundreds of telegrams or a television channel—using high frequency radio transmissions. The promise was not realized. In 1944–45, several corporations, spearheaded by Philco and Raytheon, applied for FCC permission to build experimental microwave transmission systems with the goal of commercial transmission of television, facsimile, telephone calls, and data. In 1945, the FCC decided against encouraging this competitive market, essentially giving AT&T a monopoly and protecting its investment in wire circuits. This restricted development of high-speed, high-capacity microwave transmission delayed an opportunity for fax.[54]

As before the war, telegraphy remained facsimile's main competition. The New York branch of the Federal Reserve, for example, connected with fifteen banks in Western Union's Intrafax service in 1952, but the Federal Reserve deployed a nationwide teletypewriter system in 1953 to handle money transfers.[55] Internationally, telex extended its reach with new services and circuits like the Associated Press's eight financial teletypesetter services, which produced copy ready for typesetting in 1961.[56] An AT&T engineer concluded in 1954 that compared with the teletype, "facsimile is inherently a less economic type of service, is less efficient in terms of the utilization of the frequency spectrum, and has other relative disadvantages."[57] Six years later, RCA still considered facsimile commercially unrealistic because of its cost and the firmly entrenched teletype.[58]

But teleprinters and teletypewriters had disadvantages. Overall, domestic American telegrams rose from 189 million in 1939 to 236 million in 1945 but declined to 124 million in 1960.[59] The message consisted of upper case characters only and allowed neither signatures nor stationery, making normal correspondence impossible. Teletype demanded trained operators; otherwise error rates were high. If transmission time included all preparation, particularly error-checking, faxing often outperformed teletype. In one study, teletype sent an order in 35 seconds compared with 2 minutes via faxing, but including all the preparation time changed the numbers to 22–25 and 10–15 minutes respectively.[60] Consequently, a teletype engineer expected in 1955 that faxing "will form an increasingly prevalent competitor."[61]

FAX NEWSPAPER BROADCASTING

Even before the war ended, a second generation of fax broadcasting began in 1944 based on FM not AM broadcasting and on wartime advances in electronics. The goals were faster transmission (fifteen minutes versus two hours for four pages) and multiplexing (transmitting sound and fax simultaneously over the same channel). The organizational foundations included private and government efforts to create standards, a consortium of radio stations and newspapers to finance development, and a new entrant with a radical conception of fax broadcasting. Missing from this renewed effort were RCA, whose commitment to television reduced its desire to champion FM radio, and William Finch, who licensed his equipment and in 1947 helped install his equipment in France but focused his attention elsewhere.[62]

The new actor was Milton Alden, who founded Alden Products Company in 1930 to manufacture and sell electrical equipment, primarily recording devices. In 1946 he established Alfax Programs, Inc. to expand his facsimile operations from his wartime manufacturing.[63] Obsessed with improving communications, Alden viewed fax broadcasting as a targeted information service, not a newspaper. Doubting the financial feasibility of transmitting newspapers, Alden envisioned a more encompassing product that would demand "new art in programming, and will engage the best imaginative brains of the type that produce women's magazines or the feature sections of papers."[64]

FM fax broadcasting offered three significant advantages over AM: static-free transmission, daytime operations, and higher transmission speeds. Its disadvantages were a commercially immature technology, an FCC requirement that fax not interfere with sound broadcasting, and a larger corporate battle among AM, FM, and television that delayed FM's commercialization.[65] The FM spectrum was largely unplumbed, meaning that, unlike AM, FM lacked a large base of stations and receivers—and thus actual listeners and potential advertisers. In 1948, FM stations reached only an estimated 1.6 million receivers compared with 37 million AM receivers.[66] Receiving FM broadcasting meant buying an FM radio as well as a fax receiver, but the incentive to buy was minor. The 1945 FCC frequency reallocation, the refusal of the American Musicians Union until 1948 to allow simultaneous AM-FM broadcasts, and industry's bias toward duplicating AM broadcasts instead of producing new programs for FM slowed FM's anticipated growth.[67]

Television attracted far more commercial interest. Since the mid-1920s, the development and growth of television absorbed significant resources that did not go into FM, let alone fax broadcasting. As early as January 1947, *Editor & Publisher* reported "advertisers are scrambling to get into television and overlooking FM altogether." From 1946 to 1950, radio revenues grew from $323 to $443 million, but television revenue soared from $1 to $106 million and surpassed radio in 1954.[68]

Multiplexing promised continuous fax and regular sound broadcasting. Edwin Armstrong, "the father of FM radio," had demonstrated multiplexing in November 1934 with the help of RCA engineers, including Charles Young, but RCA tests in 1941 from the Empire State Building unambiguously rejected it from both an engineering and commercial perspective, an opinion unchanged after the war.[69] Instead of waiting for multiplexing to mature, the few FM stations faxing alternated with sound to broadcast daytime "flash news."

Market demand came from radio stations and newspapers. In September 1944, Theodore C. Streibert, president of WOR in New York, and John R. Poppele, vice president for broadcasting, asked John Hogan's Radio Inventions, Inc. to design scanning equipment and receivers. Two months later, Hogan and Streibert organized Broadcasters' Faximile Analysis (BFA) with six radio stations and newspapers to develop a commercial fax broadcasting system. By September 1945, BFA and its subsidiary, the Newspaper Publishers Facsimile Service, had twenty-five members who subscribed $125,000, matched by Hogan. The money was needed because development proved far more demanding and costly than Hogan initially envisioned. The initial $50,000 budget for two scanners and five recorders grew fivefold, and the schedule for full-fledged tests was extended from 1945–46 to 1947.[70]

General Electric's Transmitter Division manufactured the equipment. Minimizing its efforts, the firm combined its standard 417-A AM-FM radio with Hogan's 8-inch-wide fax recorder, which occupied the space for the phonograph. The complex receivers necessitated so many changes that newspaper engineers had to virtually rebuild them.[71] Nor was reengineering the only problem caused by General Electric. The firm's commitment to BFA recorders did not run deep; their production was a low priority amidst hundreds of internally developed products and a corporate reorganization.[72] Nonetheless, by spring 1948 stations had purchased 12 transmitters and 300 receivers, and electrical manufacturer Stewart-Warner offered to fill orders when demand warranted.[73]

In 1947–48, several newspapers demonstrated fax broadcasting to the public and to important groups like the National Association of Broadcasters and the Advertising Club.[74] The two strongest newspaper supporters were the *Miami Herald* and the *Philadelphia Inquirer*. Under manager Lee Hills and facsimile editor Timothy Sullivan, the *Herald* invested nearly $150,000 in promoting fax broadcasting, including the world's first facsimile journalism course at the University of Miami.[75]

The *Herald*'s enthusiasm and commitment partly came from the newspaper's admitted lag in adopting other technologies like color printing. Fax broadcasting was an opportunity for Hills, who had joined the paper in 1942, to build on his reputation as an innovator. Less-heralded but important corporate considerations were delaying an expensive expansion of the printing facilities and circumventing its typesetters union.[76] Postwar technological enthusiasm contributed too. Edgar H. Felix, who had directed AT&T's first public fax demonstration in 1924, reported that "the public viewed facsimile as one of the new wonders of the world and as evidence of the magic of radio progress."[77]

The *Herald* and its WQAM-FM began promoting fax broadcasting aggressively in March 1947, publishing thirty-seven editions in two weeks. A standard performance was to photograph a luncheon, race to the paper to develop the photos, and then transmit them to the wonderment and approval of the audience.[78] The paper performed before business, civic, and educational groups, including Booker T. Washington High School, where it was "arranged in answer to requests from leaders in the negro section that their people be given an opportunity to see these initial demonstrations of 'Tomorrow's Newspaper Today.'"[79] Tests by the *Inquirer* in Philadelphia sparked similar attention.[80] Emboldened, the *Inquirer, New York Times*, and *Herald* began broadcasting multiple editions daily in late 1947 and early 1948 to hotels, banks, restaurants, and other businesses with a large clientele for public viewing.[81] Two decades after the first demonstrations, fax broadcasting finally seemed commercially viable.

Meanwhile WPEN, owned by the *Philadelphia Bulletin*, conducted a more unconventional experiment with Alden's 4-inch-wide recorder. Believing fax broadcasting was far more versatile, Alden sacrificed a newspaper appearance for a recorder that was much less expensive ($100) than the $800 General Electric machines.[82] In spring 1947, the *Bulletin* demonstrated Alden's system in a window of Gimbel's department store with a printer that enlarged the 4-inch paper to 16 inches for reading from the street.[83] In an example of centralized broadcasting sending information to many geographically dispersed users, WPEN broadcast weather reports to local airports and produce prices to farmers' markets. Engineer Robin D. Compton considered that the tests' high point was an airport in a Mennonite area where "we would find horses and buggies and horses and wagons and sun bonnets and little flat hats tied up alongside airplanes."[84]

In March 1948, the FCC held hearings to determine whether to allow commercial fax broadcasting. Advocates claimed that fax broadcasting had sufficiently matured technologically to produce profits while providing a public service. The 1947–48 experiments had demonstrated public interest, and advertisers seemed willing. Elliott Crooks, director of Radio Inventions' sales and service, predicted that a year after FCC authorization would find 100 fax publications, ranging from small audiences with less than 10 receivers to stations serving 10,000 receivers.[85]

The main dispute concerned whether to mandate a specific page width or a more flexible index. The BFA, following the interests of the radio stations and newspapers, wanted a fixed width of 8 inches for a 4-column newspaper. The forty-seven members of the panel on facsimile of the Radio Technical Planning Board (modeled in 1942 after the prewar National Television Systems Committee to create a postwar vision and framework for radio) preferred a more flexible index. Using the analogy of television, chair John Hogan noted that programs were not transmitted specifically for an 8-inch or 10-inch tube, but for any size. The issue was technical, but the consequences strategic: mandating the 8-inch width would foreclose Alden's 4-inch broadcasts.[86]

Effective July 15, 1948, the FCC authorized commercial fax broadcasting, but only for the 8-inch width, not the more flexible standard. Its rationale was ensuring full interchangeability of all types of receivers and broadcasting a newspaper not a bulletin.[87] Fax broadcasting had received the official go-ahead. And—in the opinion of Milton B. Sleeper, the editor of FM and Television—its death warrant. Sleeper had warned that the 8-inch system was too expensive, not only compared with the 4-inch system but also with television.[88] He was right. Within months, the dream of commercial fax broadcasting had died. The hundred publications had failed to materialize; indeed, newspapers soon donated their expensive equipment to journalism schools.[89]

People had flocked to demonstrations, but they refused to follow with their money. Contrary to expectations, news proved a fungible, cost-sensitive commodity, adequately served by newspapers and AM radio. Unfavorable finances overwhelmed everything else. Hogan estimated that production runs of 100–200 would reduce the $800 recorder cost to $250 and that mass production would further drop the cost to $100.[90] By comparison, $445 bought a Du Mont television with a FM radio; $60, a FM radio; and a few cents, a newspaper.[91]

The emphasis on reducing receiver cost was not misplaced. An Inquirer poll of 945 people who viewed its public demonstrations found that 74 percent expressed interest in a home receiver, but their commitment faded rapidly with price: 64 percent would pay $100, 31 percent would pay $200, but only 6 percent would pay $500.[92] The mass production needed to establish a market never occurred because the market did not exist.

Broadcasting facsimile newspapers failed, but radiofaxing did find one commercial use: notifying newspapers in 1946 of incoming shipping off the New Jersey shore. Instead of observations telegraphed from a Sandy Hook tower, pilot boats faxed the name and time of arriving ships.[93] However, while this was a successful niche market, it offered only very limited opportunities for growth.

ULTRAFAX

Far more spectacular than fax broadcasting—and equally unsuccessful commercially—was Ultrafax, unveiled by RCA on October 21, 1948, at the Library of Congress before several hundred guests. In 141 seconds, Ultrafax sent and received the 475,000 words of Gone with the Wind. More practically, RCA also transmitted

military, government, and business documents. The huge information capability of Ultrafax and television stirred RCA president David Sarnoff to suggest an
Atlantic "'airlift'" of 12–14 aircraft to relay television and Ultrafax signals between
America and Europe as well as more likely markets of military communications
and a network of high-speed circuits linking cities.[94]

Part of a larger RCA development of high-speed transmission and reproduction of information, and utilizing RCA's television expertise, Ultrafax scanned by
cathode ray tube (CRT), a gigantic step beyond the photocell. Despite the impressive technology and numerous demonstrations to government agencies and
corporations, Ultrafax, like the four other RCA CRT fax projects, failed to find
applications. Even a Soviet request to license Ultrafax, rejected by Secretary of
Defense James Forrestal on grounds of national security, failed to convince the
American military to deploy Ultrafax.[95] Despite high hopes, none of the RCA
systems evolved beyond the prototype stage.[96]

The machines suffered from a mismatch of subsystems because some components proved too advanced compared with others. Ultrafax and its offshoots
scanned quickly, indeed, too quickly for the preparation, transmission, and receiver technology available. To use the 1949 Ultrafax, the material had to be
photographed first. The other CRT models of the early 1950s either required
preprocessing or, in the case of the standard-speed fax, complex corrective equipment to keep the CRT focused at such a low speed. Not by chance, Ultrafax's
sponsors were the military and the Atomic Energy Commission, whose needs
and resources allowed more speculative experimentation with less attention to
economics.[97]

The failure of Ultrafax and its other CRT machines ended RCA's active fax
program by 1952, reducing its efforts to "theoretical and study work" for commercial applications, according to Charles J. Young.[98] RCA Laboratories did develop
a method of electrostatic printing in 1951, Electrofax, which the firm successfully
licensed for photocopying and other reproductive applications, the only profitable result of its decades of fax efforts.[99]

MANUFACTURERS AND MARKETS

In one important way, faxing became a normal business as firms entered and left
this market. In 1952, Winchester Industries acquired Finch Telecommunications'
facsimile activities, only to sell them a few years later to Air Associates, an aircraft
and communications manufacturing firm. Air Associates changed its name to
Electronic Communications, Inc., developed the "Electronic Messenger" transceiver (marketed by Western Union after AT&T declined), and itself was bought

by National Cash Register in 1968.[100] Other firms considered entering facsimile but declined, fearing—accurately, as it turned out—the lack of a market.[101] The fundamental business problem was that the postwar fax market had too many firms with too little money chasing too few orders from too few customers.[102]

Times Facsimile Corporation and Hogan exemplified the challenge of creating and maintaining markets. Signifying its desire to expand beyond weather and wirephoto systems, the Times Telephoto Equipment Company became the Times Facsimile Corporation (TFC) in 1947. TFC explored projects ranging from a portable X-ray scanner to converting railroad telegraph terminals to facsimile, but the military remained its main market.[103] So close were its relations with the Air Force that the service wrote its 1958 specifications for a new weather facsimile system around a TFC machine.[104]

The company spread itself too thinly, developing a range of equipment from a 2-inch-wide recorder to a high-resolution 12 × 18 inch weather recorder. The last began production in 1954, seven years behind schedule.[105] Such delays were largely self-inflicted. Its manufacturing had not improved greatly since World War II. Inefficient production was worsened by bidding below costs for initial production contracts, hoping for more profitable follow-on contracts. The potential for disaster was great because manufacturing a new product became, to some extent, a leap into the unknown. As Cooley lamented in 1957, "We are not smart enough to avoid all the bugs in a new design."[106]

As early as 1952, the *New York Times* considered selling TFC to avoid the responsibility for its management and financing. An important consideration was finding an arrangement acceptable for Cooley, the engineer who had put the newspaper into the forefront of wirephoto.[107] Between 1952 and February 1959, more than a dozen firms considered buying TFC before Litton Industries purchased it for $1.25 million.[108]

John Hogan's Radio Inventions, Inc. became Hogan Laboratories, Inc. after the war. Fax research continued under his son, John Hogan Jr., but shifted into instruments and materials for more profitable high-speed data plotting and recording. Its FAXpaper enabled wire services to directly print wirephotos instead of using a darkroom. In 1956, Hogan Labs licensed its technology to Stewart-Warner, which also acquired the fax business of the Allen D. Cardwell Electronics Productions Corporation (a Hogan licensee) and, with its new Datafax systems, became a major fax manufacturer. By 1960, Stewart-Warner had a quarter of the fax market, while Hogan had left the field.[109]

Even as the number of fax machines grew into the thousands, the main postwar markets increasingly became discrete specialized sectors whose requirements increasingly diverged from each other and from the slowly emerging business market. Building on wartime demand, the Weather Bureau inaugurated its national faxing system in 1946, while a separate Air Force fax network linked approximately one hundred domestic bases with 15,000 miles of rented circuits.[110] Over time, these networks grew, reflecting upgrades, expansion, and greater need. The Strategic Air Command's Strategic Facsimile Network leased 12,000 miles to connect fifty-seven air bases in 1959, while the Air Force installed 217 additional recorders at 141 stations to receive Weather Bureau maps in 1960.[111]

Weather maps expanded beyond the government, thanks to an Air Force and Weather Bureau policy allowing other users, such as airlines, to tap into its networks. But here, as elsewhere, teletypes dominated. Even while accounting for 75 of the 270 fax machines receiving Weather Bureau faxes in 1955, nongovernment meteorologists also employed 1,100 teletypes to receive Bureau reports. Faxed weather maps attracted airline attention: United Airlines employed its own fax system in 1946 to connect its dispatch center in Cheyenne, Wyoming, with Lincoln, Nebraska, and Denver, Colorado, its two transcontinental refueling stops. By 1951, Weather Bureau faxes allowed ten airlines to centralize their weather staffs, improve flight plans, and brief pilots.[112] A decade later, space exploration created a market for photofacsimile recorders to receive images from Tiros and later weather satellites. Even more than weather receivers, these very expensive receivers diverged from document faxing.[113]

For newspapers, the major developments were electrolytic recording as a lower cost (and lower quality) alternative to wirephoto and transmitting layouts for quicker and more distant publishing. The lower cost increased the number of papers that could afford to receive photographs by fax instead of by mail and thus publish the latest photos together with the latest news. Television stations, also anxious for visual material, bought wirephoto and regular fax machines to receive photographs, starting with NBC's "Today" show in January 1953.[114] By 1955, 128 television stations and 81 newspapers operated facsimile recorders, a number that increased to 500 in 1956.[115]

Radio-transmitted wirephotos failed to find a niche. Press Wireless started developing the technology in 1953 to reach small newspapers. It actually tested transmissions from New York City to the Rocky Mount, North Carolina, *Telegram*. By 1955, the firm had placed twenty-five radio receivers in Latin America. The North American market remained closed since the wirephoto services proved reluctant to lose telephone-networked customers to radio.[116]

As early as 1945, saving two to three days of delivery time by faxing newspapers and magazines from New York to Chicago and the West Coast had attracted the interest of publishers.[117] While this proved impractical, faxing copy to printing plants promised to save time and extend a paper's geographic reach. *Time* and *Life* worked with RCA in 1947 to test transmitting layouts for their magazines from New York to Chicago. Nothing came from those tests or the 1956–57 AT&T survey of possible demand and simulation via a wideband New York–Dallas–New York circuit of sending a *New York Times* layout to London.[118]

Although American newspapers experimented with the concept, the *Asahi Shinbum* introduced the world's first operational service with British Muirhead equipment in June 1959, transmitting pages from Tokyo to Sapporo. The 1964 Tokyo Olympics spurred other Japanese newspapers to follow.[119] Faxing these complete layouts demanded expensive high-resolution scanning and wideband circuits to transmit the huge amounts of data, but it enabled same-day publishing of a newspaper across the country. Less spectacularly, the *New Yorker* began faxing copy for editing and checking between its New York City and Old Greenwich, Connecticut, offices. The venerable magazine's system of messengers riding the railroads had experienced so many delays that facsimile seemed a more reliable alternative.[120]

WESTERN UNION

If one firm embodied postwar facsimile, it was Western Union, which became the world's largest manufacturer and operator of fax machines until eclipsed by telephone-based fax machines in the 1960–70s. Western Union maximized the telegraphic potential of facsimile but viewed the technology primarily, in the words of Ivan S. Coggeshall, General Traffic Manager of its International Communications Department, as "a valuable adjunct to printing telegraphy in speeding the pick-up and delivery of telegrams" and not as a promising communication format by itself.[121]

The most visible symbol was Desk-Fax, visualized by its inventor, Garvice H. Ridings, as "sort of a self-contained telegraph office" for the office desk.[122] A Western Union engineer since 1926, Ridings had been part of the original 1934 facsimile team and had, over four decades with Western Union, received fifty-four patents, 80 percent of them for faxing. In 1947, Ridings received the task of developing a smaller, less expensive version of Telefax. Building a model at home with an empty frozen lemon juice can as the cylinder, Ridings tested his ideas and ultimately lowered costs in several ways.[123] Transmitting only a third of a page decreased the machine size, yet the blank space allowed 150 typed words, more than adequate for most messages. A stylus, not a photocell, scanned the message whose typed or written characters differed in resistance from the electrically conductive paper, just like Bakewell's machine a century earlier.

Finally, Ridings simplified manufacturing. Designed without exactingly close tolerances for easy assembly line construction, Desk-Fax contained stamped and cast metal parts instead of precisely machined pieces and used pinball machine control relays. Since the transmissions were all local, synchronization was not a problem. Concentrators in main offices enabled one operator to handle many Desk-Faxes, thus saving labor.[124]

Such measures reduced Desk-Fax to a third of the cost and size of Telefax. At 7 × 10 × 11 inches and 18 pounds, the transceiver comfortably fit on a desk within easy reach of a businessman or his secretary. Costing nearly an order of magnitude less than conventional fax machines, a Desk-Fax demanded only $300 to build and install compared with $700 for a teleprinter in 1959, making direct connections to smaller users (defined as handling 5 to 20 telegrams daily) economical. Western Union benefitted through lower operating costs by eliminating messengers as well as the greater ease for customers to send and receive telegrams.[125]

Simple procedures contributed to Desk-Fax's acceptance by users. Instead of phoning for a Western Union messenger, the customer placed the telegram in the Desk-Fax drum and pressed the outgoing button. After seven seconds for the vacuum tubes to warm up, the machine sent the message, and then a timestamp inked the time for a record. Because the original remained, no carbon copies were needed. Receiving a message proved equally convenient. A buzzer alerted the recipient to insert a blank form and press a button to signal that the machine was ready. The buzzer sounded again after the message arrived for the recipient to press the acknowledgment button, creating a record of receipt.[126] Compared with telex, Desk-Fax required no special preparations or specially trained typists.

Formally introduced in New York City in mid-February 1949, Desk-Fax was hailed by Western Union president Joseph L. Egan as "one of the most important developments in this streamlined era of rapid communications."[127] Good publicity, such as placing a Desk-Fax on Jack Parr's television show to receive telegrams from viewers, was an integral part of Western Union's promotion of Desk-Fax.[128] Nonetheless, Desk-Fax spread more slowly than planned, reaching 850 sets in nine cities by September 1949, and 6,900 by December 1952, well short of its 1949 goal of 10,000. By mid-1956, however, over 26,000 units sat on desks, and 35,000 units in January 1959, compared with approximately 22,000 teleprinter lines.[129]

By the time production stopped in 1960, approximately 40,000 Desk-Fax transceivers graced offices in more than 150 cities in the United States, Canada, Cuba,

Western Union manufactured tens of thousands of Desk-Fax machines, a quantity unequaled by telephone-based fax machines until the 1970s. Western Union Telegraph Company Collection, Archives Center, National Museum of American History

and Britain. Illustrating its economic importance, New York City accounted for nearly 10 percent of the machines.[130] Western Union introduced Desk-Fax into smaller cities, such as Colorado Springs, Colorado, through the early 1960s.[131] In 1968, when less than 18,000 fax machines existed in the United States, businesses still operated 35,000 Desk-Faxes. By 1977, attrition reduced that number to 2,000 as telegrams faded from the business world.[132]

Emulating its telex service, Western Union organized automatic switchboards, monitoring, and billing equipment to enable subscribers to directly fax each other in a private network instead of going through branch and central offices. In May 1952, the firm introduced Intrafax (for Intra-company fax) for a large firm's internal communications, especially between branch offices and headquarters, and Letterfax, which sent letter-size documents in three minutes.[133]

These leased private networks averaged four units but ranged up to thirty. The cost was not trivial. An Intrafax transceiver cost $25 a month if used less than 72 hours a week or $39 if used more. The Letterfax transceiver cost $125 a month plus $4 per mile of connecting wire.[134] By late 1959, Western Union earned over $2 million from renting 1,200 Intrafax machines in 324 networks, while all fax operations brought in $4 million in 1961, nearly 2 percent of the firm's revenue. Intrafax subscribers peaked in the early 1960s at approximately 1,600.[135]

Other fax services proved less successful. For faster delivery of telegrams in residential and suburban areas, Western Union created the Telecar, a station wagon equipped with a fax receiver and radio. Tests in Baltimore in 1951 concluded the high maintenance costs and technical problems outweighed the benefits.[136] In December 1957, the self-service Autofax replaced special telephones at New York's West Side Terminal where customers called in their telegrams. The easy accessibility, complained developer Garvice Ridings, allowed "hoodlums to shock our central office personnel by sending sketches and poetry of the type found on public washroom walls" while attracting "generous quantities of gum and candy wrappers, hairpins, rubber bands, and other refuse." Consequently, Western Union restricted these machines to its offices in transportation terminals and hotels.[137]

TELEPHONE COMPANIES AND BUSINESS FAXING

The most important postwar shift was the widening government and corporate interest in faxing information and documents, not just photographs and maps. By 1954, business documents comprised one-third of RCA international radiophoto traffic, a percentage that grew as American businesses expanded abroad.[138] Domestic faxing proved more problematic.

Although facsimile did not rate mention in its 1952 perspective on future developments, AT&T interest and activities grew, stimulated by technological excitement and financial potential.[139] As one BTL engineer enthused in 1956, "It is probably no overstatement to say that a revolution in the pattern of written communications would take place if facsimile service could be made available on a basis, say, comparable to telephone calls or telegrams."[140] Faxing also attracted the Bell companies because greater use of their circuits, especially after business hours, increased their overall load factor—and income. In 1954, AT&T proceeded in its usual cautious and deliberate way to consider a fax service and nearly introduced one in 1957.

Market demand for fax applications began to shape AT&T's actions, though still significantly less than developer push. The Air Force envisioned facsimile playing a large role in its worldwide communications, distributing orders, information, and pictorial intelligence if the necessary machines, including cryptographic equipment for security, proved technically and economically feasible.[141]

More importantly, an increasing number of firms took the initiative to experiment with fax and to urge telephone companies to provide service. These early users by themselves did not provide a profitable market, but they indicated that an affordable, reliable service might attract other customers. Such interest encouraged local Bell companies to gain faxing experience and determine if a larger market existed.[142]

The wide range of experiments and trials demonstrated the growing interest in faxing documents and data to save time and improve accuracy. With the exception of libraries, almost all experiments transferred information inside an organization. Results often, but not always, discouraged applications due to faxing's high cost, slow speed, and poor reproductive quality. The cost and complexity of the equipment still compared poorly to other forms of rapid communication, including Desk-Fax, airmail, and long-distance telephone calls.

Experiments and applications "in the march toward automation and centralized control" appealed to large organizations that distributed complex data for decentralized application.[143] A major impetus for the 100-recorder Weatherfax System, opened in 1953 to supply Royal Canadian Air Force bases with daily weather maps, was that there were too few meteorologists to give each base its own specialist.[144] Another data-intensive task, transmitting utility load schedules, saw Detroit Edison test facsimile in 1952 but stay with telephone dictation.[145] Two years later, the Tennessee Valley Authority faxed a larger schedule between its Chattanooga headquarters and Muscle Shoals dam. By eliminating the tedious chore of checking for errors, faxing took only one hour compared with six hours for teletype and fourteen for telephoning.[146]

Similarly, libraries envisioned faxing to easily and quickly share materials, thus eliminating duplicate copies of journals and books, or, as the New York Public Library envisioned, storing materials in a low-rent building and faxing them to its main building.[147] Tests with RCA equipment by the Atomic Energy Commission at Oak Ridge National Laboratory in 1951 and the Library of Congress with the National Institutes of Health in 1954 validated the concept of library faxing. No applications followed because of the high cost, demanding equipment, and lack of a transmitter that safely scanned an open book.[148]

Automobile manufacturers also considered facsimile.[149] A wirephoto service linked three Chrysler plants, which allowed the firm to keep papers in its central record department, save money by reducing inter-plant messenger service, and provide rapid, error-free communications during "emergencies," a surprisingly common activity that constituted much of Chrysler's initial facsimile use. The assembly line, however, had already discovered the teletype.[150]

More successfully, the New York Telephone Company in 1951 tested faxing to reduce error rates of one-third on teletyped service orders. Not only did faxing reduce errors and save time, but the Hogan electrolytic recorders machines proved quieter than teletypes, a major attraction for an office. The test turned into normal operations, and the fifteen sets increased to thirty-four by 1957.[151]

From AT&T's perspective, the 1954–55 Strategic Air Command (SAC) and 1955 Long Lines tests showed the potential and challenges of fax service. Long Lines introduced an internal fax service in 1955. Four sets of TFC and two sets of Hogan equipment enabled direct comparison of the different systems as well as testing the concept of business faxing. The familiar problem of circuit quality again resurfaced as a formidable obstacle. Reliable, high-quality transmission demanded expensive dedicated telephoto circuits, not regular long-distance lines.[152]

In February 1954, the SAC asked AT&T to develop a Graphics Communications System. Lacking wartime fax experience, AT&T had viewed the postwar military market without excitement. SAC's request changed its mind. As well as patriotic responsiveness, a SAC facsimile network offered "very considerable additional revenue."[153]

SAC's existing teleprinter network had two major weaknesses: long transmission times, and "the man behind the machine" whose "alarming number of errors" could have dangerous consequences. Sending its ten-page strike order to domestic bomber bases took ten hours. The actual transmission took only three hours; the additional time came from preparing and verifying the tapes. Faxing promised to reduce the total time to two hours with no errors.[154]

An initial trial exposed flaws in the TFC equipment design, materials, manufacturing, operation, and maintenance. Dirt on the styli or the bushing froze recording, as did broken stylus springs. Recording on the electrosensitive Timesfax paper generated copious quantities of smoke and soot. The solutions were better training of the Northwestern Bell technicians and a redesigned recorder.[155] AT&T also developed new circuit filters and equalizers because the filters designed for the mid-1930s wirephoto equipment performed poorly with the 1950s telephone circuits and the new facsimile systems. Pleased, SAC began normal operations between its headquarters in Omaha, Nebraska, and domestic bases in July 1955.[156]

In 1955, AT&T began "serious consideration" of establishing "a complete facsimile service" using outside equipment over AT&T circuits.[157] Resolution would be 96 lpi, sharp enough to retransmit documents. Six minutes would transmit a page over any circuit with three minutes desired for local transmissions. Foreshadowing battles of the 1960s, BTL stated that FM provided higher quality transmission but preferred AM because it was easier to develop and needed less band-width.[158]

AT&T expected a large, expanding market. An April 1955 estimate predicted $1.6 million of revenue in 1956, two-thirds from SAC's one hundred stations and 28,000-mile network. Other users would rent an additional one hundred stations and 15,000 miles. By 1960, an additional 245 stations and 45,000 miles of circuits would bring AT&T a minimum of $3.2 million.[159] AT&T planned to offer Stewart-Warner and TFC terminals, the chief difference being between Hogan electrolytic paper and the more fragile, costly, but permanent TFC electrosensitive paper.[160]

Yet the telecommunications firm decided against its service in 1957. The dawning realization that leasing circuits would generate nearly 90 percent of the anticipated revenue and not the equipment, supplied by outside firms, probably cooled AT&T's ardor. And the equipment remained expensive. AT&T wanted a transmitter and receiver for $1000-$1500, but the firm initially expected to pay TFC $4,500 in 1955 and $2000 by 1957. The $1000-$1500 goal equaled the approximate cost of a teletypewriter and thus supplied the goal of a comparable monthly rent of $37. In contrast, $4,500 meant a monthly charge of $168.[161]

A second reason was the experience of the New York Telephone Company serving the heart of America's business community, which introduced its own $45–$80 monthly service in March 1957. Like fax broadcasting, this fax service produced great interest by potential customers but few actual orders and was soon dropped.[162] The third strike was the 1956 Consent Decree with the Department of Justice, which limited AT&T to "common carrier communications services."[163] While not explicitly prohibiting a fax service, the decree made AT&T very cautious about offering anything beyond basic telephone service.

RETHINKING COMMUNICATIONS

In the 1950s, a wide range of experiments by military and civilian organizations tried to expand faxing into regular administrative communications. Most tests concentrated on increasing the speed of information whose delivery had to be accurate. A few organizations used faxing to explore different ways of communicating.

The Signal Corps also tried to extend faxing to regular text communications, the sinews of a massive organization that depended on the orderly flow of paper. Envisioning high-speed faxing for "mass transmission of data . . . such as depot requisition, embarkation or de-embarkation lists, etc.," the Signal Corps in 1954–55 established experimental services to test facsimile for routine office correspondence and to interest other army corps.[164] Faxing saved two to four days over mail but failed to win acceptance.[165] Proponents had focused on saving time in transmission but did not appreciate that receiving a faxed document marked only one step in its use. Processing it into the information flow of an organization often demanded making copies or marking up the original. Although fax manufacturers developed papers for ozalid, hectographic, and stenographic duplication, none produced a high-resolution, easily copied recording.[166] This lack of reproducibility significantly reduced the attraction of faxing for normal administrative communications in military and civilian spheres.

Not all attempts at new markets failed. Alden and Western Union developed profitable niches providing systems for organizations to send short messages locally. Because the machines sent only a small piece of paper instead of a full page, transmission was relatively fast—less than two minutes. Further simplicity and lower equipment cost came from using local telephone or telegraph circuits (and thus the same power supply), so synchronization was not a problem.

Verifying signatures for banks provided one such niche, ideal because no viable competitors existed.[167] Before ATMs, people visited their bank to obtain cash. Tellers verified customers' signatures with their signature on file. If a bank only had one office, verification was not a problem. The teller simply checked the file. If banks merged or created branch offices, banks had to create duplicate files for each branch, an expensive administrative burden.

Faxing quickly linked the master file to branch offices. In 1957, Alden installed a verification system for Boston's Provident Institution for Savings, consolidating over 150,000 signatures. The fax machine only needed to send a signature, minimizing transmission time. More than three hundred banks employed thousands of Alden machines through the 1980s. Alden looked forward to bank mergers, because the new bank usually ordered a fax system to avoid the immense task of unifying the signature offices.[168] This proved to be an excellent market for Alden—small enough to discourage competitors but large enough to make a profit. Alden also sold numerous 3-inch dispatch and 5-inch Order-Write systems but never focused on faxing standard sheets of paper.[169]

Intrafax embodied similar thinking but had the advantage of being from Western Union. In an example of user-driven innovation, many customers created their own networks like WeatherFax and BrokerFax, whose names demonstrated the potential of message facsimile.[170] Several newspapers followed the *Charlotte News* in installing Intrafaxes at the local courthouse and city hall to speed reporters' copy to newsrooms while eliminating the transcription errors inherent in telephoning stories.[171]

In a major step beyond the confined structure of telegrams, Letterfax allowed businesses to send whatever they wanted on a full-sized sheet of paper, such as invoices and signed letters on company letterhead.[172] Significantly, requests from businesses, not internal thinking, pushed Western Union to develop Letterfax. Users organized the transceivers to duplicate their centralized mailrooms, broadcast from one location to many, and create other arrangements, including receiving messages without an attendant to reduce labor costs.[173] In 1954, County Trust Company in Westchester, New York, pioneered its intercity use, connecting twenty bank branches in twelve cities and towns with twenty-seven Letterfaxes in an early online system.[174]

Such applications remained rare because the high cost of renting Letterfax transceivers—five times more than Intrafax equipment—discouraged use. Nonetheless, a growing number of businesses employed Letterfax, Intrafax, and other systems, such as A. B. Dick's expensive but long-distance Videograph, to connect factories to headquarters, warehouses to department stores, and other applications where saving time and ensuring accuracy outweighed faxing's high costs and often demanding equipment maintenance.[175]

Yet even Desk-Fax equipment allowed firms to rethink their communications if they wanted. The Pennsylvania Railroad created Ticketfax as part of a larger reconceptualization of selling train tickets. Coordinating a Pullman ticket purchase was an error-prone, time-consuming frustration for railroads and customers. A ticket clerk had to phone the request to a group of clerks stationed around reservation charts. If a space existed, the customer was offered the spot, then that affirmation was relayed to the chart clerks and reconfirmed before issuing the ticket.

Starting in 1953, the Pennsylvania Railroad reorganized its ticketing process by integrating different forms of communication. A station Ready Board told customers and ticket clerks the availability of Pullman space on the twenty-three daily trains for the next two weeks. The clerk gave the passenger's request to the "space dispatcher," who took a master template with the appropriate information and faxed it to the clerk. Intrafax machines connected this operation to other stations, ticket offices, and customers equipped to have tickets faxed to them.[176] Cooley was sufficiently impressed to offer a competing system to AT&T, which declined the suggestion. Other railroads also coordinated reservations centrally via faxing.[177]

In one of many false dawns, the expectation of AT&T's introducing fax services and the establishment of an interstate Letterfax system by Reynolds Metal with Western Union encouraged manufacturers to "feel 1957 will at least see the start of a big push." Overly optimistic as always, Elliott Crooks of Hogan Labs estimated that the $3–5-million 1957 market would expand tenfold by 1962. More cautious, Milton Alden estimated a $25-million market in 1962, still a fivefold increase. Both assumed significant military demand.[178] Improved equipment and the increasing pace of technological change in electronics added to the excitement. Cooley declared the technological state of facsimile as "at this stage 'if it works, it is obsolete.'"[179]

Even the idea of the faxed newspaper resurfaced. Science fiction publisher Hugo Gernsback proposed his RAFAR (Radio Automated Facsimile And Reproduction) newspaper. Transmission would still take an hour but would include an entire normal newspaper. Like a microfilm reader, a large screen on the home receiver would display each page as the reader flicked a button. Gernsback optimistically and erroneously predicted organized labor would be the largest obstacle, not the actual technology or its economics.[180]

Although AT&T did not introduce a service, its explorations, together with the growing data communications market, created new business interest and involvement in facsimile. By 1959, fax manufacturers estimated the fax market at $10–25 million, indicative of both growth and major differences in measurements. Proponents of increased business and industrial use noted that Western Union Intrafax income doubled from $1 million in 1957 to over $2 million in 1959. In reality, two-thirds of that growth came from revenue from the Air Force's new weather map fax network.[181]

SPEED MAIL

Facsimile suffered a public setback when the Post Office suffered a humiliating political and technical defeat in 1959–60 over Speed Mail, an effort to merge facsimile and mail to deliver letters nationwide within hours. What was surprising was not Speed Mail's failure, the victim of technological overreach and effective opposition from Western Union, but that the Post Office combined such far-reaching vision and political ineptitude in one project.

President Dwight Eisenhower inaugurated the semi-accidental public debut of Speed Mail by telling the National Association of Postmasters in October 1959 about "projects in prospect for getting letters into your hands in a matter almost of moments." His Postmaster General, Arthur E. Summerfield, promised, "The next step in mail service will be as revolutionary and as progressive as anything yet achieved."[182] A 1958 paper by RCA researcher Sidney Metzger proposing faxing letters between Europe and the United States by satellite might have triggered Summerfield's interest.[183] The justifications came from a combination of challenges and opportunities. The challenges were an expected doubling of mail to 61 billion pieces by 1985 pushing against limits to mechanizing the mail process and congressionally restricted mail rates.[184]

Internally, the opportunity was "one of the most far reaching [sic] improvements in the history of the US Post Office Department with implications of cost saving, improved efficiency and faster methods of communication by mail that literally stagger the imagination."[185] Just before the project's cancellation, one official envisioned an expanded market of "items not presently considered as mail," such as bills of lading carried in trucks and internal corporate communications — fax markets that indeed developed three decades later.[186] The Post Office also justified Speed Mail as serving military needs and transferring technology from the military to civilian sector.[187] Postal officials envisioned Speed Mail equipment installed in every post office and ultimately in every business and home. If implemented, Speed Mail threatened the revenue not just of Western Union but also of the railroads and airlines that carried mail. Western Union was justly worried.[188]

Speed Mail lacked support of the telegraph and fax industries, which, in addition to seeing an economic threat, were surprised by the announcement and slightly upset that they were not consulted.[189] Western Union proved an effective and cunning opponent, pledging cooperation while covertly enlisting newspapers, the telegraphers' union, railroads, airlines and other firms in attacking Speed Mail as a "socialistic scheme" threatening private enterprise.[190]

Western Union's assault did not remain only rhetorical and political. Showing a prior awareness of Speed Mail, Western Union in late October filed rates and regulations for its new Wirefax public fax service, forcing the Post Office to battle an existing, not a hypothetical, competing system operated by the private sector.[191] Wirefax began service in Washington, D.C. and four other cities on December 1, 1959. Unsurprisingly, Wirefax never became economically important for Western Union. In 1961, its third year of operation, Wirefax earned $9,097 for handling fewer than 3,500 messages.[192] Politically, however, Wirefax served Western Union well.

In December 1959, Summerfield pledged to Western Union President Walter P. Marshall that the Post Office did not plan to compete with private enterprise and that businesses would handle any actual operation. Western Union withdrew its open opposition. One senior postal official remarked, "The cloud that once might have been known as Western Union has lifted, and with that lifting has come a different attitude on the part of many segments of the industry."[193] Any lifting of clouds over Speed Mail proved illusionary, however.

Speed Mail experienced significant financial and technological troubles. Lead contractor ITT, a firm with no fax experience but nonetheless receiving a noncompetitive contract, greatly underestimated the technical challenges, especially integrating components into a working system. In theory, a machine would open the envelope, tag the enclosed letter with a unique number, then send it to the facsimile machine and afterward destroy the original. The receiver reproduced, folded, and sealed the letter. Such was the theory. In reality, tests with forty government agencies communicating between Washington, D.C. and Chicago and Battle Creek, Michigan (for the Office of Civil Defense Mobilization) in 1959 and 1960 revealed numerous equipment flaws and poor reliability. Never solved were the major technical challenges of ensuring the privacy of documents and reliably opening and faxing a letter, not to mention the social challenges of teaching people to properly prepare the envelope for faxing.[194]

Not surprisingly, Speed Mail overran its $3 million development budget by almost 50 percent and threatened operating costs higher than regular mail.[195] Nonetheless, on November 1, 1960, Summerfield announced he would seek Congressional approval for a nationwide fax system with nearly eight hundred terminals.[196] This announcement provoked Western Union, which generated, the Postmaster General discovered, "an avalanche of letters" to Congress and scores of newspaper editorials attacking Speed Mail.[197]

The Kennedy administration had other plans. The new Postmaster General, Edward Day, formally ended the program in August 1961 because it was "a major departure from traditional methods of handling mail and threatened competition with private enterprise."[198] As an internal report wryly noted, "It is safe to conclude, however, that Mr. Day's decision was based on more than simply economic considerations."[199] Never again would the Post Office be so bold.

MICROWAVE TRANSMISSION

Speed Mail left not just a greater appreciation of the political power of Western Union, but increased business interest in faxing, an interest heightened by the expansion of microwave networks and lower cost long-distance communications.[200] The FCC had initially restricted microwave channels to common carriers and regulated utilities such as a railroad that built and operated its own private lines over its own right-of-way (Sprint originated as the Southern Pacific Railroad Intelligent Network of Telecommunications). In 1959, with the "Above 890 Megacycles" case, the FCC allowed private microwave systems despite AT&T's opposition. This regulatory action significantly reshaped the economic environment for faxing.

To keep large customers from building their own networks, AT&T in 1961 responded with Telpak, a discounted broadband service, and increasing its own microwave capabilities. Telpak reduced the cost of long-distance calling so much that Alden marketing director George F. Stafford proclaimed that Telpak put facsimile on "the threshold of a commercial breakthrough." By 1962, nine firms were developing fax equipment for point-to-point, high-volume services.[201]

Bureaucratic initiatives also signaled a growing interest in business fax. In 1954, the ITU's International Consultative Committee on Telegraphy and Telephony (CCITT) studied if it should distinguish between phototelegraphy and document transmission (also called direct recording). The goal was not to encourage a new service but to prevent "unfavourable competition" for telegraphy.[202] Two years later in a reversal of purpose, a CCITT study group stated that direct recording offered "fresh fields for operational purposes," such as sending faxes via regular telephone lines.[203] In 1959, the West German PTT declared that faxing "business papers, etc., between telephone subscribers will become fairly widespread in the near future, even in the international service."[204] In October 1962, the PTT argued that fax needed specific recommendations to avoid adversely affecting telephone service.[205] In January 1964, the CCITT agreed to develop a standard, which became the G1 in 1968.

Across the ocean, the Electronic Industries Association created the Engineering Committee (TR-29) in 1963. In the competitive and rapidly evolving electronics community, this industry-wide group became the main American organization for fax standards. One of its first activities was to establish standards for faxing by Telpak and Western Union's competing BEX.[206] The section's membership indicated widening institutional interest in facsimile. That Western Union, BTL, Stewart-Warner, and the Army's Electronics Laboratories joined was not surprising. The presence of Xerox was a harbinger of that corporation's imminent introduction of facsimile equipment.

Significantly, TR-29 did not promote facsimile or conduct market research. Possibly this was because of the competition among its members or because advocates did not think an industry-wide organization worth the expense. The lack of such a promotional body hurt fax's diffusion. Unlike in Japan, the industry lacked a unified voice and forum for manufacturers and promoters.

JAPAN

The most important Japanese event was the creation of a community of fax engineers. Starting in November 1943, several engineers at the Electrotechnical Lab of the Ministry of Communications formed a group to research sending photographs electronically. They disbanded at war's end but restarted their meetings informally in 1947 and in 1952 elected Yasujiro Niwa their first president. In 1954, the group began publishing the world's first technical journal devoted to faxing out of Niwa's Tokyo Electrical Engineering College. These men served as the nucleus for the growing Japanese research and application of facsimile. In 1957, Japan's 2,700 fax machines were second only to the United States.[207]

Like the United States, the Japanese government became the country's largest postwar user of fax technology, transmitting weather maps, police data, and supplementing the railroads' internal telegraph system. Unlike the United States, the military played a minor role, reflecting its diminished postwar role. Japan's national police, which both had a need for long-distance communications and the ability to standardize equipment, played a major role in promoting faxing. In 1950, the Ministry of Police began experimenting with faxing fingerprints and photographs. By 1960, the National Police Agency had 120 machines, a number that passed 1000 within two years, as district stations installed a new generation of transistorized equipment. In 1967, the police operated nearly one-fifth — 2,461 — of all Japanese fax machines, second only to the railroads with 3,250 machines.[208] Significantly, the police had nearly equal numbers of transmitters and receivers, implying two-way communications.

Like the United States, large organizations with many geographically separated units constituted most of the fax users. Moving beyond these applications, Nippon Telephone and Telegraph (NTT), the domestic telephone monopoly, introduced a 16-minute-per-page service between its Tokyo headquarters to eleven cities in 1958.[209] By the early 1960s, fax machines transmitted internal government documents and exchange rate information among banks.[210]

The 12,925 Japanese fax machines (and 1,661 wirephoto machines) in 1967 marked nearly a fivefold increase over a decade. A quarter were transmitters; the rest were receivers, reflecting the dominant use by railroads, media, government bodies, and others in sending information one-way from central offices. Reflecting high costs and the limits of the equipment, news generators and distributers — newspapers, news agencies, and radio and television broadcasters — and not actual consumers of news accounted for a third of all fax machines.[211] Japanese machines, developed by government and industry researchers, supplied these applications. In 1960, Japanese manufacturers built nearly 1,600 machines.[212]

Reflecting the geographic shift of innovation and application, fax development and applications were comparatively minor outside the United States and Japan. In Western Europe the British Muirhead and German Hell firms dominated the specialized (weather, wirephoto, and police) and smaller general markets, but at a much lower level of activity.[213] Faxing in the Soviet Union paralleled, if mostly trailed, Western developments, including, in the mid-1950s, a Desk-Fax-like system, albeit using telephone instead of telegraph lines. As in the West, economics played a determining role in the diffusion of fax equipment.[214]

⌒

By 1965, what had been the fax market in 1945—wirephoto, weather, war, and Western Union—had become special-purpose markets, separate from the promising but infant general-purpose business market. Significant obstacles stood before that potential. The formidable bottleneck of affordable production discouraged business faxing. The price of equipment had not, despite numerous efforts, decreased significantly and remained well above what advocates and potential users considered economically viable. These high costs reflected the demanding manufacturing (and maintenance) of electromechanical equipment and the low production runs, partly because of the high cost.

High long-distance telephone rates and poor circuit quality also continued to discourage faxing, especially for firms unable to take advantage of discounted rates available only to the largest customers. Even as faxing technology improved in performance if not price, proponents failed to fully realize that faxing's faster speed was only one criterion by which prospective users compared competing technologies. Fitting into office procedures became an increasingly important obstacle, primarily due to fax's poor reproducibility.

The contrasting success of Desk-Fax stemmed from its much lower cost—almost an order of magnitude less than a conventional fax machine—as well as its promotion as an extension of regular telegrams by Western Union and the firm's promotion of dedicated networks for users. Desk-Fax succeeded with customers because it was convenient, easy to use, and fit seamlessly into office operations. Most impressively, Western Union placed nearly 40,000 Desk-Fax machines in offices, showing that demand existed for an easily operated, inexpensive message system.

Conventional fax equipment failed the Desk-Fax test. Just operating and maintaining the equipment remained more demanding than competing services. Facsimile advanced further in the military than civilian sphere because the military, using the rubric of national security, was able to justify and dedicate the resources that a profit-based organization could not. Without the military's promotion and utilization, the facsimile market would have been much smaller and slower to develop. Certainly, much military research and development focused on capabilities such as very high speed, oversized pages, and ruggedness that business faxing did not need. Military funding, however, also promoted research in data compression and other areas that proved very important in future decades. As important, military orders provided fax manufacturers with a significant percentage of their demand.

Technology push still dominated faxing's evolution, but a much broader spectrum of market demand began to appear as ideas and applications percolated among manufacturers, telephone companies, and self-identified possible users. The thousands of fax machines—tens of thousands counting Desk-Fax—had moved beyond transmitting pictures and maps to data in many written or printed forms. Faxing data and documents had become essential for some large businesses and governments, which had multiple offices and discounted telecommunication services. The customer-driven innovation of Western Union's Letterfax demonstrated faxing's potential to transform written communications. Increasingly, at least to proponents, faxing's economic and technological challenges appeared ready to fall before a major expansion of business faxing. When that expansion began in the mid-1960s, however, its evolution was not what its early 1960s promoters expected.

The Sleeping Giant Stirs, 1965–1980

> Some day the use of facsimile will approach the point of federal use
> and will be widely used as a business communications tool to send
> documents to other companies and organizations in direct competi-
> tion with the use of U.S. mail. At that time, facsimile will not be a
> special business tool to be justified by special application, but rather
> a device required in a typical office much like a telephone, typ-
> writer [sic], postage stamp, and perhaps an office copier.
>
> —*Frost & Sullivan, 1977*

On January 22, 1973, Texas lawyer Sarah Weddington started receiving calls from newspaper reporters asking for her reaction to the U.S. Supreme Court's 7-2 *Roe v. Wade* abortion decision. Not knowing the Court's reasoning, Weddington, the plaintiff's lawyer, telephoned a colleague in Washington, D.C., and asked him to go to the Court to get a copy of its opinion and read her the main sections. Later that day, Weddington received a telegram officially informing her that she had won her case and that the Court had airmailed its full decision to her.[1] Such was the state of communications in the early 1970s: the telephone for immediate information, telegraphy for quick written information, and the mail for lengthy communications. Facsimile advocates hoped to replace or at least augment all three.

Their fondest expectations were not realized, but from 1965 to 1980 faxing changed decisively from specialized machines for specific markets to increasingly essential general-purpose machines for government and businesses. The most important development was not technological but economic, the liberalization of telephone networks in the United States and Japan. Technology push remained more important than market pull, but general-purpose fax machines became the fax market. While installed machines in the United States grew from 20,000 in 1970 to over 250,000 in 1980, technological leadership had passed to Japan, which manufactured more than 100,000 machines annually. Europe, home of faxing, trailed far behind.

When did this modern world of facsimile, a world of machines designed for offices instead of specialized applications like maps, begin? An obvious date is November 11, 1965, when Xerox and Magnavox announced their six-minute per page Telecopier. Less visible but just as important was telecommunications deregulation. In 1968, the Federal Communications Commission's Carterfone decision began the slow journey to open access to the American telephone network, which was essential for widespread faxing. In contrast, the quick 1972 deregulation of the Japanese public telephone network ignited an explosion of fax diffusion, propelling Japanese manufacturing and applications.

This chapter charts the changing regulatory and technological environment for facsimile, the importance of standards and compatibility, the rise of general-purpose fax machines by new entrants like Xerox, the decline of older fax firms, and the failure of most startups. In an excellent example of technological prematurity, subminute machines in the early 1970s fizzled as engineers failed to reconcile cost and capability. A second generation of less technically challenging but more affordable three-minute per page machines fared better.

Despite the initial Wall Street excitement about facsimile in the late 1960s, actual placements and accomplishments remained disappointingly modest. The ideal fax machine, according to a 1971 corporate survey, transmitted a page reliably and inexpensively in less than a minute over a flawless telephone network to a compatible machine recording on bond-like paper.[2] In contrast, the costly first- and second-generation machines needed three to six minutes to transmit a page over telephone circuits of varying quality and record on low-quality paper.

Although the equipment improved greatly in performance and price by 1980, faxing encountered its familiar problems of poor economics, poor reproductive quality, and the gap between desired and actual performance. A new obstacle, incompatibility among competing manufacturers and generations of equipment, partly due to business strategy and partly to technological differences, further dimmed the attraction of faxing. The new audience of office workers, which was less technically skilled and had more options for communicating than earlier niche markets, proved less willing to put up with what they perceived as problems or inferior quality. These shortcomings initially limited faxing to internal faxing within large organizations, especially the government. Even so, by 1980 fax had become an increasingly important communications tool, especially in Japan.

REGULATIONS AND STANDARDS

A huge shift separated the regulatory worlds of 1965 and 1980 from restricted to more competitive, open, and flexible markets. The changes occurred first domestically in the American and Japanese markets and then internationally as part of a larger deregulation of telecommunications that created a more fractured, dynamic industry that did, as proponents predicted, spur technological innovation and reduce long-distance telephone charges.

Before June 26, 1968, the FCC required an expensive physical interface, like the Data-Phone, between the telephone network and any attachment. That day, the seven commissioners ruled unanimously that the Carterfone, an acoustic-inductive device that connected private communication systems to the telephone network, was lawful and that telephone companies must allow foreign attachments, destroying AT&T's monopoly on connections to the telephone system.[3]

Regulatory change was not a genteel, obscure activity but a battlefield where firms fought each other and the FCC for competitive advantage. Petitioning for new services, appealing decisions, pointing regulators to allegedly improper conduct by rivals, and other legal maneuvers shaped this regulatory environment. So important was the FCC that AT&T devoted a high-capacity Xerox LDX fax system with a dedicated Telpak broadband circuit to link its New York corporate headquarters with its FCC lobbyists in Washington, D.C.[4] Not surprisingly, change occurred slowly, measured in years from when a firm applied for an FCC decision through the inevitable challenges from losing parties. Such protests might ultimately fail, but the delays provided time for the losing parties to develop their own services, maximize monopolistic profits, or otherwise benefit.[5]

Indeed, AT&T successfully delayed efforts by manufacturers and federal and state regulators for full open access until 1976, when the FCC finally allowed the direct attachment of modems and other equipment to the telephone network.[6] Eliminating the Bell-mandated protective coupler greatly aided faxing by reducing the cost and effort to install a fax machine. The acoustic coupler and Data-Phone became relics, quickly replaced by the universal RJ-11 plug-in jack.

The international regulatory environment remained predictable until the 1970s. Voice circuits only carried conversation at set tariffs; acoustic couplers were prohibited. Leased circuits provided high-quality transmission, but these dedicated lines were expensive. In 1972 the FCC received requests from Xerox to allow acoustic coupling for overseas transmission and from Western Union International (the international cables section spun from Western Union Telegraph Company in 1963) to provide unrestricted overseas voice, data, and fax communications. In 1976, the FCC allowed overseas data transmission over the telephone network, effectively enabling direct faxing. To an unknown degree (uncertainty about its legality meant firms did not publicize their faxing), many corporations began faxing overseas directly.[7]

Even as economic deregulation encouraged faxing, incompatibility caused by poor technological regulation and deliberate corporate strategy fragmented and frustrated the fax market. As the CCITT (International Consultative Committee on Telegraphy and Telephony of the ITU) gloomily stated in 1979, "One of the biggest handicaps in the development of document facsimile services was incompatibility within Group 1, then between Groups 1 and 2 and now between Groups 1, 2, and 3."[8] The inability to create and enforce compatibility among competing manufacturers' machines (and sometimes within the same manufacturer's equipment) demonstrated the importance of effective standards for a network technology and how their absence delayed the diffusion of faxing.

Technical standards were not neutral; they embodied their creators' desired norms and goals to the exclusion or detriment of other possibilities. Conflict and arduous negotiations proved to be a normal part of the standards process as vendors, companies, and countries maneuvered for individual as well as collective advantage. Standards benefitted producers, sellers, and users by providing economies of scale, decreasing technical uncertainty, and reducing transaction costs. Making competing systems compatible increased network benefits while reducing uncertainty and risk for consumers. To some extent, developing a communication technology was a game of standards. Unlike a stand-alone technology like a photocopier, faxing worked only by connecting with other machines.

Standards emerged either by stakeholders negotiating a common standard or a firm imposing its standard through market supremacy. Because of the time needed to reach agreement and the political need to compromise, communal standard setting could limit instead of encourage a technology's potential. The four-year CCITT process did not reward rapid development of a technology. Market-based standards did not necessarily produce the "best" technology, but one that succeeded—often for reasons unrelated to the standard.

In 1964, the CCITT began developing a standard for business faxing, which became the G1 standard in 1968.[9] Loosely written and essentially *ex post facto*, issued after many six-minute machines entered operation, this G1 standard proved unsuccessful as manufacturers marketed incompatible equipment.[10] Nor did G1 affect the faxing in the United States, where the large domestic market made international standards less important—no official American delegation participated in Study Group XIV fax standard development until 1975. In 1966, the

TR-29 fax committee of the Electronics Industry Association had created the first American fax standard, RS-328, which differed significantly from the G1 standard but proved no more successful.[11] Faced with similar but incompatible equipment, many potential customers decided to wait until one standard triumphed, justly afraid of being unable to communicate if they chose wrong.

Despite G1's failure, efforts began on a new standard. Developed simultaneously with the new three-minute equipment, the G2 standard proved "fairly well observed."[12] As important, the G2 process developed procedures for the critical and contentious role of testing to verify the feasibility and effectiveness of proposed algorithms. Key to their acceptance were the creation of the British Facsimile Industries Compatibility Committee to conduct tests in a neutral environment and prior agreements about appropriate criteria, their measurement, and interpreting the test data. Those negotiations greatly smoothed the path for G3 testing a few years later.[13]

TECHNOLOGICAL PROGRESS

The technologies underlying faxing—and expectations of future technologies—evolved greatly, as did their patrons. Civilian research outpaced its increasingly specialized military counterpart, proof of the maturing technology and rising expectations of a big business market. The biggest single factor, advances in electronics and computing, comprised part of a larger explosion in microelectronics. Transistors, integrated circuits, large-scale integrated circuits (LSI), very large-scale integrated circuits (VLSI), and the growing ability to process and store data increasingly shaped the fate of faxing.

Until the Xerox Telecopier, a fax system usually consisted of a separate transmitter and receiver. Solid state electronics greatly reduced equipment size, enabling both functions to be packaged into a single transceiver (*trans*mitter-re*ceiver*). The Telecopier and its analog competitors were one-piece desktop models, though some needed a large desk. By 1980, single units were so common that "fax machine" had replaced "transceiver."

The business market began to pull fax development when customers asked for specific features as they gained experience with their equipment. Most important were automatic (or unattended) operations. A machine that transmitted or received without anyone standing over it greatly enhanced the convenience of faxing, also reducing labor costs. Other desired features were automatic document feed, sheets of plain paper instead of rolls of electrochemical paper, automatic dialing, and delayed transmission (also called store-and-forward). The last option stored the fax in a computer memory until a certain time, such as after 11 p.m.

to take advantage of lower long-distance rates, and then transmitted the document. Manufacturers offered these features by the mid-1970s, though at a price: the 25-number automatic dialer for the Xerox Telecopier 200 cost $75 a month, a substantial addition to the basic $195 rental.[14]

Each aspect of faxing involved technical choices; seldom were they clear-cut. For scanning, the major decisions were choosing between a flatbed or cylindrical scanner and deciding on the type of scanning mechanism. Despite the theoretical advantages of flatbeds, the more technically mature cylinders comprised the majority of models until the late 1970s.

Similarly, the gap between electromechanical and electronic scanning was increasingly obvious — the former was slower, required frequent maintenance, and demanded precision engineering such as synchronizing the clutch on the Telecopier I's 233-teeth gear. Nonetheless, electromechanical systems maintained their dominance into the 1970s because they cost less. Nor did electromechanical scanning remain stagnant as fiber optics allowed simpler, more compact designs with less movement. The few electronic systems in the 1960s, such as Xerox's LDX and A.B. Dick's Videograph, were so expensive that only high-volume faxing justified their cost and the necessary wideband circuit.[15]

By reducing long-distance rates and improving the quality of its circuits, AT&T did far more to advance facsimile than if it had offered machines or services. By electrically shielding components, installing better filters, and moving to digital trunking, the Bell network shrank incidence noise an order of magnitude from the mid-1960s to mid-1970s. Eliminating larger problems, however, made previously minor problems, like phase jitter, more visible.[16] While the quality of circuits improved, greater resolution and data compression made transmissions more sensitive to the environment. What a 1960s Stewart-Warner Datafax might have ignored, a 1970s 3M 9600 highlighted. As an AT&T history dryly reported, "By the mid-1970s, it was apparent to everyone that the game consisted of a constant sequence of improvements and more stringent objectives for data transmission, only to have the limits tested by more demanding data services."[17]

If developers had to make hard choices, at least they were familiar with the technologies. Not so prospective users, who confronted a bewildering range of options about image quality, cost, speed, circuit type, and other factors. The most variable option — and the one linked most directly to cost — was speed. In 1968, the choices ranged from one page in thirteen minutes to ten pages in one minute. For a regular telephone circuit using an acoustic coupler, the standard speed was six minutes with a resolution of 96 or 100 lines per inch (lpi). Many machines differed only in their type of connection and circuit. Of seventeen models offered in

1968, most operated over higher quality leased lines instead of the public network for better quality and faster transmission.[18]

Pushed by military need and funding in the mid-1950s and later by commercial need to transmit data electronically over telephone lines, the modem (modulator-demodulator) proved an essential technology for faxing.[19] In the late 1950s, AT&T introduced the first modems with 300 bits per second (bps) and 1200 bps speeds on conditioned circuits. By the early 1960s, commercial modems reached 2400 bps, but speed was not inexpensive: The Codex 9600 bps modem in 1968 cost $23,000. A decade later, 9600 bps remained the maximum practical speed, but the cost had dropped dramatically to $3,500 (compared with $500 for 2400 bps).[20]

Most important for faxing was the AT&T Data-Phone, introduced as a "so-called 'modem'" in 1965.[21] Customers physically connected their fax machines permanently to a Data-Phone, which converted the analog signal into telephone frequencies and, at the receiving end, reconverted back. By opening faxing to smaller users, the Data-Phone promised to be the "real key to the future business success" and "the biggest boon in years."[22] It served a growing demand for data and fax communications, placing 132,000 sets in 1970 and 200,000 in 1972.[23]

Expanding greatly on previous research, data compression developers explored several approaches, including redundancy reduction and run-length encoding. Commercial applications suffered from the high cost and limited capability of electronic components to compress and store data, the limited efficiency and incompatibility of compression algorithms, and the lack of efficient error-correcting methods.[24] Even though significant advances occurred, the high costs deterred most potential customers, the military excepted.

Recording remained the weakest component of faxing. Users faced what Telikon developer Robert Wernikoff defined as a Hobson's choice between excellent copy if cost was no object and reasonable cost but poor quality.[25] The main methods—electrosensitive, electrolytic, and pressure-sensitive paper—all had drawbacks, primarily poor resolution but also an odd feel, high cost, and fading. Paper quality actually worsened with electrosensitive paper, which was easily markable and, most obnoxiously, smelled bad as the white zinc layer burnt off. Filters and scents helped, but good room ventilation was essential. The paper's advantages were its easy storage and lack of processing.[26]

The 1960 introduction of the Xerox 914 photocopier, which had "perhaps a greater and more instantaneous impact on offices than any device before or since," reduced one of fax's greatest handicaps, its lack of reproducibility.[27] Photocopying greatly facilitated office acceptance of faxing by providing an easy way of duplicating received faxes. Eager offices rented over 200,000 Xerox 914s, far exceeding the estimate of 5,000 machines by consulting firm Arthur D. Little, as the number of Xeroxed photocopies soared from 50 million in September 1961 to 490 million in March 1966.[28] This explosion of an office copying culture increased expectations of easy circulation of original documents and raised interest and expectations in erasing the physical barrier of the office by reproducing papers at a distance.

The historically competitive environment of mail started to shift in faxing's favor as costs increased (a first class letter doubled from 5 to 10 cents between 1963 and 1974 and rose to 20 cents by 1981) and the quality of service decreased. As memories of Speed Mail faded and the political power of Western Union declined, interest in faxing revived at the Post Office (renamed the Postal Service in 1971). Its Office of Advanced Mail Systems Development conducted and funded research that manufacturers, sensing major equipment sales, supported.[29] Into the early 1980s, fax services feared the establishment by the Postal Service of a competing domestic service. One threat never materialized. In 1979, the Postal Service proposed requiring stamps on every hand-delivered electronic communication, telegrams excepted, but the FCC successfully defended its electromagnetic turf.[30]

Even as Western Union suffered as domestic telegram delivery dropped from 94 million in 1965 to 55 million in 1980, telex remained the technology of choice to rapidly transmit a written message for business and governments. As computerized equipment and gradual deregulation provided faster, better service (including lower-case letters and different alphabets), British users rose from 14,000 in 1965 to 40,000 in 1972 and 80,000 in 1980. In 1978, telex had over 110,000 American and 230,000 European users. Worldwide, the number of subscribers peaked at over 1.7 million in the mid-1980s.[31]

In 1968, compared with the young but growing competition of digital data transmission, faxing had disadvantages of cost, incompatibility, and output that was not machine-readable; but it had advantages of simpler operations and handling graphics.[32] Over the decades these advantages grew, even as the field of computer-based message systems advanced. Even in the early 1960s, some engineers integrated facsimile with computers and other information and communications systems.[33] By the late 1970s, some firms employed computers to coordinate networks of fax machines and link faxing to other forms of electronic communication and parts of the office. Comfax's $50,000 Facsimile Communications Controller enabled a network of fax machines to perform tasks most machines could not emulate for nearly a decade and a half, such as store-and-forward. Not surprisingly, the very high cost limited those markets.[34]

FAX BOOM AND BUST

More than any other event, Xerox's entry into facsimile signaled a new age, conferring legitimacy on a field of small, specialized firms with little public recognition. Rival Alden Electronics president John Alden hailed Xerox's entry as "the greatest thing that could have happened. A great marketing and merchandising organization has recognized that facsimile has a future."[35] Xerox had the resources—technical expertise, manufacturing capability, sales agents, service technicians, and advertising dollars—to make facsimile a serious business product. Xerox also demonstrated how hard, even with its resources and reputation, it was to profit from facsimile.

In the mid-1960s, Xerox was a highly respected, highly successful corporation whose 914 photocopier had revolutionized business by reorganizing office operations around multiple copies of documents. Copying at a distance seemed the "next natural step" beyond photocopying when Xerox formed a facsimile group in 1962.[36] Significantly, the corporation named its first fax product the LDX, for Long-Distance Xerography, and its general-purpose machines Telecopiers.

The Rolls Royce of fax systems when introduced in 1964, the LDX was one of the first civilian fax machines to graduate from vacuum tubes. Despite its transistors, the LDX consisted of a 650-pound Scanner (only slightly more than the 648-pound 914 copier) and a 425-pound Printer. The LDX found far fewer customers than expected due to its $800 monthly lease fee (including 10,000 copies—500 pages per business day), the cost of the Telpak or other broadband circuit, and the growing availability of less capable but far less expensive machines.[37] Only a few organizations needed such high volume (eight pages a minute) and point-to-point transmissions. Southern Railway operated the country's largest LDX network, thirty-three scanners at yard stations and ten receivers at its Atlanta communications center, handled by the firm's own microwave system.[38]

Although an economic failure, the LDX established Xerox in facsimile. Ironically, Xerox never again was the technological leader, forgoing multiple opportunities to acquire leading-edge technology and patents. The LDX taught Xerox that a market for a low-cost facsimile existed, but the firm lacked a machine. Fortunately for Xerox, Magnavox had one. In 1961, building on research to transmit electrocardiograms over the telephone, engineers at Magnavox Research Laboratories in Torrance, California, began investigating facsimile, and in 1962 management approved commercial development. This "plug and put" system worked by plugging the electrical cord into a power socket and putting a telephone receiver into an acoustic coupler. By mid-1965, the machine was ready to market.[39]

Magnavox historically sold to consumers. It lacked the sales force and strategy to sell to businesses, but its officers were smart enough to look for a firm that did. After unsuccessful negotiations with Western Union, Magnavox agreed with Xerox to market the Magnafax 840.[40] This 46-pound machine, delivered in July 1966, began Xerox's commercial domination of the fax market. Although later dismissed—accurately—as "large, noisy, slow, [and] unreliable," the Xerox Magnafax Telecopier I marked the start of successful mass production of fax machines.[41] Possibly embarrassingly for the company that pioneered xerography, the 840 recorded by pressing a stylus against carbon paper, the same system RCA had used three decades earlier. Rent was $50 monthly, $35 for the unit and a minimum charge of $15 for 100 pages (5 pages per business day).[42]

Apparently stressed by reliability problems and a poor relationship with Magnavox, Xerox exercised its option in 1967 to build the machine under license and introduced the slightly modified Telecopier II in mid-1968 for $75 a month.[43] Unwisely believing its own publicity about facsimile, Magnavox responded by creating a sales force to market the Magnavox 850, essentially a more reliable Telecopier I. After losing several million dollars, Magnavox sold its fax operations in 1973 to 3M, despite efforts by an independent group, including Magnavox Systems general manager, Peter Philippi, to buy the operation and a Justice Department antitrust inquiry into whether the sale would give 3M a dominating position in the fax market.[44] The second-largest American photocopier firm in the mid-1960s, 3M became the country's third-largest facsimile distributor by the mid-1970s by buying outside technology and customers, unusually acquisitive actions for a fax firm but normal for a large corporation.[45]

Did facsimile have a future? In 1967–68, Wall Street thought so, believing that facsimile "could become the next phase of the office-machine boom."[46] The investment firm Harris, Upham & Company, which specialized in finding profitable new technologies and companies, proved particularly influential. Stating that the resolution of technical problems, the entrance of Xerox, the "monstrous pile-up of paper in offices and the growing log-jam in post offices" provided both problem and answer, the firm's April 1968 report concluded optimistically, "The technology is here now; the need seems apparent. The sleeping giant has indeed begun to stir."[47]

After all, if businesses used nearly a million photocopiers, then what could the demand be for a long-distance photocopier? In 1967, Telikon developer Robert

The two-piece 1964 Xerox LDX weighed nearly 1,100 pounds but could transmit eight pages a minute. Xerox Historical Archives

Wernikoff proclaimed facsimile "will soon be as familiar as office copiers."[48] One "common calculation" luring investors to "this marketing man's dream" was that leasing fax machines to only 1 percent of America's 30 million businesses meant a $300 million market—no small potatoes when Xerox's 1968 revenue was $896 million.[49] Harris, Upham estimated that the fax market could grow from 18,000 machines in 1968 to 500,000 machines and $500 million in 1978.[50]

The importance of such optimistic estimates of present and future markets lay not in their accuracy or exact numbers but in shaping expectations about the future and reducing perceived technological and market uncertainty. Estimating the present size and shape of the fax market was not easy. American manufacturers did not disclose detailed data about placements and numbers, correctly

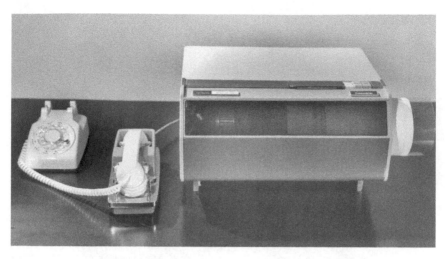

The introduction of the Xerox-Magnafax Telecopier I with its acoustic phone coupler in 1967 began faxing's long road toward revolutionizing communications. Xerox Historical Archives

viewing this information as valuable to competitors. Unlike in Japan, no neutral industry-wide association collected data, nor did the federal government gather detailed information. While filling this void, market research firms measured fax diffusion and usage differently. Varying assumptions made predictions of the future even more wide-ranging. While Samson Trends and SRI in 1971 estimated the 1970 market at $18 million and $25–29 million respectively, 1970–71 analysts' predictions of the 1975 market ranged from $100 to 200 million.[51]

Some fax firms benefitted immensely from a broader Wall Street excitement about those "tiny scientific companies put together by little clutches of glittery-eyed young PhDs, their company names ending in 'onics.'"[52] In November 1971, Comfax Communications Industries' $7 stock offering soared to $26 its first day, very impressive for a firm that had no earnings, no developed product, and definite financial losses.[53] As an internal 1970 Addressograph-Multigraph report stated, the enthusiastic "reaction of venture capital to any company simply claiming to have such an entry" illustrated the uncritical excitement about facsimile.[54] By 1973, according to fax industry analyst Howard M. Anderson, 126 companies investigated manufacturing fax equipment and 32 actually tried it.[55]

Graphic Sciences appeared to justify the excitement. Its meteoric appearance in 1967, raising nearly $40 million in stocks and bonds, presaged the dot-com boom three decades later. Anticipating that Graphic Sciences' technology and patents made it the next Xerox, investors and speculators bid Graphic Sciences' stock from the initial $10 to $28 within a month of the offering and to $120 in 1969 before falling back to the $30s, a level maintained until the 1973–74 recession knocked it below $10.[56]

Without Xerox, Graphic Sciences never would have existed. If Xerox had heeded the advice of Robert K. Dombrowski, its Information Systems Division Vice President for Marketing and Planning, Xerox would have acquired Milton S. Cohen's patent for a graphic transceiver. Instead, and not for the last time, Xerox balked, a victim of a self-inflicted Not-Invented-Here complex, and Dombrowski and three other Xerox executives left the company to form Graphic Sciences and buy the patent.[57]

Graphic Sciences delivered its first units in 1969, a year later than planned due to extensive redesign.[58] While technically superior to Xerox's Telecopier, that decision made the dex (decision expeditor) deliberately incompatible. Graphic Sciences believed incompatibility did not matter since, as CEO Sullivan G. Campbell stated, "many users really don't want to communicate with outside companies."[59] The executives' experience with IBM and Xerox, unlike that of an AT&T manager, did not inculcate the importance for compatibility in communications.

The consequences greatly damaged the new firm and the field of business facsimile. Incompatibility cut Graphic Sciences off from Telecopiers and forced potential users to choose between an established and an unknown firm. That Graphic Sciences had nearly a third of the market after a year of sales, a share it never surpassed, indicated how successful direct competition might have been against Xerox.[60] Not until 1973 did Graphic Sciences introduce a Xerox-compatible machine. By then, the damage had been done. The market fractured and remained smaller than expected as incompatibility frustrated actual users and deterred potential users.

Nonetheless, Graphic Sciences maintained its technological leadership, offering options for speed, resolution, and type of telephone connection—seventeen models by 1975, including the first three-minute machine, the dex 180. Financially, however, Graphic Sciences suffered from price wars with Xerox and Magnavox and an attempted takeover by the firm distributing dex equipment in Canada and the United Kingdom.[61] Facing "rapid technological changes in the industry and substantial competition from other major companies," the directors decided to be bought. After discussions with more than twenty firms, Graphic Sciences early in 1975 became part of the Burroughs Office Product Group.[62]

While not technically superior to Graphic Sciences and other competitors, Xerox's reputation, sales force, and service combined with machines that performed sufficiently well at a reasonable price enabled the firm to seize and hold a large chunk of the market. Although Xerox remained the market leader and also introduced new machines, as early as 1971 market analysts wondered whether "gimmickry, price cutting, and the sheer power of a numerically superior sales staff to move an unexceptional transceiver line" would suffice or if the firm should withdraw completely.[63] One glaring shortcoming was the inability of some of its machines to handle legal documents, essential for law firms. The problem was not technological—competitors offered such machines—but a failure to appreciate what the market could be.

Xerox at least had the troubles of the leader. What had been the fax market—photographs, maps, military needs, and specialized applications like signature verification—increasingly became cloistered niches. Of twelve fax manufacturers in 1961, only a few entered business faxing, and none became major players.[64] Instead of developing attributes like affordability to survive in a more demanding, competitive and uncertain, albeit much larger general market, these niches encouraged other criteria, such as high resolution and high speed. The result was small numbers of high-cost, high-performance equipment. As late as 1981, specialized machines comprised 2 percent of the approximately 100,000 machines produced, but over 11 percent ($34 million) of their $305-million value.[65]

The largest American fax manufacturer since 1956, Stewart-Warner perceived its market narrowly as internal communications for high-volume users like corporate mailrooms, so its equipment, like the 1968 Datafax Electronic Mailbox, was costlier and larger than Telecopiers. Although it developed a digital subminute prototype in the early 1970s, Stewart-Warner decided against competing in the general business market. In 1982, Sanyo bought its Datafax operations.[66]

Alden and Litton stayed in their niche markets, not always by choice. In 1966, Alden had about 10 percent ($2.4 million) of the market and hoped to ride Xerox's entry to 20 percent of the mid-1970s expected $500 million market.[67] Alden emulated Xerox's "Mail Letters by Phone" marketing with its "Mail by Telephone" advertisements, but it did not offer a full-page machine until the early 1970s, too late to make an impact. Instead, Alden retained its focus on special-purpose equipment ranging from extra-wide transmitters for blueprints to thousands of small signature verification machines.[68]

Since purchasing the Times Facsimile Corporation, Litton Industries had become a wonderful example of how acquisitions can fail in their new corporate environment. Although Litton marketed a Japanese bandwidth compression system for newspapers to fax pages to printing plants, the firm decided in 1969 to remain outside the business market by neither acquiring the American rights to sell a Japanese machine nor developing a solid state version of its Messagefax. Instead, Litton offered its major client, the military, the $75,000, thirteen-page per minute Fastfax. But even the military shifted to less expensive equipment.[69]

With its Desk-Fax machines, growing networks, and profitable service, Western Union appeared to have the basis for widespread fax operations in the early 1960s. A decade later, surveys discussed Western Union in the past tense. The company faced three challenges—competition, technology, and management—and failed all three. Western Union was not oblivious, promoting a vision of a unified "recorded message" communications, investing over $500 million in 1966–70 in computer-based services like Mailgram and EasyLink, building a microwave network to deliver "data power" as easily as businesses obtained electricity, and expanding its 28,000-customer telex network by buying AT&T's 42,000-customer TWX network.[70]

But Western Union also provided wake-up calls, delivered candy, and operated in supermarkets, among other activities. The result, the FCC stated, was deteriorating telegram services, partly due to "investing too much time and money in other types of wire communications and not enough in its telegram business." In a self-reinforcing downward spiral, demand dropped as prices rose, and service worsened while new services demanded money to implement, yet failed to replace the lost revenue.[71]

Western Union's focus on fax as an adjunct to the telegram represented a sad case of managerial narrow vision that failed to take advantage of the widest and deepest facsimile technology base through the mid-1960s. As Harris, Upham noted in 1967, "As a communication and service company searching for identity, they seem uncertain about their facsimile future."[72] Instead of building on Letterfax, upper management ignored it.[73] Like Xerox, the firm rejected opportunities to acquire the rights to outsiders' fax technology, possibly because its joint forays, like the 1968 Stewart-Warner-Western Union Info-fax 100, fared poorly.[74] Thus, facsimile's leading developer and user in the 1940s–50s vanished from the growing fax market in the 1970s.

In fairness to Western Union, business fax had failed to catch on. In early 1971, the United States had approximately 20,000 general- and special-purpose fax machines. Xerox, Magnavox, and Graphic Sciences accounted for 13,000–15,000 machines. Stewart-Warner was a distant fourth with 2,000 machines.[75] As the Electronic Industries Association laconically reported, facsimile "was once hoped to be the ultimate communications medium, serving industry, commerce, and consumers alike. Time has tempered this optimism."[76] The problem was not a lack of supply but a lack of demand. Over a dozen firms offered nearly two dozen fax machines renting for $40–175 monthly.[77]

The perennial problems of performance and economics frustrated faxing, albeit for a much larger, less technically sophisticated audience. What would have thrilled a 1950s audience did not impress a new generation and new market of office workers. Although unparalleled for transmitting documents, faxing suffered from slow speed, high costs, unreliable and incompatible equipment, few published directories, poor print quality, and noxious odors.

Slow speed disrupted office routines and kept telephone bills high. Executives did not relish waiting for the six minutes for one page, let alone for a longer document. Faxing a three-page letter either tied up a local telephone for eighteen minutes or, if the fax machine was in a central location, kept a secretary from her desk. By training staff to type during transmissions, the University of Nevada library recovered nearly 75 percent of this dead time — when the noise and odors did not distract the workers.[78] The high cost of long-distance charges was harder to overcome. A daytime three-page fax between San Francisco and New York cost $8.45 in 1970 compared with 6 cents for a letter and $7.35 for a 100-word telegram.[79] Options for reducing these charges were limited. Delaying transmission for the drop in telephone rates at 5 p.m. and 11 p.m. reduced costs but also the utility of faxing. If large enough, the firm might employ a WATS line, which enabled unlimited long-distance calls within a region, but faxing had to compete with regular voice conversations.

Poor reliability hurt early analog and digital fax machines to the point of recalling systems.[80] The inability of EG&G's Bandcom 1000 fax machine to reliably maintain a stable transmission forced the firm's Graphic Transmission Systems subsidiary to withdraw from the business market in 1970 and instead focus on specialized systems for the government.[81]

Fax machines still consisted of very complex, delicate electromechanical equipment — like Xerox photocopiers, reliability problems and the counterbalancing service contracts were normal. Worse was incompatibility. As analysts accurately warned, "Without such compatibility, facsimile will at best be confined to a small corner of the communications world, playing the role of the in-house telephone" instead of becoming an "essential part of every home or office communications center."[82] More than one commentator unfavorably compared the universal service provided by telephone with the incompatible service offered by fax machines.[83]

The business model of closed networks and intra-organization faxing did not need compatible machines. By locking a customer into a firm's machines, incompatibility benefitted that manufacturer by reducing competition. Rarely was the argument so open. Instead, Graphic Sciences claimed compatibility would stifle technological development by reducing a firm's incentive to upgrade its equipment. Developing their own technologies also enabled companies to avoid licensing or infringing patents.[84] In reality, deliberate incompatibility fragmented the market and scared away potential users fearful of choosing the wrong system.

Incompatibility did generate innovative "work-arounds." Companies offered "gateways, adopters, converters, or similar technologies" and services to receive, convert, and transmit faxes between incompatible machines or even between different types of telecommunications such as telex and fax.[85] Other work-arounds included acquiring different fax machines to ensure that faxing could always occur and creating directories of compatible machines.

Ignorance added to incompatibility. Decentralized acquisition meant large organizations rarely knew how many and what kind of fax machines they had. Often it took an inventory ordered by top management to discover, for example, that DuPont employed thirty-four models from seven vendors. Even the military never knew the exact number of its machines.[86] Nor did ignorance stop there. A continual challenge was discovering whether someone had a fax machine, whether it was compatible, and if so, its number. The answer, a directory, suffered from an obvious problem of obsolescence and a less obvious problem of providing prospective clients to competitors: Graphic Sciences and Magnavox/3M sales agents used Xerox's Teledirectory to identify potential prospects.[87]

Ignorance also extended to faxing in general and a firm's products in particular. Education meant telling customers about the existence of faxing, which meant marketing, which took nationally known firms with large advertising budgets—Xerox in the mid-1960s, Exxon a decade later, and Federal Express in the mid-1980s. For the overwhelming majority of potential users, facsimile was an unfamiliar concept. Indeed, in a 1967 *New Yorker* article about Xerox, John Brooks described the firm's two fax machines without mentioning the "f" word: "the LDX, by which documents can be transmitted over telephone wires, microwave radio, or coaxial cable; and the Telecopier, a non-xerographic device, designed and manufactured by Magnavox but sold by Xerox, which is a sort of junior version of the LDX and is essentially interesting to a layman because it consists of a small box that, when attached to an ordinary telephone, permits the user to rapidly transmit a small picture (with a good deal of squeaking and clicking, to be sure) to anyone equipped with a telephone and a similar small box."[88]

For actual selling, firms depended on hundreds of sales agents who proselytized for faxing, educating the ignorant and convincing the skeptical to invest in this new product. Simply explaining to potential customers what a fax machine was and how it could benefit that firm comprised a major part of selling. As, in Hugh Aitken's words, "the individuals who moved information from one system to another, interpreting it and changing it in creative ways as they did so," sales agents played an essential role in the commercialization and diffusion of faxing.[89] Sales agents proved so important that when a new firm entered the fax market, one indication of its seriousness was whether it hired leading agents from existing firms.[90]

Judging by sales literature and internal manuals, agents focused on price, performance, and productivity, and viewed their competition as telex and other firms' facsimile offerings.[91] Good agents not only sold machines, which produced their commissions, but provided their firms with feedback about how customers actually used their machines and what they wanted. Very good agents demonstrated their worth with quick thinking. When negotiating a contract for twenty-six machines with Ford, a 3M sales agent was asked if he had ever owned a Ford. The salesman promptly responded that "by tomorrow my wife Barbara will be the proud owner of a Ford product." He drove her car to the contract signing.[92]

Companies tried to distinguish themselves by name as well as product. Just as promoters at the turn of the twentieth century employed high-tech names like the Electrograph, so too did manufacturers choose names to link their fax machines with desired attributes, such as Telautograph's Quick-fax and 3M's Versatile Remote Copier (VRC). The goal, never met, was, as Qwip president John Cochran stated, to "make Qwip as synonymous with the concept as Xerox is with copiers."[93] That no firm's name ever became a generic noun or verb for the industry, unlike Xerox in photocopying, reflected the inability of one firm to dominate the fax market.

Xerox transferred to faxing its highly profitable photocopier leasing and metering policy, which rented each photocopier with a set amount of paper and a charge for each additional sheet.[94] Unfortunately for Xerox, faxing suffered from the low consumption of paper and the ease of breaking leases. Per page charges did not produce profits because consumption proved so small compared to the amount used for photocopying. The Telecopier's 200 monthly rent included 300 pages (approximately 15 pages/business day), with a charge of 15 cents for the next 400 pages and 5 cents a page thereafter. Yet when "serious business users" consumed 10–15 documents daily, most renters did not even meet their monthly quota.[95]

The high cost of fax machines meant most users leased, rather than bought, their equipment. Leasing benefitted users by reducing their risks of choosing the wrong equipment and avoiding a costly purchase in their budgets. As an operational and not a capital expense, leasing did not distort budgets or attract supervisors' attention, even though leasing was more expensive than buying after as few as two years.[96] Leasing usually included service, relieving users of the worry of repairing their equipment.

Leasing benefitted manufacturers by providing a steady and profitable revenue stream after recouping the equipment cost and by encouraging new users to experiment with a new machine. The risk was users renting for a short time and then cancelling the lease. Fax manufacturers justly feared this churning: customers ended half of their Telecopier leases in 1971 and a third of Qwip leases in 1975.[97] The reasons varied—unrealistic expectations (or unrealistic sales pitches), switching to a competitor, corporate cost cutting, inadequate use, high monthly costs—but the result was very high return rates.[98]

USERS

The federal government remained the single largest user, comprising in 1971 approximately one-third of the fax market, a share that dropped to a tenth in 1977 as the new technology diffused to wider audiences. Usage differed markedly by agency and agenda. Those dealing with national security were more likely to fax: the military accounted for over half the government's share in 1971 and a third in 1977.[99] It had the financial and human resources, less concern with cost, worked with complex technologies daily, and had the demand. The Air Force F-15 flight testing program alone transmitted more than 5,000 pages monthly (250 pages a business day) via eight Dacom terminals in 1974.[100] In contrast, 60 percent of all government users in 1976 sent two or fewer faxes daily.[101] Indicative of faxing's slow progress, a quarter-century earlier Western Union had defined a small user as generating five to twenty telegrams daily.[102]

The Cold War arms and space races encouraged faxing. Their urgency and cost-plus contracts meant that contractors valued time over money. The thousands of engineering changes in development and construction of large-scale projects demanded the precise and coordinated exchange of information. NASA engineers joked that their most valuable communications tool was the T-38 jet, flying revised engineering drawings around the country.[103] NASA also employed fifteen LDXs for high-speed, high-quality transmission in 1971 as well as approximately two hundred slower Magnafax 850s. By 1981, NASA employees faxed over 60,000 pages monthly.[104]

Like their major customer, aerospace firms employed faxing extensively.

Lockheed had been an early adopter, faxing engineering drawings between its Sunnyvale, California, headquarters and Cape Canaveral since the late 1950s. As its projects grew in complexity and geographic dispersion, so did its faxing. For Lockheed's C-5A airplane, a nationwide network of fax machines reduced coordinating an engineering change from four days to four hours. In 1966–67, the firm faxed over 5000 changes.[105]

The Boeing Space Division installed six LDXs to produce graphics for teleconferences (a word Boeing invented) for its 1960s Apollo work. Although originally done to save time and money by reducing travel, the project's manager considered the improved information flow "probably of much greater value." In addition to teleconferences, which generated up to 10,000 pages monthly in 1968, the LDXs faxed so much information that the Division's office in Washington, D.C. cancelled its teletype service, a forerunner of what would happened when faxing became popular.[106] In the 1970s, Boeing was an early and large customer of the expensive subminute Rapifax, ordering fifty initially and doubling that order within a few years as the space shuttle program grew. By using fax machines to coordinate activities, rush orders to the shop floor, and keep administrators informed, Boeing saved transaction time and telephone costs, and thus money.[107]

Outside aerospace, faxing made fewer inroads. Based on the responses of 232 (of 1,250) firms in 1971, SRI concluded that American businesses remained "essentially virgin territory." Over 90 percent of faxing was internal to a firm or linked specific contacts, such as an architectural firm to a construction site. Articles and advertising reinforced this intra-organizational perspective of closed networks of communications.[108] Railroads employed approximately 400 machines, benefitting from fewer errors in transmission and easier tracking of them when they occurred because faxing left a paper trail.[109] Southern Railway operated the country's largest LDX network and also used Stewart-Warner Datafax machines to fax reports, switching orders and other urgent operating data among its fifteen divisional offices.[110]

The few inter-company networks had specific needs like speed and accuracy.[111] Because faxing eliminated transmission errors and reduced fraud by providing the authenticity and authority of letterheads and signatures, financial transactions comprised a large part of international faxing.[112]

Facsimile served as a centralizing technology to speed internal correspondence from mailroom to mailroom. Installing an inter-mailroom fax system was organizationally easy. Communications managers preferred placing the expensive equipment in a centralized location or mailroom to maximize use instead of departments operating their own machines. The General Service Administration implemented this approach at its sixty-six Federal Telecommunication Record Centers. Unfortunately, as the Air Force had learned in the 1950s and countless organizations relearned, centralized faxing deprived users of its major benefit, immediate and convenient access.[113]

New applications usually faced challenges more social and political than technological. Users had to be willing to fax. Not all were. English Ford Motor executives, for example, proved remarkably resistant to their communications department's inducements to fax, which included a portable trolley-mounted fax machine. Not even the 1971 British postal strike could induce them to use the new mode of communication.[114]

Pushed by fax manufacturers, a few firms linked machines with different capabilities into internal networks to promote horizontal as well as vertical communications.[115] In some areas, faxing began to change the nature of work and not just accelerate the speed of communication. Reducing labor costs by replacing a dedicated teletype operator with a less-expensive clerical worker attracted some firms.[116] Faxing saved the time of more skilled (and expensive) workers. When Magnafax transmissions replaced engineers' telephone calls from International Harvester's tractor testing grounds to its headquarters, errors decreased and draftsmen did not have to duplicate work.[117] Faxing enabled closer coordination and faster business procedures. RCA's Distributor and Special Products Division installed more than 200 3M fax machines at its largest distributors to handle thousands of monthly orders. Tellingly, some distributors then installed their own equipment to accelerate their own data flows.[118]

Not all institutional challenges concerned the actual fax technology itself. In 1975, the New Orleans Security Association for Financial Institutions established a fax network to send urgent alerts about con artists, forged checks, and other bank crimes. The promoters had to convince bankers to take cooperative security seriously and realize that a crime against one bank was easily perpetrated against others unless they shared information. The effectiveness of the system, including helping to arrest a woman with four hundred stolen travelers' checks, eventually swayed doubters.[119]

A few journals, like *Business Week*, faxed to speed writing and editing between branch offices and headquarters.[120] Newspapers, facing daily deadlines, adopted faxing as "a sort of error-free teletype network" between domestic offices and for reporters to send their copy to editors instead of telephoning or telegraphing their stories.[121] As gonzo journalist Hunter S. Thompson raved, "This is a fantastic machine, and I carry it with me at all times. All I need is the Mojo wire and a working telephone to send perfect Xerox copies of anything I've written to anybody else with a Mojo Wire receiver . . . and anybody with $50 a month can lease one of these things. Incredible. What will they think of next?"[122]

Faxing helped create the national and worldwide newspaper. Pioneered by Japanese papers in the 1960s, faxing full-size layouts of pages enabled printing newspapers anywhere a satellite printing plant existed, whether a few miles, a state, or a continent away. Because of the large quantities of information transmitted, data compression proved economically essential to reduce the cost of renting wideband circuits. Indeed, Litcom in 1968 became the first American firm to market a Japanese fax system, a bandwidth compression system for newspaper transmission from Toho Denki, the forerunner of Matsushita Graphics, and newspapers gave Dacom its first major non-military sales.[123] The *Wall Street Journal* used the equivalent of a television channel (3 megacycles, equal to 800 telephone circuits) to transmit pages from San Francisco to Riverside. The extremely wide bandwidth allowed the $48,000 Westrex Pressfax system to send a newspaper negative at 800 lpi resolution in only 4.5 minutes.[124] Slower systems used less bandwidth (48 kc for the Japanese systems) for less cost but more time.

By 1980, faxing had made national versions of *Pravda*, the *Wall Street Journal*, and other newspapers unexceptional. The *International Herald Tribune*, which labeled itself a global newspaper, began faxing its layout from Paris to London in 1974, to Zurich in 1977, and to Hong Kong in 1980. Instead of flying copies to Asia for delayed distribution, Hong Kong printing enabled same-day distribution, lower costs, and greater circulation.[125]

Libraries experimented again with interlibrary faxing. The expansion of universities in the 1960s, the rising costs of magazine subscriptions, and tight budgets pressured libraries to coordinate, not duplicate their holdings. Faxing seemed a solution that would also reduce the time to send an article from days or weeks to hours. Tests discovered unexpected benefits from the first systemic analyses of interlibrary loan operations, conducted to compare the cost of faxing, and improving administrative procedures. For the Smithsonian, faxing meant "no records to be filed or pulled from our circulation records, no overdue notices to borrowers, no lost pieces to track down, and no handling and packaging to return the loans."[126] The tradeoffs of poor image quality and high costs (up to $2.50 per page) due to long-distance charges and retransmission, however, outweighed benefits for most libraries. Only Hawaiian libraries, which already used WATS lines and suffered from slow mail from the mainland, found faxing unambiguously positive.[127]

New technologies offered opportunities for entrepreneurs to provide new services and products. For facsimile, firms attempted to profit by providing transmission, compatibility, and information services. Most impressive were the large number of companies entering what they perceived as a promising market, the wide and evolving range of services offered, and their rapid winnowing. The first wave in the late 1960s and early 1970s consisted of public fax services. The basic principle was simple. The document was delivered to the store, transmitted to the closest receiving station, and then picked up, delivered, or, like Picture Telegraphy systems four decades earlier, mailed.[128]

Typical was Facsimile Transmission Network, which introduced its Faxmail in 1970 with an impressive network of more than two hundred cities. A walk-in customer paid $4.50 a page and the long-distance charges, while subscribers paid a third less.[129] Unfortunately, public faxing suffered from high costs, poor publicity, unreliable pick-up and delivery, and too many firms dividing a small pie. Despite the opportunity provided by the March 1970 postal strike, sales were poor. Of eight firms in 1969, one had gone bankrupt, six had severely retrenched, and only Vistagram Centers, backed by the resources of parent Graphic Sciences, remained in 1971.[130]

Not all services failed. The most successful applications occurred where people willingly paid for speed and convenience, such as by obtaining commercial truck permits. To avoid high annual fees, truckers preferred to buy state permits valid only for a day or week. Before faxing, they had two options. If they knew their schedule, they could apply by mail. More realistically, drivers stopped at the first town in a state to telegraph their fee to the state license bureau and waited for the approval. Sensing opportunity, American Facsimile Services (AFS) and Transceiver East offered their respective Insta-Com and Instacom permits. The driver stopped at a truck stop to fill an application, which a service-provided machine faxed and replied with the permit within 20 minutes. By 1976, these firms filed over a million permits annually, earning $7.5 million.[131] By 1985, AFS had 18 franchises, 3,000 registered trucking firms and truckers, and fax machines in 800 truck stops.[132]

STARTUPS AND SUBMINUTE FAXING

The continuing flow of startups indicated that many entrepreneurial engineers and investors saw potential in facsimile. Their near-universal failure to translate innovative technologies and promising ideas into commercial success also demonstrated that potential remained in the future. Often a patent was the firm's only or main asset. Faxon, which failed in the early 1970s, recouped part of its investors' funds by licensing its major patent in the late 1980s to fax manufacturers.[133] Small size meant startups lacked the financial resources to survive if their products suffered delays or problems. Nor could they easily afford to market their equipment to customers justifiably skeptical about the new firm's prospects.

Access to funding was essential. Graphic Sciences's ability to raise $40 million set it apart. Their inability to raise enough capital meant Electronic Image Systems (EIS) and Visual Sciences were purchased by Addressograph-Multigraph and Matsushita Graphics respectively. Faxon's inability to secure funding or a partner to manufacture its analog white-space skipping technology in the early 1970s meant its window of opportunity vanished as development shifted to digital faxing and Japan. Only infusions of money from CBS, Savin, and Ricoh and government contracts enabled Dacom to morph into Rapifax.

Nowhere were startups more important than in subminute faxing, the most technologically challenging sector. The strategic question facing manufacturers in the early 1970s was whether to leap from analog six-minute to digital subminute machines or to pursue the less-challenging three-minute analog machine. The main digital promoters were startups and their government patrons, primarily the national security community and NASA. The more conservative analog proponents were the firms actually producing machines, which had both experience making and selling equipment and a vested interest in maintaining their existing products.

In some ways, the dichotomy was false. Digital's advantages had been recognized since the mid-1950s. What was new was the growing sense that, as EG&G engineers proclaimed in 1965, existing equipment was "generally two decades out of date" and harnessing contemporary technology offered a daring firm a profitable market.[134] Digital faxing also promised compatibility with computer-based communications, an increasingly influential vision of the future.[135] The battle revolved around the classic dilemma of technological evolution: Should firms invest in incremental innovation or develop more promising but more demanding radical invention?

Three-minute analog proved the more successful path in the mid-1970s. Easier to develop and manufacture, these G2 systems proved that the technically good is often a commercially wiser choice than the best technology. Reducing vertical resolution to 64 lpi from 96 lpi enabled a six-minute machine to transmit a page in four minutes and a three-minute machine in two minutes (hence, the terms 6/4 and 3/2 machines).[136] As discussed below, some manufacturers even offered analog data compression.

In contrast, advanced digital technology proved a recipe for frustration and failure. Many were the Loch Ness machines, whose existence was rumored but never surfaced. Only a few subminute machines moved beyond prototyping, and only Rapifax survived commercially before the late 1970s. The technology and economics simply proved too daunting. As *Fortune* predicted in 1973, the necessary sophisticated electronics, "similar to that of a small digital computer," cost too much.[137] Subminute was a speed too fast unless users had heavy fax loads and deep pockets, like the military.

Institutional and broader economic forces hurt too. Large firms spurned ventures with startups. Most importantly, Xerox did not take advantage of several opportunities to buy, invest, or form ventures with small firms offering digital and other advanced technologies.[138] The 1973–74 recession forced the cancellation of thousands of leases, pushing manufacturers to focus more on maintaining existing markets than introducing newer, more expensive machines.[139]

Subminute machines combined multiple promising technologies. The components in Graphic Sciences' 1973 feasibility model — "laser technology, solid state imaging devices, complex MOS and bipolar LSI circuit technology, data compression techniques, and flatbed scanning" — evolved into standard features in the G3 machines that revolutionized faxing after 1980.[140] Hope and expectation again outran economic and technological reality, but this time by more firms on a larger scale. In 1971–72, the challenges did not seem insurmountable save to Xerox, which dropped its subminute project due to its high cost. Several firms planned to introduce subminute machines in 1973 for $150–300 a month.[141] Instead, aborted machines littered the digital landscape. Sometimes the corpses were only projects; other times, the failed efforts also destroyed the firm, or at least its independence, as EIS and Comfax illustrated.

Robert E. Wernikoff and EIS demonstrated not only the pitfalls that startups faced in transforming their ideas into products but also the indispensable role played by the federal government, MIT, and firms around Boston's Route 128 in the American electronics industry in the 1960s. In the 1950s, Wernikoff received three MIT degrees while studying with information theory pioneers Claude Shannon, Norbert Weiner, and Y. W. Lee. MIT also trained his EIS colleagues, Paul Epstein and William Schreiber, who also had expertise in information and coding theories. The Air Force funded Wernikoff's research on data reduction of images.[142] After working for a contractor to MIT's Lincoln Laboratory and the federal government, Wernikoff became convinced — and convinced several investors — that digital faxing promised faster transmission and higher resolution than analog. Founded in September 1963 and initially operating out of the basement of defense contractor EG&G, EIS developed a fax machine, the Telikon (from the Greek *telis*, 'convey or transmit,' and *ikon*, 'image').

Turning its patent into a product proved much harder than anticipated. As EIS discovered, a complex technology consisted of many other technologies whose development and integration proved to be major challenges.[143] Just as Telikon testing began in late 1964, financial survival dictated that EIS shift to building four photofacsimile recorders to turn NASA Nimbus weather satellite data into negatives or prints and digitize it for computer processing. While it was a significant achievement, the contract monopolized the firm's main engineer, James E. Cunningham. Moreover, an endless flow of demanding documentation, quite undesired by a struggling small firm, accompanied the NASA contract.[144]

Lacking the resources to complete and market its Telikon, EIS sought a larger partner. However, in 1963 and 1965 Wernikoff failed to convince Xerox to incorporate his data compression process into the LDX.[145] Demonstrations and discussions with firms ranging from diazotype copying firm Bruning and US Steel to the Wall Street Journal generated interest but not investment (save for a small contract with Bruning). For Wernikoff, life was schizophrenic, alternating between euphoric demonstrations that would have sold a hundred machines "if we had any to sell" and the depression of "endless visits, telephone conversations, and strategic sessions" over contracts. Negotiations concentrated as much on the exclusivity of licenses, advance payments, and royalties as on the actual technologies.[146]

In late 1964, their "cash position, which is despairingly thin at best, curled up and died" when NASA and Bruning delayed reimbursing EIS. Only a personal loan of $20,000 from Wernikoff and Epstein, neither of whom had taken a salary for months, maintained the firm.[147] After inconclusive discussions with Xerox and Western Union, in 1966 EIS became a division of Addressograph-Multigraph, an office equipment manufacturer.[148] The new owner heralded Telikon as EIS's "most spectacular new development," but ominously that praise appeared only after mentioning the three EIS products with customers.[149]

Despite outside interest—the firm repeatedly rebuffed AT&T requests to rent systems—Addressograph-Multigraph slowed Telikon development because its estimates of demand declined from 4,800 machines in 1967 to 1,800 machines in 1970 and of monthly rent rose from $600 to $1650. The slow acceptance of analog faxing shaped this growing pessimism.[150] EIS found its resources increasingly diverted by its parent company into areas promising larger markets, such as converting pictures into electronic data.[151]

The Telikon II proved to be an over-the-horizon product, always due to enter the market but not quite yet. In 1967, EIS predicted its first delivery in 1968 and demonstrated the Telikon II to hundreds of military, government, and corporate officials. The demonstrations continued in 1968, but expected delivery slipped to 1970.[152] As late as mid-1970, the business press and Wernikoff expected a 1971 introduction.[153] Instead, Addressograph-Multigraph downgraded EIS organizationally in October 1970, prompting the three founders to leave. Full-scale introduction never occurred, although the National Security Agency bought two units for Richard Nixon's San Clemente "Western White House" to receive encrypted material from Washington, D.C.[154]

Addressograph-Multigraph's poor finances also hurt Telikon. Its new copying machine, a higher priority than Telikon, had required an expensive reengineering effort that absorbed company resources. The 1970 recession had reduced demand for office products, so the firm was firing employees and delaying new offerings. The economy recovered, but Addressograph-Multigraph's commitment to facsimile did not.[155]

Electronics Associates, Inc. (EAI), formed after World War II to produce analog and hybrid computers, had entered facsimile in 1969, understandably thinking its computer expertise would enable it to create the next generation of fax machines. Finding that developing a subminute machine was more difficult than expected, EAI joined fellow startup Comfax in producing prototypes of the 250-pound subminute Comfax/EAI Fax-1, the "most advanced machine available anywhere" in late 1972.[156] Low reliability and high cost—$335 a month excluding the automatic document fielder ($30–40), the $600 annual maintenance contract, a leased line, and, of course, a second machine at the other end of the line—meant that the seventy machines produced in 1973–75 were soon abandoned.[157]

Not all new firms failed. Access to capital and corporate support enabled Dacom/Rapicom and Visual Sciences to turn their patents into products. Established in 1966 by two Lockheed engineers and funded by the military, Dacom specialized in data compression.[158] The firm expanded to fax systems tailored for niches where speed was more important than cost, such as encrypted communications and full layout pages for newspapers, placing 150 machines by 1975. The government, primarily the Defense Department but also the White House, was its largest client.[159]

Dacom's significance lay in its successful expansion from government markets to sell the first commercially profitable civilian subminute machine. This process took nearly a decade and repeated injections of capital. In December 1970, a partnership between CBS subsidiary Colfax Inc. and Savin Business Machines Corporation bought 60 percent of Dacom to develop a high-speed, low-cost business fax machine. With Ricoh of America, Savin's photocopier supplier, they formed the Rapifax Corporation. Four years later, Ricoh bought out the other parties despite a lawsuit from Dacom.[160]

Ricoh's investment produced its first visible fruit in 1974, two years behind schedule. The Rapifax 100, essentially a less costly and less capable Dacom, was designed in the United States and manufactured in Japan. Very reliable, it transmitted one page in 35, 50, or 90 seconds, depending on the resolution (67, 100, or 200 lpi). Rapifax targeted heavy users (defined as more than 10 pages daily), exchanging high rentals for lower long-distance charges. The *Chicago Tribune* saved more than $3000 annually between its Washington bureau and Chicago newsroom. More importantly, the savings in time—6 minutes for a 6-page story compared with 36 minutes for their Telecopiers—eased deadline pressures. Unsurprisingly, the federal government was a major customer.[161]

High speed had been achieved, but not low cost. Though half the cost of its Dacom sibling, Rapifax's rent of $300–350 contrasted with $60 for the six-minute Telecopier 400 and $80–90 for the three-minute dex 180. Newer three-minute machines slashed Rapifax's potential market. Even though its 1,500 machines monopolized high-speed faxing until 1977, Rapifax never had more than 4 percent of the total market and made a profit only in 1979.[162]

The most successful small firm was Visual Sciences because it forged advantageous international alliances to manufacture and market its machines with three large firms, Matsushita Graphics in Japan, Plessey in the United Kingdom, and 3M in the United States. Like Graphic Sciences, Visual Sciences emerged from a corporate disagreement. After failing to convince Litton to expand into business faxing, chief engineer Frank DiSanto, financial vice president Walter Scheer, and some of the Times Facsimile staff who joined Litton quit to form Visual Sciences in 1969. Financing came from Dennis Krusos, a small integrated circuit manufacturer.[163]

Friends since 1954, DiSanto and Krusos were convinced that the Carterfone decision had opened a new market. Their assets were DiSanto's design for a machine and good relations with Toho Denki, a Japanese manufacturer of fax machines. Those relations had developed when Litton acquired the American rights to Toho Denki's bandwidth compression system for transmitting newspapers.[164] Krusos and DiSanto and Konosuke Matsushita, the firm's founder, signed an agreement that gave Visual Sciences the world market outside Japanese, Taiwan, and Korea for any desktop transceiver Toho Denki manufactured, a very advantageous position—if Visual Sciences could exploit it.[165]

Visual Sciences tried to find a large firm to market its machine in North America. After approaching nearly forty electronics and telecommunications companies, it was still looking. The risk was perceived as too high—after all, as a Western Union official told them, nothing could replace the teletype. Turning to

Europe, DiSanto and Krusos signed an agreement with the British telecommunications firm Plessey in December 1971 for a thousand machines. This agreement convinced 3M in 1972 to buy 16,000 machines over four years. Years later, Frost & Sullivan concluded that the high targets "can only be credited to 3M's enthusiasm for a new market, and the sales skills of Visual Sciences."[166]

Disappointing sales prompted Plessy and 3M to decline to develop a new machine with Visual Sciences. After unsuccessful negotiations with Xerox, Visual Sciences and Matsushita Graphics (Matsushita Electric Industry purchased Toho Denki in 1972 and renamed it Matsushita Graphics Communications System) agreed in 1977 to develop a three-minute machine and establish an American marketing firm, Panafax. Aggressively marketed as the first machine to meet the new G2 standard, the MV (Matsushita Visual) 1200 desktop proved very successful.[167]

What Panafax did not do was make money. When Visual Sciences could not provide new capital, Matsushita Graphic's share grew from 51 percent to 75.5 percent as did the firm's frustration with the restrictive agreement. In May 1982, Matsushita Graphics bought out Visual Sciences, paying $31 million for its quarter-share in a firm with a negative worth of $11 million. The result was a revitalized Panafax, fully controlled by Matsushita Graphics, just as the American market began to explode.[168]

LATE 1970S GROWTH

By the late 1970s, fax finally generated profits for the major manufacturers—Xerox, Graphic Sciences, and 3M—but these profits were low compared with their other operations. Its $50 million investment had not produced profits for ten of the twelve years Xerox sold facsimile. One problem that facsimile, earning well under $25 million annually, had within the multi-billion dollar firm was its relative unimportance even as Xerox faced an increasingly competitive photocopier market and hostile legal environment. Symbolizing the low priority, Xerox had to deny a rumor in 1977 that it intended to sell its fax division.[169]

Like many market leaders, Xerox moved conservatively and cautiously, as demonstrated by the 1975 three-minute Telecopier 200, which scanned by laser and recorded xerographically but required a hardwired connection. Its $195 rent was substantively more expensive than earlier Telecopiers but not significantly less than the Rapifax.[170] As Frost & Sullivan scathingly reported, "given an excellent market opportunity to dominate a new segment of a growing market which it had created, Xerox, instead, successfully resisted the temptation to step into the jet age and created the Xerox Telecopier 200, which is the equivalent to the finest and last, high performance, propeller driven airplane."[171] Yet the Telecopier 200 proved commercially successful, demonstrating again that an adequate product marketed and supported by a major firm often succeeded better than technically superior competitors from less-known firms.[172]

The lost potential was even greater than it appeared. In 1971, Comfax and EAI had negotiated manufacturing their Fax 1 as a Xerox product. If the proposal had been realized, Xerox would have combined a subminute machine with excellent manufacturing, sales, and service capability. Xerox's corporate staff recommended buying the rights but was overruled by its Office Systems Division. Instead, the division introduced—four years later—the slower Telecopier 200.[173]

Why did Xerox, which created the business fax market by licensing its Telecopier technology from Magnavox, decide against licensing the EAI Fax 1? The apparent cause was internal corporate dynamics. Newly formed in Dallas, the Office Systems Division considered that licensing a machine was not the most impressive way to demonstrate its technological competence to headquarters in Rochester. The Not-Invented-Here factor, an arrogance that implied a technology not emerging from Xerox's research and development was not worthy of Xerox, clearly played a role. Technically, the Telecopier 200 was also an optically challenging machine, and Xerox was an optically oriented corporation. This focus on optics may explain why Xerox also dropped its research with Fairchild Camera & Instrument on charge-coupled devices (CCDs), which transformed fax scanning in the 1980s—in Japan.[174]

Slow internal development cost Xerox time and technological leadership to other firms, none of which had Xerox's finances, reputation, or marketing. The greater loser was the fax market, deprived of impressive digital technology for several years. The fax market had been Xerox's to lose, and lose it did. From more than three-quarters of the market in 1971–72 to two-fifths only five years later, Xerox's share steadily declined as competitors introduced newer, more capable equipment.[175]

Just as the entry of Canon, Kodak, and IBM devastated Xerox's leading position in photocopiers, so too did the entry of Qwip, 3M, and Rapifax provide serious competition at the low, middle, and high ends of the facsimile market. Exxon Enterprises' twenty-two startup companies absorbed some of Exxon's profits from the post-1973 growth of oil prices and produced some fascinating office equipment. Taking liberties with the English language, Exxon (itself an artificial word) introduced the Vydec word processor, the Qyx electronic typewriter, and the Qwip facsimile transceiver.[176]

In 1974, Quip Systems introduced the simplest and least expensive fax machine yet, breaking from the network model of use. With its low $29 rent, half the rate of the Telecopier 400, the six-minute Qwip 1000 was, to use president Richard Nelson's analogy, the Kodak Instamatic camera compared with the competition's Nikons. The price attracted thousands of new users and existing users switching from other machines.[177]

Like Graphic Sciences, Qwip deliberately chose to be incompatible with competitors, this time resulting from the decision to cut costs. And Qwip did this by eliminating features previously considered essential, including framing and synchronization. No framing meant the received document could begin anywhere on the page. Eliminating synchronization was not a problem if both machines used the same power grid. If not, however, the copy could be unreadable.[178]

Analysts questioned how Qwip could profit by renting a $29 machine.[179] A partial answer was Qwip's policy of returning machines to central depots instead of fixing them on site (which proved disastrous when the depots proved incapable of keeping track of machines sent and received). A more complete answer was the deep pockets of Exxon absorbing Qwip's losses. Qwip moved upmarket with more costly, capable, and compatible new machines, growing from 1 percent of the market in 1977 to nearly a quarter five years later. By 1980, Qwip trailed only Xerox in installed units (50,000 versus 64,000, twice the number of 3M or Graphic Sciences), and it had placed more machines than anyone else since 1977.[180]

Qwip contributed significantly to facsimile. Its low cost made facsimile economically attractive for tens of thousands of users, if only to experiment. Low cost also necessitated new ways of marketing that went beyond sales agents. Selling through United Telephone System, the second-largest independent telephone company, began the slow shift to fax machines as a commodity.[181] Finally, Qwip's heavy advertising, like Federal Express's ZapMail a decade later, popularized the concept of facsimile.

By the late 1970s, a sense — or hope — emerged that facsimile was finally moving from "communications' stepchild . . . to forming the cornerstone of electronic mail systems."[182] *Business Week* announced in February 1978, in a statement off by less than a decade, "This may finally be the year when the facsimile industry starts living up to the glowing forecasts long made for it."[183] There were reasons for renewed optimism. Diffusion was wide if not deep in corporate America. A 1977 survey found 97 percent of the 900 largest U.S. corporations used facsimile, a major improvement from the 1971 SRI survey. Several firms operated more than a hundred machines, and two-thirds of the respondents planned to expand their applications and would increase their usage further if the cost of high-speed equipment dropped.[184]

Relative costs had decreased significantly. In 1968, a Telecopier II cost $75 monthly, and sending one page across America cost $4.05. By 1978, when that $75 was worth $130 adjusted for inflation, $65 leased a 3-minute Qwip Two, and the telephone call cost $1.37.[185] The breakeven for renting a subminute machine compared with a G1 or G2 machine was fifteen pages a business day in 1978, but only twelve pages, 20 percent less, three years later.[186]

Expectations of major technological advances encouraged fax's prospects. Satellite-based wideband transmission networks promised corporate faxing at one page per second, while the convergence of the fax and photocopier would make the fax machine both a local and long-distance duplicator.[187] White-space-skipping for analog machines to transmit a page in only one or two minutes, unsuccessfully advanced by Faxon in the early 1970s, finally matured in the late 1970s in Japan.[188] More concretely, faxing generated an estimated $200–300 million for long-distance carriers annually in the late 1970s, attracting specialized common carriers such as Graphnet, which leased AT&T lines for fax traffic.[189]

New services offered compatibility among competing machines as well as store-and-forward. The expectations proved, again, overly optimistic: Faxpak, part of ITT's digital Com-Pak communications network, began operations in 1979 two years late, delayed by designing software to connect incompatible machines. That compatibility proved to be important: a quarter of Faxpak's 4,000 subscribers in 1981 linked their Rapifax machines with more than sixty different G1 and G2 machines.[190] Faxpak and competitors like Sprint's Speedfax lost tens of millions of dollars as businesses failed to subscribe in the expected numbers. Nonetheless, to maintain their technological credibility and larger competitiveness, firms expanded these services to include electronic mail text-to-fax conversion and fax gateways.[191]

Businesses used compatibility services because new analog and digital machines remained incompatible and expensive, restricting their market. To gain a competitive edge, the Japanese firms kept their white-space-skipping algorithms incompatible, ensuring that the most advanced as well as the oldest generations of fax machines remained unable to communicate. As troubling as the continued incompatibility, continuing over-competition threatened to continue, as probable supply outstripped predicted demand. Frost & Sullivan warned in 1977 that "barring a technological breakthrough, which is quite possible," demand was growing too slowly for the large pool of current and planned manufacturers to profit.[192]

As part of the larger information revolution (a concept becoming popular), faxing was caught in conflicting visions of the future. These visions were important because they shaped popular, government, and corporate perceptions, thus affecting research priorities, equipment purchases, and other decisions. An optimistic future for faxing collided with visions of "office automation" (OA), "the highest profile and most hyped development innovation of corporate computing during the second half of the 1970s and the first few years of the 1980s."[193] The goal envisioned seamlessly connecting electronically hitherto independent equipment for the effortless, paperless flow of ideas and information for "the office of the future."[194]

A movement promoting OA developed, complete with conferences sponsored by professional societies (e.g., the 1984 Institute of Electrical and Electronic Engineers International Conference on Office Automation), extensive press coverage, much corporate research, and the inevitable consultants. On a practical level, a continual flow of new products entered the office. Typing moved from manual to electric typewriters and stand-alone word processors, and then jumped to personal computers, which themselves became increasingly connected and multifunctional.

Consequently, the main market for facsimile, the business office, itself underwent considerable theoretical and actual changes in the decades that fax machines tried to become standard equipment. Underlying these changes were the drive for faster, more efficient communication and efforts by manufacturers and firms to sell their equipment and services.

Because G1-G2 facsimile was analog, faxing seemed to fit poorly into OA assumptions of word processors, personal computers, workstations, and other equipment linked into a digital network. As *EMMS* paraphrased the dominant industry view in mid-1983, "The technology is often considered to be a bit dated, out-of-step with the technologies now being integrated in the office of the future. Thus stand-alone facsimile is often looked at as a technology whose days are numbered."[195]

JAPAN

By the late 1970s, Americans increasingly acquired fax machines manufactured in Japan, reflecting lower production costs but also such advanced capabilities that a NEC executive labeled fax the "fifth generation" of high technology exports, following cameras, radios, televisions, and cars.[196] The contrast with struggling American startups half a decade earlier could not have been greater. That surge in exports reflected the dynamic growth of facsimile in Japan, growth that Japanese fax manufacturers expected to increase.[197]

Japanese society was based on handwritten, not typed, communication. Its written language of 92 phonetic *hiragana* and *katakana* characters and 2,000 basic (and over 50,000 total) ideographic *kanji* characters fit poorly into the regime imposed by Western telegraphy. *Hiragana* words can have different *kanji*. For example, 'bridge' and 'chopsticks' are both phonetically *hashi*, but are written differently in *kanji* because a simple phonetic structure means the language has many homonyms. In addition, a *kanji* can have multiple pronunciations and meanings.[198] The complexity of *kanji* typewriters—one for newspaper printing had 192 keys with 13 shifts—made typing a highly skilled profession limited to newspapers, government offices, and a few other areas.[199]

Transcribing Japanese into English characters (*romanji*) to send by telex and retranslating the received message were laborious, error-prone tasks due to the many possible meanings of a phonetic character. Furthermore, the desire to save money by minimizing the number of words sent meant the text might consist of abbreviations, code words, and other shortcuts common to telegrams but not part of everyday language. Intrafirm telegrams often included additional jargon and abbreviations specific to that firm, if not opaque to outsiders. Thus, a telex message was potentially much more difficult to decipher and understand than a regular handwritten message in Japanese.

Faxing seemed ideal for this "nation of *kanji*."[200] Indeed, a 1977 American analysis concluded, "The Japanese character set is probably the prime driving force now for facsimile in Japan."[201] Like the original document, a fax easily conveyed handwritten comments and the imprint of a *hanko*, an individual's personal seal, both standard ways of sharing information and authorization within an office. Japanese communicating favored the personal and close, not the impersonal and far. Just as a face-to-face conversation was socially superior to a telephone conversation, so too was a faxed handwritten note superior to a typed missive. Into the late 1980s, the assumption was that a typed letter had been sent to others also.[202] Yet until the 1972 deregulation of the telephone network, Japanese fax technology, applications, and diffusion trailed the United States. Deregulation sparked an explosion of development and demand. By 1977, R&D and sales surpassed American levels.

A more open market, however, was not the only factor promoting faxing. The Japanese government became a major shaper and accelerator of faxing. Of the five major state roles in shaping the evolution of new technologies—promoter, buyer, operator, regulator, and standard setter—the first four have received the most attention because they are most visible. For Japanese faxing in the late 1970s, the last two categories were more important because the government shaped both the macro-environment for telecommunications and a micro-environment for fax technology that encouraged domestic cooperation and competition, following its basic information technology policy.

The main state actors were the Ministry of Posts and Telecommunications (MPT), responsible for telephone regulation; the semi-governmental Nippon Telephone and Telegraph (NTT), which operated the domestic telephone system; and Kokusai Denshin Denwa (KDD), which controlled international telephone connections. The Ministry of International Trade and Industry (MITI), an avid promoter of industrial development, played only a minor role. The main private actors were the manufacturers and their trade group, the Communications Industry Association of Japan (CIAJ).

Government visions of a grand telecommunications future and telephone deregulation shaped the legal, economic, and intellectual environment for faxing. Since 1968 and especially after the 1973 oil shock, parts of the Japanese government, including the Economic Planning Agency, MITI, and MPT offered competing visions of a computer-based "information society" (*joohookai shakai*). MPT eagerly heralded new communications technologies, including fax, email, videotex, and cable TV, while promoting a more distant integrated world of digitized communications. The government did not have a vision monopoly. NEC president and chairman Koji Kobayashi avidly promoted the integration of computers and communication, while a Japanese version of OA embraced, rather than shunned, faxing.[203]

It is important to recognize that fax never received pride of promotional place in Japan as well as the United States. Government white papers of the present and visions of the future viewed facsimile as a sideshow to other computer-based information technologies like multimedia. In this sense, fax evolved as a "normal" technology, not benefitting from special state attention like the much-ballyhooed fifth generation strategic computing initiative or CAPTAIN videotex.

More direct in its commercial impact was the 1972 liberalization of the telephone system. By eliminating strict NTT restrictions on attaching equipment to its telephones, the Public Electric Communications Law (*kaisho kaeno*) enticed companies to expand or enter faxing.[204] Its prompt implementation in comparison with the Carterfone decision gave Japan a lead of four years of accelerated technological development. In the decade after deregulation, the fax market grew from an annual production of 10,000 machines worth 5 billion yen to 360,000 machines worth 190 billion yen.[205]

Decisions about the future of faxing devolved from NTT and KDD to individual manufacturers, increasing experimentation and innovation. As attention shifted from wideband circuits to standard telephone lines with their high long-distance charges, research focused on data compression to reduce transmission times.[206] By 1980, nearly all faxing occurred over the public telephone network instead of dedicated lines, increasing NTT's revenues.[207]

Manufacturers concentrated not only on reducing the costs and improving the capabilities of their machines but also on developing and blackboxing the complex electronics into well-designed packages requiring minimum training and skills.[208] NTT's facsimile laboratories, the world's largest, served as an incubator of fax technology with their long-term approach, growing investment in R&D, and cultivation of young engineers and scientists who often moved to industry and academia.[209]

Before deregulation, manufacturers received orders from NTT and customers to build machines. After deregulation, a growing number of firms manufactured and then marketed equipment. Deregulation also meant no single firm dominated the market due to growing competition, low barriers to entry, and growing segmentation of demand. Before 1972, the top four firms had 84 percent of the market, with Matsushita at 61 percent. In 1975–79, the top four firms only had 66 percent and Matsushita only 29 percent.[210]

To accelerate the development of the fax industry, the government encouraged R&D, bought equipment and subsidized its use, reduced taxes on fax machines, and established networks to encourage faxing, all standard methods of promoting specific technologies. MITI promulgated its "Elevation Plan of the Facsimile Equipment Manufacturing Industry" in December 1978.[211] Despite the bold wording, it is doubtful whether this plan did more than raise the visibility of fax. More meaningful was MITI's 1977 promotion with MPT of a tax measure that halved the time to deduct equipment to five years, encouraging businesses to buy expensive machines.[212] In 1983, the government offered a 7 percent tax relief on purchases of fax machines, a move the CIAJ considered very important for the vitality of the industry.[213]

Near-term networks and long-term computer-fax links comprised two goals in Japanese visions of fax's future. By centralizing expensive switching and service equipment, a network allowed users to acquire less capable and thus less costly fax machines while still accessing sophisticated services like broadcasting. Further in the future was linking fax machines with other equipment as part of a larger rationalization and increased capability of the office.[214]

To turn these visions into reality, NTT promoted a low-cost, low-speed "Minifax" and provided an increasingly capable fax network, F-Net, discussed in the next chapter. NTT's mid-1970s goal of 40 million sets in twenty years attracted manufacturers. Even 5 percent of that goal still meant two million sets, nearly a hundredfold increase over the existing base.[215] Despite its 100,000 yen goal, the Mini-fax transceiver cost a million yen ($3,800) when introduced in 1981. NTT's subsidized 15,000 yen ($57) monthly rent attracted over 50,000 small and medium businesses as well as homes by 1983.[216] MPT developed its Administrative Information and Communications Network to provide both a government information infrastructure and an initial market for digital fax technology.[217] These initiatives helped lay the foundations for G3's rapid acceptance by stimulating the development and diffusion of faxing.

Before deregulation, governments and large businesses comprised 90 percent of Japanese faxing.[218] In 1980, governments and large firms still dominated faxing, but the number of machines had soared, as had the range of applications and users (especially in banking and insurance). More than three-quarters of new users introduced faxing to replace or reduce telexes. Also attracting new users was a greatly expanded world of transmittable material: government communications consisting primarily of handwritten documents, followed by typed documents, official notifications, and newspaper clippings.[219] Like their American counterparts, many firms internally faxed via high-capability, high-cost equipment to support centralized communications and control. Other uses proved uniquely Japanese. To eliminate errors in telephoned orders (such as sending wedding roses instead of white lilies to a funeral), the Florist Telegraph Delivery Association bought 360 machines.[220]

Fax production increased tenfold in value and eightfold in numbers from 1975 to 1981. The larger increase in value was caused by a shift to more expensive, capable, and faster machines.[221] Until 1978, G1 comprised the majority of installed machines, but the growing market rapidly switched to faster machines: A 1980 survey found 22 low-speed, 48 medium-speed, and 61 high-speed models. By then, Japan had not only greater per capita diffusion of fax machines than the United States but greater use per machine.[222]

EUROPE

Four different markets existed in 1980: Japan, the United States, the much smaller Western European market, and everywhere else. The first two had large, unified markets and rapid growth. Japan and the United States had approximately the same number of fax machines—250,000, compared to 100,000 for Europe or respective ratios of fax machines per thousand citizens of 2.0, 1.1, and 0.3.[223]

The slower European growth reflected the abundance of alternative communications, especially telex; the division of the continent into separate, tightly regulated national markets; equipment restrictions; and the high cost of telephone calls. PTT political and physical protection of national telephone systems and manufacturers slowed the diffusion of faxing.[224] Equipment providers had to modify and certify their offerings for each PTT, an administrative and engineering challenge.[225] Absent were the push of telephone deregulation, American startups, and the pull of the ideographic language of Japan.

Other priorities ranked higher than facsimile for European firms and governments. Combining limited vision with pragmatism, Siemens Telecommunication Group announced 1975 goals of providing every household a telephone and every business a telex, establishing a new service of "office teleprinters" to link desktop to desktop, and, finally, offering a facsimile service for exchanging documents. In Siemen's future, facsimile and desktop communications did not mix, a future still unchanged in 1981.[226]

The French government boldly promoted *télématiques*, the "growing interconnection between computers and telecommunications," to modernize French society, guided by the 1978 bestseller *L'informatisation de la société* [The Computerization of Society], which placed the information revolution within a larger "struggle between knowledge and tradition."[227] Most successful was Minitel, a videotex online service from the French PTT. Beginning in 1983 as an alternative to telephone directories, Minitel grew to 2.5 million terminals in French homes and businesses in 1987 and 6 million in 1991.[228] The government's Mass-fax program's goal of domestically building one million inexpensive G3 machines by 1990 proved far less successful.[229]

Domestic PTT public fax services with G1 and G2 machines also proved disappointing. In 1975, the British service transmitted only 300 pages among eleven cities in its first six months. A Swiss service, started in 1976, handled only 600 pages monthly, or one page per its eighteen offices daily. In comparison, airliners transported four tons of mail daily to the United States from Switzerland.[230] Yet PTTs remained willing to experiment with new services that competed against their telephone and letter services—fifteen national PTTs in Western Europe operated public fax services in 1986. Private faxing elicited less support due to concerns about potential damage to their telephone networks and whether PTTs had responsibilities to assure quality and compatibility.[231]

While concentrating on their home markets, manufacturers in Japan and the United States also sought international sales through subsidiaries, joint ventures, and other arrangements. Although small in number, these sales produced helpful revenue, boosted the value of faxing for multinationals, and raised faxing's image internationally. Different national PTT standards usually meant incompatibility of foreign versions of American and Japanese equipment. In 1977, only 1 percent of American machines were compatible with their European counterparts.[232]

Such incompatibility, coupled with regulatory restrictions and high cross-border telephone charges, hindered international faxing. This glum picture altered economically as regulatory worlds shifted toward more openness and competition, and technologically when Ricoh introduced its subminute Japanese (Rifax 600), European (Kalle Infotec 6000), and American (Rapifax 100) machines, which enabled corporations and governments to fax directly internationally.[233]

Growing international demand stimulated services. In March 1978, RCA Global Communications and KDD introduced Quick-Fax, charging $10 per page from the United States to Japan. To their surprise, the firms discovered that less than 10 percent of the traffic came from telex. Nearly half the transmissions consisted of legal documents in English with handwritten comments. Financial data, engineering drawings, and other graphic content comprised most of the other transmissions. Clearly, faxing had benefitted from the growing globalization of business by offering new opportunities for the rapid transmission of images, expanding beyond the limited framework of telex. By 1980, Quick-Fax served sixteen other countries.[234]

◌

With the Telecopier and its successors, mass production of fax machines finally began. The numbers of installed machines climbed from thousands to tens of thousands to hundreds of thousands. Increased supply meant nothing without increased demand. Decisively outstripping the specialized markets that had previously defined facsimile, the market for general purpose faxing finally appeared in the late 1960s as faxing slowly became an essential component of communications for large corporations and the government. Despite the addition of Telecopiers, Magnafaxes and dexes to the pool of Datafaxes and other machines, market demand remained below expectations, leaving many disappointed entrepreneurs and investors in the large gap between predictions and reality. Technology push still dominated over market pull.

Why had facsimile fared so poorly? Why had customers not flocked to fax? Although the numbers of users grew greatly, to numbers unthinkable in the 1950s, the self-reinforcing combination of incompatibility, poor economics, and technical limitations meant slower growth than advocates had predicted, and manufacturers needed to profit. Manufacturers deliberately introduced proprietary machines incompatible with competitors both to lock customers into their systems and because their business model of closed networks and intra-organization faxing did not need compatible machines. In contrast, open networks and inter-organization faxing depended critically on compatibility. While intra-organizational faxing did dominate initially, the incompatibility in the early generations of general-purpose fax machines scared away potential users fearful of choosing the wrong system and fragmented the market.

Liberalization of telecommunications changed the macro-environment, but although it was dropping in cost, faxing remained expensive. The biggest technical failure was the inability to create an affordable machine that could transmit a page in less than a minute. This was not for lack of effort, but, to paraphrase Cornelius Ryan, subminute faxing in the 1960s–70s remained "a technology too far." However, impending changes in compatibility, cost, and capability in Japan seemed poised to change that pessimism.

CHAPTER 5

The Giant Awakes, 1980–1995

It is as though starship technology has arrived at last, and it is possible to "beam" a document to any part of the globe in seconds.

—D. G. *Elliman, 1991*

Perhaps nowhere was the fax gap between Japan and America better displayed than in academia in the late 1980s, when fax machines proliferated in Japanese offices and homes, while an American professor "spent countless hours" trying to find a fax machine on his campus to fax a reply to a faxed invitation from Japan. Indeed, when an American professor admitted that she did not have a fax machine, her Japanese colleague humorously responded, "No fax! Do you have indoor plumbing?" Similarly, after stockbroker Jim MacKintosh acknowledged he lacked a fax machine, a client asked, "How the f*** do you operate in the dark ages?"[1] The shock that an otherwise educated, respectable adult did not have a fax machine demonstrated how rapidly faxing became an essential part of daily life for tens of millions of Americans and Japanese.

Starting in the early 1980s, the combination of increasing deregulation, true compatibility, quickly dropping costs, greater competition, and rapid technological change created a blossoming of new machines, applications, and services even as competing technologies became less attractive. This easier access, easier use, and attractive economics generated an explosion of faxing in Japan that quickly spread to the United States and the rest of the world.

Part of the larger "information revolution" based on increasing access to faster and fuller communications at lower cost, faxing rode this wave but also swelled it, making communicating much easier for anyone at any distance, whether down the hall or around the world. Most significantly, faxing helped change expectations of information—increasingly including images—from rare to abundant, from hard to find to easy to acquire, from difficult to obtain to simple to create and disseminate. Faxing benefitted the growth of visual culture by making the sending, receiving, and alteration of images easier and less expensive.

As fax machines changed from a "technology of the future" to the "technology of the now," market pull finally surpassed technology push as millions of new users adopted faxing, creating unprecedented and rapidly increasing demand. The multiplication of commercial and private users diminished the importance of the government as a consumer and patron, just like computing. In Japan, annual production grew from 104,000 fax machines in 1980 to 4.9 million in 1990 and 7.7 million in 1995. Total production increased from 250,000 machines in 1973–79 to 18 million in 1980–89, and 77 million machines from 1990 to 1999. The United States had approximately 250,000 fax machines in 1980, 500,000 in 1985 — and 5 million in 1990.[2] Each new user increased the utility of faxing for earlier users, creating a virtuous circle of network externalities of increasing opportunity and value, an increase popularized as Metcalfe's Law.[3]

In 1988 faxing, "the biggest technological explosion since the personal computer," became, *New York* definitively stated, "America's hottest new verb."[4] Though lagging behind Japanese usage and diffusion, a fax machine had become an essential component of doing business in the United States, from local deli orders to international diplomacy. While faxing sandwich orders to a deli sounds trivial, it embodied the virtues of fast, error-free transmission of information with a printed record for both parties that sparked faxing's rapid acceptance and diffusion into the international business community.

This chapter begins with the complicated and often bitter international struggles to develop standards. Not until the 1980 G3 standard did competing manufacturers offer truly compatible equipment. Coupled with decreasing cost and growing capabilities of electronics and the mounting cost and decreasing quality of competing methods of communication, the sleeping giant finally awoke. This rapid rise surprised American manufacturers and market analysts, who viewed G3 in the past tense and as about to fall before the theoretically superior, all-digital world of G4 and the Office Automation movement. The expensive failure of Federal Express's ZapMail, one of the many unexpected difficulties in the evolution of computer-based communication, demonstrated again that the "best" technology was often what met users' actual needs and desires, not the most technologically advanced approach.

Faxing finally fulfilled its promoters' promises to become an essential new technology in the evolving global telecommunications networks as tens of millions of businesses and people purchased machines. The rapid diffusion of faxing in Japan and the United States reshaped "business as usual" everywhere as demonstrated by the worlds of art, law, and politics. Less publicized were the negative aspects of faxing, primarily the lack of privacy, government eavesdropping, and proliferation of junk fax. Two of the more surprising aspects of fax's acceptance were how readily people valued speed and ease of operations over authentication and security.

Fax's acceptance proved even more penetrating and profound than advocates anticipated. Faxing's instancy and imagery became a vital part of daily institutional and personal life. The widespread adoption of fax machines created two forms of public spectating for audiences of onlookers. In the first, artists knowingly presented creations in the new medium of fax art. Far more widespread were the spontaneous audiences of office workers who read faxes designated for others.

Most excitingly, thousands of unexpected applications appeared as people experimented and innovated. This tinkering demonstrated on a grand scale a democratic innovation from below of creativity in endeavors ranging from the frivolous to the vital. Like any tool, people applied it for their own ends, whether centralizing control, forging college transcripts, intercepting faxes from revolutionary groups, or ordering pizza.

G3 STANDARD AND COMPATIBILITY

The most important event in fax history since 1843 occurred in a Tokyo conference room in 1977. The ultimate result was G3, one of the most successful standards in telecommunications history. G3 fax machines scanned and compressed images digitally, converted them to voiceband analog signals for transmission over regular telephone circuits, and reconverted the signals into digital data for recording. G3 facsimile became a formal CCITT Study Group XIV question in 1972.[5] Pushing the participants was the failure of G1 and G2 standards to create a unified, compatible market. Nonetheless, expectations remained low. A 1972 industry survey predicted de facto compatibility from the dominance of a single manufacturer, who ideally would license its algorithms and thus ensure its profits while still allowing compatibility and competition.[6] The appearance of commercial subminute machines with proprietary algorithms, however, seemed to render the CCITT effort futile.

The primary standards battle concerned whether to use 1-dimensional (1-d) or 2-dimensional (2-d) coding for data compression.[7] A 1-d code scanned one line at a time. A 2-d code used statistical probability to anticipate the next line. The tradeoff was less efficient, less complex coding and thus less demanding electronic components versus more efficient coding that demanded more complex software and hardware. Japanese engineers favored the more challenging 2-d codes, objecting to the more mature 1-d Huffman code's lower efficiency and inability to handle the high resolution required for the intricate *kanji*.[8] Not by chance, this advocacy of more complicated algorithms demanded the microelectronics that Japanese industry led the world in manufacturing. Yet a compromise appeared in 1975. Although most CCITT delegates favored the simpler 1-d Huffman code, they allowed the option of a 2-d system.[9]

The creation of a CCITT 2-d code was not assured, however, because several Japanese manufacturers had developed their own incompatible algorithms and remained reluctant to give them up. At this point, the state played a crucial role. In 1977 the Ministry of Posts and Telecommunications (MPT) provided guiding momentum by establishing a committee with representatives from NTT and KDD, the Communications Industry Association of Japan (CIAJ), and major fax manufacturers. By creating this neutral arena, the MPT pushed these different actors to cooperate to create a single national standard.

Consensus did not come easily. The committee narrowed six competing algorithms to NTT's Edge Difference Coding (EDIC) and KDD's Relative Address Coding (RAC), reflecting both the power of the telephone services and the efficacy of their codes.[10] While some differences between the two were technical, the underlying challenge proved the organizations' longstanding rivalry. After high-level negotiations failed, lower level talks between laboratory engineers and managers eventually succeeded. The psychological turning point came when one NTT manager wrote "READ" on a blackboard. He explained that the name contained both codes (*RAC* and *EDIC*), described what one did with a faxed message, and, when pronounced in Japanese, was "lead," which was what NTT and KDD would do — once they thought in terms of combining forces, not conceding.[11]

After the creation of the Relative Element Address Designate (READ) code in 1977, the MPT campaigned for its CCITT adoption. Befitting a major standard, its acceptance was a contested political process because of the high economic stakes. To Japanese surprise, CCITT acceptance of READ was not automatic but required much negotiation, testing, modifications, and concessions, including making READ royalty-free like the Huffman code.[12] The insistence on royalty-free standards was not unreasonable for acceptance by potential licensees: Iowa State University earned $36 million in royalties from twenty-four fax manufacturers from David C. Nicholas's 1973 patent for variable-length coding of digital data after the university threatened to sue for patent infringement for machines made before 1991.[13]

By 1979, Study Group XIV had received algorithms from Europe and America as well as Japan. The increased acceptance of 2-d coding came from the sharply decreasing cost of electronic components and critical experiments conducted in neutral environments. Months of testing competing algorithms by manufacturers, PTTs, and governments preceded the decisive November 1979 meeting in Kyoto. There, delegates' "frank and useful discussion" battled with the "spirit of compromise and a desire to reach a conclusion" before a Modified READ, submitted by British Telecom, became the G3 international standard, a decision affirmed by the CCITT plenary assembly in 1980.[14]

The G3 standard provided stabilization but not full closure for compatibility. Over the next two decades G3 evolved significantly, as did the CCITT (renamed the ITU Telecommunications Standardization Sector), which streamlined its procedures to study and approve a new feature in a year instead of the previous four years, thus maintaining its relevance versus competing standard-setting organizations.[15] Critically, the G3 standard served as a floor and not a ceiling on innovation. The Non-Standard Facility (NSF) option allowed manufacturers to include proprietary features, thus giving incentives for buyers to stay with one firm's equipment and for manufacturers to innovate. Several proprietary features, such as faster transmission, later became standard features.[16]

By 1999, the G3 protocol supported 1200 × 1200 pels (picture elements, also called pixels) resolution (compared with 100 × 200 pels in 1980), secure internet transmission, and color images. Modems transmitted at speeds inconceivable in the 1980s, reaching 56,000 bps (5 seconds per page).[17] The 1990s G3 machines were thus far more capable and faster while still compatible with earlier G3 machines.

Actually achieving compatibility was neither automatic nor easy. As well as imposing strict internal quality control, Japanese manufacturers discovered they had to share information. Out of public view, the CIAJ hosted meetings where competitors tested the compatibility of their machines in the same room. As companies introduced more proprietary features and more complex software, this testing became more important. Confidentiality agreements ensured that proprietary information would not aid rivals. This testing proved so successful that the CIAJ institutionalized it as its Harmonization of Advanced Telecommunication Systems (HATS) program.[18]

TECHNOLOGY

Compatibility alone would not have created the fax boom. The machines had to improve. And they did, becoming faster, cheaper, smaller, and better. What Dennis M. Roney, the president of Pitney Bowes Facsimile Systems, called "the happy results of microcomputer technology applied to facsimile" understated the changes.[19] What was economically or technically prohibitive in the 1970s proved affordable a decade later. Without the massive increases of computing power and memory at rapidly decreasing cost, the ambitious 2-d code of G3 fax would not have succeeded.

Shifting from electromechanical to electronic components reduced cost, size, and noise while improving capability and reliability. Nowhere was this more apparent than the modem. In 1980, a fax modem required 400 square centimeters for 10 LSI circuits and 150 other discrete components. By 1984, a faster modem demanded only 65 square centimeters with 30 components. Six years later, a single chip provided even more capability. Prices dropped too. In 1977, a 1200/2400 bps modem cost $530, and a 9600 bps modem, $3500. In 1990, Rockwell International sold 9600 bps modems for $48.50 in lots of 10,000.[20] Modems remained an American quasi-monopoly due to Rockwell International, which used its military expertise and attention to clients' needs to develop an almost impregnable position protected by patents, continual innovation, and low prices.[21]

New illumination and sensors accompanied the switch to digital scanning. Instead of powerful lamps and photocells, the light from low-power helium-neon lasers or light-emitting diodes (LEDs) reflected from the document into a bar of solid-state photodiodes or charge-coupled devices (CCDs).[22] The mechanisms of scanning changed accordingly. Gone were the curved platens, revolving drums, rotating scanner heads, and other precisely machined components. Instead, flatbed rollers smoothly passed pages under a stationary scanning bar.

The switch to digital brought new metrics. Resolution, formerly measured in lines per inch (lpi), now was in pels, which quantified how much information a page contained. The G1-G2 96 lpi resolution had 96 dots per inch, or 861,696 pels for a 8.5×11-inch page. The G3 100×200 lpi resolution meant 1.87 million pels to transmit and the G4 400×400 lpi 14.96 million pels. These huge amounts of data demanded data compression that the G3 and G4 standards provided.

The improved quality of domestic and international telephone circuits also contributed to fax's success.[23] Together with the development of efficient data compression and error-coding algorithms, the bottleneck of poor transmission shrank significantly. G3 transmitted not in minutes per page but in pages per minute. These faster speeds combined with the sharply dropping cost of domestic and international long-distance calls radically transformed the competitive economics of faxing.[24]

Faster scanning and transmitting would have been for naught without a revolution in recording. By itself, thermal paper appeared an unlikely choice. It felt flimsy, darkened on exposure to heat (as many a person who placed a hot cup down discovered), and faded. Compared with the alternatives, however, thermal paper proved outstanding. Recording did not generate obnoxious smells, smoke, or noise, greatly easing its acceptance in an office. The sharper resolution (6 lines/ mm compared with 4–5 for electrostatic paper) improved reading of Japanese *kanji*. Most importantly, the reliable recording mechanism reduced the cost of a fax machine by thousands of dollars.

The two key elements were heat-sensitive paper and an array of 1,728 film resistors, one per pixel, that selectively heated the paper, then quickly cooled. Thermal printing emerged in the early 1970s for strip chart recorders, but the promise of the fax market accelerated its development. Overcoming problems like abrasion from contact with the thermal paper and quickly cooling the heating elements so their residual heat would not smear the image, Japanese firms developed commercially viable printers and paper.[25] Thermal machines comprised 4 percent of Japanese fax machines sold in 1977—and 54 percent in 1982. Of thirty-three new machines in 1981–82, twenty-six of them (three quarters) had thermal paper compared with six electrostatic and one plain paper.[26]

Though essential to the economics of the early fax boom, thermal printing existed as a transitional technology. Pulled by the desire for higher quality and pushed by the transfer of technologies from computer printers, fax manufacturers fiercely competed to offer alternatives. Its replacement occurred in a classic trickle-down evolution as more expensive machines adopted laser, thermal transfer, and inkjet recording, followed by less expensive models. By 1991, new plain paper models outnumbered thermal models. Sharply dropping prices closed the multi-thousand dollar gap with thermal printing. For example, Sharp's least-expensive laser fax machine sold for $4000 in 1989 but half that a year later, while $1000 bought an inkjet in late 1992. By the late 1990s thermal printing had retreated to very inexpensive machines.[27]

The rapidly evolving array of printing technologies reflected competitive market pressures as firms introduced features to differentiate themselves, fill (and often create) niches, and gain market share. This shotgun approach, common in Japanese marketing, encouraged experimentation and innovation among manufacturers as well as consumers. The continuing profusion of new models—average life was less than a year by 1989—meant quick copying of a popular feature.[28] Improving convenience and operations proved as important competitively as speed. Automatic document feeders and dialers reduced repetition. Automatic paper cutters and paper trays holding hundreds of pages helped handle the increasing volume. Available first as options, such features increasingly became standard.

Packaging these improved features and complex technologies into an easily understandable and usable machine proved to be essential for widespread acceptance. As a result, a user had only to plug in the telephone and electric cords, load the paper, and the machine was ready. Basic office skills—punching a telephone number and inserting paper into a photocopier—sufficed to fax. Small size meant easy placement almost anywhere in an office.

JAPAN

Duplicating a pattern in other business and consumer electronics like the LED watch, once a few Japanese firms established the viability of fax manufacturing, larger firms quickly entered the market.[29] Exports of fax machines increasingly fueled production, rising from one-fifth in 1981 to over half in 1987.[30] These rapidly expanding exports seemed paradigmatic of Japan's high technology strengths of evolutionary product development, mass production of sophisticated electronics, and marketing. For most Japanese, however, the increased domestic diffusion of fax machines outweighed the economic benefits of growing exports.

Before the fax boom began, Japanese fax manufacturers viewed the fax market as bifurcating into high-cost, high-capability equipment for large firms and more numerous, less-costly, and lower-capability machines acting as a "complement to the telephone."[31] Feature-seeking users and newly entering manufacturers thought differently. The fierce competition among manufacturers meant advanced functions quickly migrated into increasingly inexpensive yet increasingly capable machines. The market actually trifurcated into high-capability business, standard business, and personal machines. By 1996, each sector comprised one-third of the value of machines produced in Japan, but the division by numbers of machines was 6 percent, 21 percent, and 71 percent, showing the huge difference in price.[32]

Faxing flowered in Japan, fueled by a more permissive regulatory environment and the increasing capabilities, shrinking size, and lower costs of machines. The 1985 Telecommunications Business Law provided a second deregulatory jolt, increasing domestic telephone competition (which lowered rates) and permitting smaller enterprises to provide value-added services. Most visible were small "pay fax" retailers, who enabled customers to fax while buying flowers and gasoline or sipping coffee.[33] The 1986 MPT Information Flow Census claimed the overall amount of information generated increased by 10 percent in one year, but faxed information increased by 38 percent.[34]

By 1988, over three-quarters of businesses with more than ten people had a fax machine, as did 55 percent of offices with five to nine people. A year later, telephone lines dedicated to fax machines comprised 6 percent of all lines and an astounding 24 percent of business lines.[35] Faxing quickly became part of Japanese work culture, including the *chuken fakkusu* ("faithful faxer" or "middle-rank faxer"), a mid-level, middle-aged man distrustful of modern office machines partly because he handles them poorly. By early 1994, fax machines were the most common office machines: 81 percent of businesses with more than five employees had a fax machine compared to 35 percent with PCs, 42 percent with word processors, and 63 percent with photocopiers.[36]

To promote business and home faxing, NTT introduced its Facsimile Communication Network (F-Net, initially called Facsimile Intelligent Communication System) in 1981. Telephone engineers and managers, whose professional experience revolved around networks they controlled, liked the idea of a centralized system that provided services like store-and-forward at a lower cost than incorporating them in individual machines. F-Net expanded in geographic coverage and technical capabilities as it grew from 800 subscribers in 1981 to 85,000 in 1985, to more than a million by the mid-1990s.[37] Emboldened by the 1985 deregulation and F-Net's growth, in August 1988 advertising and human resources company Recruit introduced its Facsimile Network Exchange (FNX), which became F-Net's main competitor. Like F-Net, Recruit continued to attract customers by adding new applications such as faxing data to a computer.[38] In the United States, firms like Xpedite offered services like blast faxing, and communication carriers offered fax networks, but none had the breadth, depth, or reach of F-Net and FNX.

Another major difference between the United States and Japan was home faxing. While most American home offices had fax machines—80 percent in 2000, or 14 million machines, personal homefax machines never became popular. In contrast, more than half of all Japanese households in 2002 had fax machines.[39] As early as a 1971 Science and Technology Agency Delphi survey of technological trends, informed opinion assumed receiving newspapers and other public information would be the main attraction for homefax. Instead, work—extending the reach of the office—drove the initial diffusion, as companies provided homefax machines for traveling salesmen and top executives.[40]

Home faxing soon became a way to actively seek information and communicate, not just to passively receive news. Japanese became avid faxers. They shopped, answered questionnaires, searched for information, sent comments to live television shows, learned about up-to-date ski conditions, and communicated with friends and family, whether separated by continents or a few apartments. Faxing enabled word processor owners to easily transmit their documents and for parent-teacher associations and other groups to organize. The benefits were not only economic: tutoring and testing at home for highly competitive college entrance exams instead of going to a class saved hours of commuting for students and teachers.[41]

Manufacturers actively encouraged the homefax market. Ricoh established correspondence courses for pre-college students with the free loan of a fax machine, hoping to entice parents as well into faxing.[42] One effective promotion was *Fax Life: Sources of Convenient Information* [Seikatsu marutoku jyohogen], which Hitachi Hiplan, a Hitachi subsidiary, started publishing in 1991 and convinced three other big fax firms, Matsushita, Sharp, and Sanyo, to join in 1992. By providing a wealth of information about hundreds of fax services and applications, *Fax Life* gave fax owners ideas about how to better employ their machines. Initially free for new buyers, *Fax Life* appeared in bookstores to attract existing users in 1996 after NTT joined the group. Circulation jumped from 100,000 to 250,000 an issue.[43]

Homefax's attractions and diffusion increased in the late 1990s and into the twenty-first century due to continuing price decreases and increasing capabilities. A constant stream of innovations kept users replacing their homefaxes with newer, more capable machines. The homefax served as a cordless telephone and answering machine, providing multiple functions in one small package, an important consideration in an apartment. Newer features included a liquid crystal screen to avoid printing, wall-mounting to save counter space, color printing, a portable hand scanner, and the capability to send written messages to cellphones.[44]

AMERICAN COMPETITION

Faxing grew partly because the "skyrocketing cost of mails, indefinite postal deliveries, traffic delays for messenger services, hang-ups in elevators and reception rooms," and other problems decreased competitors' attractiveness. Inside a corporation, a message might need a day and a letter two days or more to reach its destination, so bypassing the company mailroom gave desk-to-desk faxing a major advantage.[45] Slower, costlier mail service also benefitted the economics of fax. In the United States, even as two-day delivery dropped from 88 percent in 1968 to 82 percent or lower in 1988, postage rose from 15 cents in 1978 to 32 cents in 1995, and from 50 yen in 1976 to 80 yen in 1994 in Japan. In 1993, faxing ten pages long-distance cost $1.25 during the day but only 65 cents at night, less than the 75 cents needed for postage.[46]

Nor did the U.S. Postal Service's ventures into advanced technology do much better. Its domestic and international electronic mail and fax services, E-COM and Intelpost, attracted disappointedly low levels of traffic in the 1980s, outpaced by organizations and individuals acquiring their own fax machines.[47] The Service's 1989 contract with a private firm to install fax machines in post offices fared equally poorly, dissolving in 1991.[48]

Overnight delivery services benefitted facsimile by drumming into people and businesses the need for and expectation of quicker delivery. Faxing did not reciprocate. Its decreasing cost kept reducing the economic advantage of overnight delivery services. In 1983, faxing cost less if sending fewer than twenty pages. Two years later that break-even point had risen to sixty-five pages.[49] While slowing Federal Express's growth—the firm estimated faxing displaced 20,000 documents daily in 1988—faxing devastated telex. As the 1988 slogan for ATT Mail stated, "I told you to send it, and that means fax, not telex." Despite computer-based, more capable equipment and services introduced by PTTs and equipment manufacturers, outgoing telex traffic and the number of subscribers in OECD countries started shrinking in 1986–87. From 1983 to 1992, worldwide telex traffic annually dropped 8 percent and subscribers 9 percent, declines that continued in the 1990s: in Hong Kong, telex minutes dropped from 95 million in 1987 to 17 million in 1997. Telex quickly disappeared as a significant means of communication.[50]

SELLING

Waves of new machines also changed sales patterns. In the United States, save for the more expensive and capable office equipment, selling shifted from the sales agent to the store as fax machines became a commodity in the late 1980s. In the mid-1980s, Canon and other firms used agents to expand the market for lower-cost machines. Like their predecessors visiting large corporations a decade earlier, these agents were proselytizers, spreading the concept of faxing, but this time to the much larger world of small businesses. Because these companies primarily communicated with other firms and not internally, sales agents had to convince multiple firms that photocopying over a telephone would benefit them. Not all successful arguments stressed better business communications. For Canon's Tony Borg, the compelling benefit of faxing that male business owners understood was quickly organizing a football pool.[51]

Japanese firms encouraged office products stores, many of which already sold

Japanese-made photocopiers and other equipment, to sell facsimile equipment. Office machine dealers found fax machines and photocopiers essentially boxes to be sold, requiring none of the post-sales support common to personal computers. Selling thermal fax paper, which had a high profit margin, added to the incentive of selling fax equipment. Fax machines appeared in "big box" stores and mass merchandisers like Wal-Mart in the late 1980s. By 1991, over half of all sales occurred in retail stores and another quarter by dealers.[52]

Putting machines into stores did not sell them. People had to know facsimile existed and be convinced to use it. Less than 20 percent of Americans knew what faxing was in 1981. Advertising, articles, demonstrations, training, and social pressure changed that. In 1984–86, Federal Express's ZapMail spent millions of dollars in advertisements, publicizing and legitimating the unfamiliar technology. Its initial advertising referred to "fax," but the quotation marks had disappeared by 1986.[53]

Journals like *Administrative Management* advised whether to introduce facsimile.[54] Until the mid-1980s, articles focused on intra-firm networks to speed transactions, reduce errors, and save money. Facsimile also required no keyboard or training, making it usable by secretaries, workers in the field, and executives who did not want to use a keyboard. In the late 1980s, however, articles began emphasizing rapid inter-firm communications.[55] Reflecting and leading the diffusion of faxing into popular culture, mainstream magazines began running articles on this exciting new technology.

The Japanese emphasis on market share instead of profits and the growing number of manufacturers—nearly thirty in 1994—created competition so vigorously intense that manufacturers worried that aggressive newcomers were pushing prices and thus profits down too fast even as production soared.[56] Total revenue peaked in 1988 at 580 billion yen ($4.5 billion), then dropped to a low of 407 billion yen ($3.2 billion) in 1992 before reaching a new peak of 609 billion yen ($5.3 billion) in 1999 and then declining slowly. The real profits accrued to telephone companies. In America, long-distance charges and additional telephone lines brought billions of dollars into telephone company coffers, an estimated $2.3 billion in 1988 alone.[57]

No one firm dominated this growing market, where leadership changed as quickly as the introduction of a new firm or machine. Prices fell far faster than analysts predicted and vendors expected. In 1983, the market research firm International Resources Development predicted a gradual decline in the cost of a G3 machine from $7,600 to $4,200 in 1992. Instead, psychological and accounting barriers—$3000, $2000, $1000—kept failing, years earlier than anticipated, the last in 1985. By late 1987, $500 bought a basic machine.[58] Once the price of machines fell below $1000, the price threshold above which many managers had to seek higher approval for purchases, the machines spread rapidly. List prices were misleading, bulked up to allow reductions that sounded good for the customer but still profiting dealers. The standard margin was 50 percent—that is, a dealer paid the manufacturer half the list price. Adding to the confusion, competing firms sometimes sold the same equipment, differing only in nameplate and price.[59]

In Japan, the magic number that transformed a machine's apparent affordability, 100,000 yen, was reached in the early 1980s. For homefax, the price point was 39,800 yen—the pleasant-sounding *san-kyu-pa*, reached in 1993–94 in Akihabara, the Tokyo electronic district.[60] As table 5.1 shows, average prices (which hid large differences among types of machines) dropped very sharply after 1980, greatly increasing the attractiveness of buying machines.

The rise of G3 shifted technological leadership and manufacturing, save for modems, from America to Japan. The increasing importance of digital electronics played to the strengths of Japanese manufacturers and disadvantage of American firms, which reduced their investment and risk by marketing equipment designed and built in Japan. By 1985, manufacturing had almost completely shifted to Japan. Indeed, when the International Trade Commission considered initiating an unfair trade investigation, it found essentially no American industry to protect.[61]

The market share of the main American firms, Xerox, Exxon, Burroughs (Graphic Sciences), and 3M, plunged from 85 percent in 1982 to 45 percent in 1985, while Japanese firms quadrupled their share from 12 percent to 48 percent.[62] Some American firms' fax efforts literally self-destructed. Exxon's Qwip, which lost over $100 million before its sale in 1984 to Harris Corporation's Lanier Business Products, literally became a case study in how not to manage entrepreneurial enterprises.[63] Not only established fax manufacturers and startups faltered: Although involved in fax standard setting, IBM introduced its Scanmaster fax system in 1984 with a fatal flaw, G3 incompatibility. Even IBM failed to use its own product, buying thousands of Pitney Bowles machines instead.[64]

Management consultant Peter Drucker blamed the voluntary abdication by American firms on their erroneous perception of the market. American firms had asked, 'What is the market for this machine?' Japanese firms instead asked 'What is the market for what it does?' and created that market.[65] While partially correct, Drucker neglected the long American experience with the fax market.

In the late 1970s, investing in a new generation of technologically more challenging fax machines was not an obvious choice. The new technologies, such

TABLE 5.1.
Japanese fax production and revenue

	Yen		
	(1 B)	Machines (1,000)	Yen/machine (1,000Y)
1975	23	20	1,170
1980	120	104	1,157
1985	312	917	340
1990	461	4,925	94
1995	483	7,684	63
2000	547	9,930	55
2005	453	11,010	41
2009	421	11,010	38

Source: CIAJ, *Tsushin kiki chuki juyo yosoku*
[Communications Equipment Demand Forecast]
(Tokyo: CIAJ, annual).

as solid-state scanning, differed sufficiently from the existing analog and elec-tromechanical systems to require American firms to hire new people or retrain researchers.[66] Profits had been thin or nonexistent for over a decade, and market researchers did not envision a large increase in demand. From this perspective, the decision not to invest seemed reasonable. Why devote resources to a slow field when other areas promised higher returns, especially if G3 became only a transitional stage due to the imminent and "inevitable merging of facsimile into the overall OA [office automation] scene"?[67]

Fax was a stealth technology, sneaking under the radar of industry assump-tions to suddenly explode in the market for electronic communications. Look-ing at the popular and more informed technical and business press and where firms, governments, and standard-setting organizations invested their resources, fax's great success in the 1980s–90s clearly was not supposed to happen. This informed consensus—a worldview based on expectations of continuing progress in computing and electronics and fax's under-fulfilled expectations for over a century—envisioned fax as at best a minor player in the all-digital, paperless of-fice of the future. Compared to the digital promise of electronic mail connecting millions of computers and videotex reaching out to tens of millions of televisions, how could a long-distance analog-digital photocopier compete?[68]

The industry also gravely misjudged the potential of G4 facsimile, whose im-minent and revolutionary arrival had been predicted and promoted since the late 1970s and which became a CCITT standard in 1984. Instead of G3's theoretically inelegant (but efficient) conversion of digital scanning data for analog transmission and reconversion for digital recording, advocates envisioned the all-digital, very-high-speed (up to 20 pages per minute) G4 would seamlessly integrate in a planned, comprehensive manner with the seven-level Open Systems Interconnection (OSI), the Integrated Service Digital Network (ISDN), and the X.400 electronic message protocol. These efforts, discussed in the next chapter, proved so complex that they never fully succeeded, instead being overtaken by more agile, less costly, and faster alternatives. The failure of these larger systems handicapped G4.[69]

The main reasons for the failure of G4, however, came from the surprising resilience and evolutionary improvements of G3, which by the early 1990s did almost everything G4 promised to do—but for far less money, without having to wait for ISDN, and maintaining full compatibility with the analog telephone network and older G3 machines.[70] The all-digital G4 machines were incompatible with G3 machines. One advantage of G4 was the introduction of Modified Modified READ (MMR), which provided greater data compression and resolution (400 lpi) than G3, but in 1988 American and British G3 advocates asked the CCITT to accept MMR as a G3 option. This proposed modification, labeled G3bis, provoked a fierce fight between countries with a substantial stake in G4 development (Germany, France, and Japan) and the rest of the world, comfortably ensconced with G3 and seeing no reason to restrict the rapidly growing technology.[71]

As a 1990 CCITT report concluded, the real technical issues of interoperability between G3 and G4 comprised only part of a "more complex, many-sided question including economical [sic], service and even political issues" revolving around essentially incompatible architectural models of telecommunications.[72] The underlying issue concerned whether G3 should receive capabilities that threatened G4 and the all-digital world of OSI. From this latter perspective, G3 was an upstart hurting its more rational and sensible sibling who was due to inherit the family estate. The G3bis problem proved so intractable that the CCITT delegates could not reach a consensus but pushed the final decision to the 1992 plenary assembly, which approved the enhanced G3 standard.[73]

The approval of G3bis meant the performance gap between G3 and G4 narrowed, while the cost gap grew with every purchase of a G3. In 1981, IRD estimated a standalone G4 machine cost $20–25,000. In 1986, the firm pegged a machine's cost at $5,000–10,000 and expected a decrease to $3,500–7000 by 1995.[74] While these were significant drops, G4 prices remained far above their G3 counterparts.

Consequently, millions of G3 machines poured off assembly lines compared with only several thousand G4 machines, far below expectations. As late as 1989, when eight manufacturers marketed seventeen models, predictions of a bright G4 future—now envisioned for the mid-1990s—still appeared.[75] By 1991, GammaLink president Hank Magnuski dismissed G4 as "past its prime," unlike G3, which still had the potential for improvement, partially by implementing features planned initially for G4.[76]

Consequently, most American analysts missed the fax surge, consistently underestimating expected growth of fax machines and overestimating their expected cost. IRD reported in 1985, "All in all, the boom in fax shipments in 1985 is unlikely to continue beyond 1986." As late as 1987, market research firm FIND/SVP predicted a total of 830,000 G3 machines in 1990 and 1.7 million in 1995, numbers that were off by millions.[77] Who can blame them? For years, fax's promises had exceeded the reality. When the sleeping giant finally awoke in America, its advocates and observers were as stunned as anyone. In Japan, the CIAJ did much better, predicting the number and value of fax machine manufacturing one and two years ahead (correlation coefficients of 0.98/0.91 and 0.86/0.81, respectively, for 1988–2008), but its longer-term estimates proved far less predictive (0.19/0.32 for five years ahead).

Like any contested technology, pre-G3 machines did not immediately disappear. Into the mid-1980s, worksheets for managers to compare communication costs included telex, G1, and G2 as well as G3 machines.[78] New G2 machines appeared through the mid-1980s, and a top-end G2 with many automatic features cost much less than a faster but less-capable G3. As late as 1983, G1 comprised 60 percent, G2 25–30 percent, and G3 only 10–15 percent of the 350,000 machines installed in America, two-thirds of which were rentals. Two years later, sales comprised nearly four-fifths, and G3 two-thirds, of over 500,000 machines.[79]

The G2 and G1 backward compatibility of early G3 machines probably hurt as much as helped G3 diffusion because of the added cost of compatibility and the decreased pressure to upgrade to G3. Including G2 compatibility added $600 to the $4600 cost of Xerox's 295 G3 machine, only $100 less than the optional computer link. A small firm with limited faxing needs could spend $500 on a G2 instead of $3000 on a G3 and still fax with G3 machines.[80] Murata's 1985 introduction of its 7100 model with no G1–G2 compatibility proved to be a turning point. A deliberate decision to reduce cost, this exclusion gambled that G3's rapid spread rendered backwards compatibility unnecessary. The gamble succeeded.[81] In 1986, a comprehensive guide listed 50 G3 models from thirteen companies. In 1988, the same guide listed 136 models from thirty-six companies.[82]

USERS

Conveying information with the authenticity, authority, permanency, and dense content of a letter but as quickly and casually as a telephone conversation, faxing's real revolution was in changing how people communicated and worked formally and, importantly, informally. Whether hailed as a "turning point for PR" or a "quiet revolution in the mortgage marketplace," faxing accelerated the work cycle from weeks or days to hours or minutes. The need-for-speed mindset wholeheartedly embraced facsimile, whose immediacy gave a "hot off the wire" urgency lacking in a letter or telex.[83]

Unlike OA, facsimile easily dropped into existing office operations, although adhering too closely to routine reduced facsimile's benefits: turning off the fax machine at the end of the business day lost any after-hour messages.[84] Integration and acceptance, however, were not automatic; they required negotiation within the office. Despite the similarity to letters, facsimile raised practical and procedural questions. Transmitting local faxes incurred no expense, but receiving faxes and sending long-distance faxes incurred definite, if small, costs. Organizations had to determine what constituted reasonable personal use. A common modus vivendi was moderate personal use, such as requests to radio shows and lunch orders. Personal long-distance faxes and running a sumo tournament, betting pool, or political campaign exemplified prohibited activities.[85]

Cover sheets quickly became standard because they provided vital identifying information about the recipient and sender (and created a niche market providing humorous and specialized coversheets).[86] Such protocols eased faxing's acceptance, as did the guidance in business manuals.[87] By the early 1990s, etiquette experts like Miss Manners had integrated faxing into the hierarchy of written and oral communications with a social status more formal than a telephone call or email but less polite than a mailed letter. The golden rule applied: "Be a gentleperson with your fax," and always remember that confidentiality was not assured.[88] Security consultants made the same point, but less eloquently.

The competence and assumptions about a person, country, or organization sometimes became reified through faxing or its absence. Artist Jane Bunnett, explaining why it took three years to organize a recording session in Cuba, rhetorically asked, "Ever try to fax Cuba? Once we couldn't get through for weeks so we flew from Toronto and here were all these people standing around the fax machine. It was like, 'Hey, guys, there's no paper in it.'"[89] At one large American corporation in 1993, management faxed the twenty-three male board members but overnighted the two female directors the same information—incorrectly assuming that they, as women, did not have fax machines.[90]

Negotiations proved excellent for faxing because its written record provided more precision than conversations while avoiding potentially embarrassing face-to-face situations, whether delicate diplomatic negotiations, breaking up a band, or an imperfect command of a language. Real-time negotiating at a distance enabled preliminary discussion—sometimes measured in rolls of fax paper—before the parties formally met. Especially in service-oriented businesses, people learned that faxing did not replace personal contact but did accelerate and improve communications.[91]

Faxing validated the declaration of Melvin Kranzberg, a founder of the history of technology, that "Technology is neither good nor bad; nor is it neutral."[92] Facsimile acted as both a centralizing and decentralizing technology, depending on who used it and how. Faxing could be enabling, such as allowing a sales agent to transmit orders from a client's office, or enslaving, such as reducing that agent's authority and autonomy. The very visible acceleration of office operations, coupled with pressure from greater competition and outsourcing, made fear often a motivator as important as opportunity.[93]

Like other technologies, faxing affected different groups of people differently. A new sense of personal and economic geography emerged as people realized they could separate themselves from their work—but usually only if they ranked sufficiently high. For others, the long arm of the office reached out to them. Reporter Jeffrey Young labeled facsimile an "electronic lasso," an apt definition since users could rope with or be roped by it.[94] Faxing enabled firms to locate backroom operations in low-rent buildings, and salespeople and executives in more expensive and prominent locations. This differed only in scale, not concept, from faxing low-wage clerical work offshore.[95] This outsourcing expanded greatly as telecommunication costs dropped.

Faxing enabled partners in law firms to read at home the drafts their associates wrote in the office. In 1986, only two partners in Paul, Weiss, Rifkind, Wharton & Garrison had fax machines. The next year eight partners did. By late 1988, twenty partners had fax machines and the firm also had one machine on each of its eleven floors as well as a dedicated fax room. The firm's computer services director, William Hunnell, considered faxing a "dramatic, important life-style change for lawyers."[96]

Faxing blurred if not destroyed the distinction between work and everything else. For workaholics, fax machines enabled them to take their work anywhere. For everyone, deadlines never ended. Instead of meeting a Friday deadline by readying a package for a Thursday overnight pickup, faxing encouraged—or de-manded—working up to the deadline as well as procrastination.

Similarly, the fax machine enabled longer weekends at beach houses and other escapes from city offices while working. Perhaps the worst houseguest was an attorney who arrived "with a box—not a roll—of fax paper; our machine was working all weekend."[97] For a traveling parent, however, faxing proved an in-expensive way to keep in touch with home: A welcome fax from a spouse and children waiting on arrival at a distant hotel triggered a fax back in response.[98]

For others, faxing meant no escape from their superiors. On the popular Japa-nese television show *Otoko no Ibasho* (A Man's Place), the gift to a newly pro-moted section chief from his superiors of a home fax machine extended the office into the sanctity of the home.[99] Diplomats, sales agents, and distant negotiators found themselves with less leeway and free time, continuing a trend that started with the telegraph. In the middle were the receptionists and others who spent hours at their fax machines handling this correspondence.

Seeing an opportunity in travelers, services quickly appeared to provide faxing in hotels, airports, and other venues. Offering a fax service was as easy as install-ing a machine (or employing one already in use) and charging a fee. Responding to demand from conference organizers and business travelers, hotels offered fax machines in business centers, then in executive suites, and, for the truly elite, in poolside cabanas.[100]

Not all services succeeded. The most spectacular G4 failure, ZapMail, popu-larized and legitimized faxing while demonstrating, albeit inadvertently, the vir-tues of compatibility and the end of the industry model of intra-company faxing. Its promoter, Federal Express, one of the most innovative and fast-growing firms of the 1970s, acted from the normal corporate motivations of fear and opportunity to embrace the wrong future. The fear came from competitors considering fax delivery services.[101] The opportunity began in the late 1970s, when Federal Ex-press investigated capturing corporate communications sent by telex and "Xerox-telephone hook-ups."[102] A 1981 study by J. Vincent Fagen, a Federal Express founder, tried to capitalize on the firm's extensive network of customers and ser-vices by expanding from the physical to the electronic delivery of documents and possibly data, voice, and video.[103]

The ambitious result, ZapMail, consisted of two tiers of service. Low-volume customers would use regular delivery and pickup, but Federal Express offices would fax the documents for door-to-door delivery within two hours. High-volume users would have their own ZapMailer fax machine. Ultimately, two Federal Express satellites would handle communications among 50,000 satellite dishes on clients' roofs. Compared with 1984 earnings of $1.4 billion, the firm anticipated ZapMail revenues at $1.3 billion in 1988 and $3.5 billion by 1993.[104]

The reality proved quite different. Four flaws marred ZapMail: high costs, technical problems, a focus on the mailroom instead of the office desk, and equipment incompatible with G3 machines. The initial charge of $35 for five pages contrasted with $14 for Federal Express's overnight delivery.[105] The "normal" types of technical problems included software difficulties and delays in receiving equipment. Although operations began on July 1, 1984, Federal Express did not install ZapMailers in large customers' premises until March, 1985.[106]

ZapMail's emphasis on centralized mailroom to mailroom delivery proved to have been fundamentally misconceived. Federal Express's success in overnight delivery came partly from bypassing mailrooms for direct desk-to-desk delivery. Employees did not like reverting to dependence on the mailroom. Equally flawed but more understandable was the decision to use the technologically advanced G4 NEC ZapMailer, which was incompatible with the slower G3 machines. Like Graphic Sciences earlier, Federal Express promoted a closed system to establish a fax monopoly equivalent to AT&T's hold on long-distance service. Realizing its error in 1986, Federal Express started retrofitting its ZapMailer II machines with G3 capacity and offering a G3 machine, but it was too late.[107]

Demand and income never reached expectations. Transmissions averaged slightly over 2,300 daily the first year, contributing less than one percent of the firm's income. By 1986, ZapMail handled 11,000 documents daily. Impressive compared with previous fax services, this paled compared with the 550,000 packages Federal Express handled daily. When Federal Express finally halted ZapMail in October 1986, losses reached $317 million.[108]

While less spectacular than ZapMail, public fax service remained problematic. Two-thirds of the 170 fax service firms established in Manhattan between 1965 and 1992, the vast majority after 1983, failed.[109] Far more successful was adding faxing to existing businesses like print and copy shops. A store just needed space for the machine and a phone line. By 1994, nearly a third of convenience stores offered faxing and other services like photocopying.[110]

The mobile fax machine proved essential for real estate agents and others whose office was their car or briefcase.[111] Linked with radios or cellphones, faxing provided information to mobile emergency crews and utility repair vans. Mobile fax, however, never became a large market due to the poor quality of transmissions and the development of alternatives like the electronic mailbox, which permitted accessing faxes anywhere.[112]

Color fax also proved unsuccessful. Although attracting interest since the 1930s, its complexity (sending multiple images and recombining the transmissions at the receiver) and expense meant color remained a curiosity despite occasional interest from manufacturers, PTTs, and newspapers.[113] In 1991, Sharp introduced a 3.5 million yen ($26,000) color fax machine, followed by other manufacturers who built on the 1994 ITU color fax standards. Although inkjet multifunctional machines included color as standard, color never became a major attraction for faxing.[114]

A third-generation effort to fax newspapers in 1989–92, part of a larger foray into electronic communications, fared slightly better than the previous pre-electronic efforts. The modern faxpaper began on April 3, 1989, when the *Hartford Courant* launched its one-page *FaxPaper*, an afternoon summary of the next day's paper for local businessmen and insurance executives. Initially priced at $2,500, the subscription soon fell steeply to $1,500, then $600. The paper had 2,000 subscribers in 1992 before ceasing publication.[115] More typical was the five-month existence in mid-1990 of the *Chicago Tribune*'s fax paper. Condensed versions of regular newspapers, lacking in graphics and depth, proved to be a recipe for failure since the lack of advertisers matched the lack of subscribers.[116]

Successful faxpapers targeted smaller audiences with specific needs and desires. Providing specific information such as stock quotations and restaurant reviews for a small fee proved more profitable. In the early 1990s, newspapers introduced fax-on-demand services, allowing a reader to call a designated telephone number, punch in the code for an article, and receive it at the reader's fax.[117] Another niche was geographic. The *New York Times*'s *TimesFax* reached over 30,000 readers in 1992 by targeting cruise ships, resorts, American executives in first-class Japanese hotels, and Navy ships. Though not adding greatly to the bottom line, *TimesFax* extended the paper and readership worldwide.[118]

More profitable was the specialized newsletter, serving narrowly defined groups willing to pay for up-to-date information. In 1993, McGraw-Hill faxed more than thirty newsletters, substituting the speed of faxing for the slowness of the mail.[119] Not all faxletters aimed at the business community: *The Missoulian*'s *Fish Fax* updated summer fishing conditions in Montana, and the *Providence Journal*'s *Friar-Fax* soothed the desire of distant fans for information about the Providence College basketball team.[120]

FAX ART

No better demonstration of how fax machines enabled people to imagine, explore, create, manipulate, and communicate was the art world, part of which enthusiastically embraced faxing. The concept of fax art (also called telecopy-art

and art(e)fax) was simple: Artists faxed images to a museum or gallery whose curators hung them on walls to flow like a "waterfall and the movement of film frames
in a projector."[121] Public participation—faxing contributions as well as viewing—
and appreciation of the art's process, variation, and randomness were important
goals. Fax art reversed the essence of faxing: providing limitless numbers of exact
replicas. Instead, fax artists intended their exhibits to change with new submissions, thus creating an ever-changing reality.

Faxing attracted artists for a range of reasons. Some, like Sue Lowenberg, a
co-organizer of the 1986 International Night of Telecopying, and Peter Max, en-
joyed fax art's "instantness." Others, like Joseph Kossuth and art critic Martin
Prinzhorn, used fax art to challenge accepted ideas of reproduction and authenticity. Weaving contemporary technologies and art together excited artists like
Eduardo Kac, who also experimented with slow-scan television, video, DNA and
telepresence.[122]

The Fax 1 production between creator Tom Klinkowstein at Amsterdam's
Mazzo art gallery and Robert Adrian at Vienna' Blitz Bar gallery using G2 machines introduced fax art on August 5, 1981. A year later, Adrian organized The
World in 24 Hours where artists in fifteen cities on three continents sent faxes,
slow-scan television, or other telephone-transmittable media to the Ars Electronica '82 show in Linz, Austria. The concept quickly spread. In 1984, for example,
Partifax at the Grimsby Public Gallery in Toronto lasted two months, with one
week devoted to faxing from four European cities (Berlin, Bristol, Pavia, and Vienna) in four countries as a component of the L'Unita festival.[123]

Fax art shows continued into the twenty-first century, part of a larger wave of
mail art, networked art, techno-art, and performance art "exploring the relationship between people and the everyday technology that they use to communicate
with each other," where artists collaborated to alter the artwork they received
before sending it on, according to the Brighton Disembodied Art Gallery. Like
redrafting a legal brief, faxing superbly suited this revisionary process. For the
1993 Telaesthesia, the gallery received one hundred faxed wallpaper designs from
twelve countries. Each new arrival was photocopied, made into strips, and hung
on the wall, changing the exhibit.[124] As well as stimulating the art world, faxing
also challenged curators charged with preserving the new art.[125]

Fax art mimicked patterns of fax usage, with participating artists and organizers located primarily in prosperous, democratic countries in North America and
Western Europe. The global art village had many blank spots. As fax machines
spread, so did public participation—a 1990 University of Illinois show received

more than 900 faxed contributions.[126] Appropriately, the first Chinese fax art show in 1996 linked the image of the fax machine as a tool for free speech and economic opportunity.[127]

EXCITEMENT OF FAXING

Modern facsimile had two faces, the public persona of machines empowering people, small businesses, and political movements, and the far less publicized corporate market of large-scale, high-volume systems. The first image captured imaginations and sparked excitement the way Apple's MacIntosh did for a generation of computer users. Faxing seemed a leveling technology allowing individuals and small firms to communicate as efficiently as larger firms. Communicating by fax meant no one knew that your mercenary company was only "a retired military guy sitting in a spare bedroom with a fax machine and a Rolodex."[128]

Faxing created excitement and empowerment as the creation of thousands of applications and businesses demonstrated. This was the true "democratization" of a technology, of innovation in action as people developed uses for faxing far beyond those promoted and imagined by fax vendors. Faxing's ability to easily transmit information in images — as opposed to the limited text of telex — contributed greatly to its success as "creative faxers . . . discovered that these machines can transmit not only sales charts and legal briefs, but pizza orders, song requests, party invitations, greeting cards, ski reports, amniocentesis results, baby footprints, children's drawings, and vows of eternal love."[129]

Decreasing costs made previously expensive actions affordable, like faxing instead of mailing minor league baseball statistics for centralized compilation, thus creating demand that fax promoters never predicted.[130] Humor quickly appeared on the faxways as people swapped jokes, including the "Guide to Safe Fax," and sent cartoons, thus enabling the rapid diffusion of visual humor often sexual in nature.[131] Children faxed letters to Santa (receiving replies that were "kept general and contain no promises") and God accepted faxes from anyone via the Wailing Wall in Jerusalem.[132] More pragmatically, God and Mammon did mix: Japanese students preparing for college entrance examinations could pay 3000 yen to receive a faxed prayer from the Dazaifu Tenmangu Shinto shrine.[133]

The diffusion of fax machines provided fertile soil for a broad, rich array of services and usage far beyond business communications. Hundreds of firms, mostly small, provided a wide range of enhanced facsimile service, a phrase encompassing any application beyond basic faxing, such as broadcasting and electronic mailboxes.[134] For firms and government agencies seeking to avoid investing in the equipment and expertise for large-scale faxing, "blast faxing" services proved ideal for rapidly disseminating press releases and alerts, such as traffic accidents or volcanic eruptions threatening air travel.[135]

In Japan innovative services offered real-time information, whether current snow conditions and parking availability at ski resorts, or daily market data for the 637 chrysanthemum growers of the Akabanemachi Japan Agricultural Co-operatives. Most of the services were new: Less than half of the hundred firms answering a 1993 survey existed before 1991.[136] The central and local governments provided fax machines to hearing-impaired people and civic groups for emergency warnings. Communications were not all one-way. Once fax machines spread widely, police stations provided their fax numbers to hearing- and speech-impaired people.[137]

Other, less publicized uses developed. For a society that highly valued politeness and attention to details, a fax discreetly obtained the correct *kanji* for a person's name before sending a more formal letter. *Fax Life* sponsored a popular eBay-like auction service, with sellers faxing their offers to Hitachi Hiplan, which collated and then faxed lists to buyers who either contacted the seller directly or via Hiplan.[138] The Japanese system of addresses also contributed to fax's popularity. Because an address indicated location but also when it was built, street numbers were not sequential. Finding a specific address often challenged Japanese as well as foreigners. Before faxing, arranging a meeting usually demanded a courier with a map, detailed telephone conversations about directions, or an agreement to meet at a prominent location. Visitors observed, "Japanese homes all seem to have fax machines, and trading maps seems to be one of their indispensable purposes."[139]

Faxing facilitated information flows of all kinds, enabling entrepreneurs to create niche opportunities. Hoping his "faxed 'tip sheet' . . . could become to tabloid news what McDonald's is to food," Tom Colbert established Industry R&D in 1992 to collect unusual local news stories and fax the collation to subscribing Hollywood filmmakers and newspapers seeking ideas. Zapnews, a fax service that provided radio stations with less extensive and less expensive news feeds than the wire services, inspired Colbert.[140]

In Japan, faxing quickly became an expected and popular way for viewers to participate in television discussion, news, and game shows. For the stations, faxing proved faster, error-free, and more efficient than writing a caller's message. As American radio stations discovered, faxing also provided a mediating layer of technological separation between irate or angry callers and radio staff.[141]

Nowhere was fax's potential better demonstrated than at Southwest Texas

Methodist Hospital in San Antonio. To reduce delays in processing prescriptions, its pharmacy tested fax machines at three nurses' stations in 1988. Turnaround dropped from four to eight hours to one hour, but the two-week trial also discovered faxing improved tracking of orders. Thirty-six fax machines soon equipped all the hospital's nurse stations.[142]

The hospital deserved attention for three reasons. First, even a few years earlier, a hospital could not afford forty fax machines. Second, the unexpected benefits of faxing, the greater control and knowledge about patients' drugs, proved almost as important as the expected benefit, faster turnaround and better use of aides. Third, this was a bottom-up experiment, conceived and implemented locally. This was the true excitement of fax, the opportunity for people to take a technology and turn it into something that they wanted.

Faxing became the greatest boon to distant communications since the airplane. International and distributed operations of any kind—legal, scientific, business, governmental, academic—became much easier, less expensive, and faster. Faxing did not cause the expansion of globalization in the 1980s–90s, but it certainly facilitated it by circumnavigating not only time zones but poor mail and telephone services, allowing exchanges of letters within minutes instead of weeks. Typical of the enthusiasm about faxing was Texas attorney Peter S. Vogel, who gathered needed signatures in five cities in one morning by fax. Sending and receiving the follow-up papers took three weeks.[143]

Faxing's ability to quickly post new findings—and thus establish priority—and to cooperate over a distance greatly attracted scientific communities. Speed did not always prove desirable, however. The rapid spread of unpublished data and ideas threatened to bypass existing procedures and organizations that reviewed and published articles in science journals. Nowhere did the dichotomy between fast transmission of new material and a fuller understanding of that material appear wider than the cold fusion episode of 1989–90, where, Science News claimed, "The term 'publication by fax' already has pejorative connotations."[144]

As important as saving time and money was the ability to fax in one's own language instead of translating and retranslating a message into a telex-compatible form with the attendant potential for error and misinterpretation. Many firms abandoned their telex machines and even overseas agents.[145] In 1987, a survey of American businessmen showed 16 percent would telex a six-page letter internationally, 18 percent would fax it, and 29 percent would use the mail. In 1990, telex had shrunk to only 2 percent and mail to 20 percent, while faxing soared to 48 percent.[146]

In Japan, faxing caused much of the growth in international telephone calls from 23 to 135 to 761 million minutes from 1980 to 1986 to 1996, while telex minutes went from 38 to 44 to 5 million. By 1988, and probably earlier, faxing surpassed conversations on overseas telephone calls. The gains came at the expense of Quick-Fax. KDD's service rose from 24,000 messages in 1982 to peak at 250,000 messages in 1985 before declining equally sharply to 60,000 in 1988. The drop came from firms buying their own machines and faxing directly, eliminating KDD as the middleman, just as European newspapers in the 1930s bypassed PTT services.[147]

POLITICS

Just as it changed the business world, so too did faxing shake the assumptions and patterns of politics. Faxing combined with existing communications technologies like television and newspapers to create a seeming revolution by providing activists, politicians, and ordinary citizens with powerful tools to influence private and public debate. In the late 1980s and early 1990s, early adopters initially employed faxing quite innovatively and successfully, enabling outside groups to organize more effectively than before, expand their base, and score some noticeable political successes. This democratization of access and effort generated much excitement about fax's revolutionary political potential.

Faxing, proclaimed University of Southern California journalism professor Joe Saltzman, enabled "freedom of the press as it was meant to be—a uniquely individual way to communicate to others. It brings back memories of 17th-century English broadsides, Revolutionary War American pamphlets and circulars, and Third World placards and posters."[148] As Eric Barnouw had warned in 1982, however, faxing also provided new possibilities for greater centralization and control from above.[149] Even as faxing's visibility made it the tool of the underdog and revolutionary, states secretly monitored such faxing, wealthier campaigns outfaxed opponents, and public relations firms quietly orchestrated "Astroturf" movements for their clients.

Like all technologies, faxing benefitted those who were the most organized, the most active, and the most energetic. The applications were not new. Unprecedented was the ability to communicate and coordinate on a larger scale. Instead of weeks or days, faxing needed only hours or minutes to inform the media, arrange meetings, receive the latest polling data, and conduct the scores of other activities essential for any political campaign. In a realm where the timely acquisition and diffusion of information could be a matter of political life or death, faxing enabled eager users to do more, stay more informed, and disseminate information far more quickly and accurately than ever before.

A universal technology equally accessible to the neophyte and the professional, faxing gave centralized organizations the ability to appear decentralized, and decentralized organizations the ability to act in a coordinated manner. Similarly, faxing allowed politicians to appear modern and open. In May 1994, Japanese Prime Minister Tsutomu Hata asked citizens to fax comments on his policies in an unsuccessful attempt to deepen his political support. Three machines received hundreds of faxes daily until his minority government collapsed after nine weeks.[150]

Activists created fax networks, their own Committees of Correspondence, to strengthen their internal cohesion and quickly communicate with thousands of people. A barrage of faxes urging members of a group to call or write letters could create a seemingly spontaneous grassroots movement literally overnight. The results could be quite effective, as discovered by the Clinton Administration during its abortive 1993 efforts to create an energy tax and by proponents of the 1995 Conference of the States. In both cases, opponents created fax networks to organize and mobilize, linking with talk radio for greater outreach. One informal fax network, Speak Out America, claimed over 100,000 members willing to fax, write, and call their representatives.[151] The mobilizations succeeded: neither energy tax nor conference occurred. Successful opponents of revising immigration law in 2007 similarly flooded Congress with over a million faxes.[152]

By 1990, faxing was an essential part of daily American political activities. From a technology of wonder, it had evolved into a tool of toil. Even as volunteers for small groups spent hours over stand-alone fax machines, better-funded candidates "blastfaxed" thousands of pages an hour via firms like Xpedite, which faxed over 750,000 pages for the Bush and Clinton 1992 presidential campaigns.[153] The revolutionary potential of faxing decreased as public relation firms and lobbyists provided people with prepared letters that they signed and then were faxed free. These "Astroturf" campaigns, which appeared local in origin but were organized professionally, diminished the impact of a fax. Politicians, valuing the easy flow of information, maintained public fax numbers but also private numbers to minimize unwanted faxes.[154]

Compressing news cycles proved to be one of faxing's most significant consequences. Until 1990, the news cycle took two days, starting with publication of a candidate's statement and ending with an opponent's reply. By 1992, rare was the story that did not carry the opponent's reply with the original statement. Faxing enabled instant detailed responses, as was sharply demonstrated when the Democratic National Committee faxed a rebuttal of George Bush's August 20, 1992, acceptance speech at the Republican National Convention to reporters covering the speech inside the Houston Astrodome.[155] By accelerating the news cycle, faxing contributed to a de-emphasis of contemplative coverage, as reporting changed to more of a "he said, she said" focus.

Internationally, faxing played an important role in advancing democracy by weakening regimes in Panama and Poland, strengthening public scrutiny of the new government in Mozambique, and resisting a coup in the Soviet Union.[156] The common denominator was the easier domestic and international diffusion of information and strengthening communication among activists.

The Soviet Union experienced the most successful application of political faxing. Not by chance, this was also the country where political activists and journalists instead of businesses had imported many of its fax machines. The easing of the Communist Party monopoly on information coincided with the rise of Solidarity in Poland, a rise based partly on communications equipment smuggled from the West (including printing presses, photocopiers, personal computers, and fax machines). The Solidarity experience provided a partial template for activists inside and outside the Soviet Union like publisher Phillip Merrill, who advocated peacefully overthrowing the Soviet Union by flooding it with fax machines, photocopiers, and other communication technologies to enable Soviet citizens to spread information and communicate with each other directly.[157]

Glasnost—the ending of censorship—and relaxed import controls allowed outside organizations to send fax machines to groups in the Soviet Union. Much of the initial equipment came from diaspora communities responding to requests, whether in the Baltic states, the Ukraine, or Armenia, for information technology, especially fax machines and photocopiers. Faxing proved invaluable in sending information to journalists—Western, Soviet, and, increasingly, former Soviet— for stories that reached a much wider audience. Interfax exemplified this new world of communications. Established in late 1989 by Radio Moscow and Interquadro, a Soviet-French-Italian venture, Interfax quickly became an indispensable source of news for Western news bureaus, diplomats, and even the office of Communist Party General Secretary Mikhail S. Gorbachev. Other fax-based news services, such as Baltfax, became important sources of regional information. By late 1988 journalists stationed in Moscow experienced a surfeit of information, delivered by overflowing fax machines.[158]

The August 1991 coup to overthrow Gorbachev vividly demonstrated the effectiveness of faxing and its ability to link with existing media. Admittedly, the unsuccessful coup was a textbook of what *not* to do. Open telephone lines allowed Soviet fax machines to transmit declarations from Boris Yeltsin's White House to hundreds of Russian and foreign destinations. Yeltsin's followers faxed statements to Radio Liberty, the Voice of America, and other Western radio stations, which rebroadcast them into the Soviet Union, thus reaching tens of millions domestically while sending information and pleas for assistance to the international community, strengthening its opposition to the ultimately unsuccessful coup.[159]

But faxing was a tool, not an automatically triumphant democratic technology. Its effectiveness reflected the local penetration of fax machines and the relative level of official ruthlessness as well as a movement's strength and support. In Myanmar possessing a fax machine meant arrest, whereas anti-Communists shipped thousands of fax machines into the Soviet Union and its satellites to weaken them.[160] Nor was faxing inherently democratic. Right-wing death squads in Guatemala faxed death threats to journalists, union leaders, and others they deemed subversive, adding to the existing arsenal of anonymous letters, funeral bouquets, and telephone calls. Extortionists and terrorists similarly faxed warnings, press releases, and threats.[161]

Similarly, Middle East dissidents and ordinary citizens faxed to promote political change, but the regimes remained in power.[162] Spurred by the 1990–91 Gulf War, Saudis used their "huge network of home fax machines" with so much commentary on previously faxed communications that "faxing a political letter has become like publishing a pamphlet." By the end of 1992, a conservative counter-reaction meant that "the fax machines have fallen silent."[163]

The limits of political faxing appeared most obviously in June 1989, when urban China seemed on the brink of a democratic revolution. Students created a domestic network of fax machines, based primarily in universities and research institutes, to communicate among themselves, which a Western observer labeled "seeking truth from fax," a play on Mao Tse Tung's "seek truth from facts" (*shishi qishi*).[164] The expansion of that informal network abroad captured public attention. The images of Chinese students in the West faxing news about the Tiananmen Square democracy movement into China to evade the government's censorship thrilled Westerners like NBC News president Michael Gartner, who wrote, "The fax machines are, in a way, the fuel of the revolution. The faxed materials inform, encourage, embolden the young revolutionaries. They have become the wall posters of this generation. Never has there been anything like it."[165]

Working journalists were equally excited. Traveling around China after the Tiananmen Square massacre, Frank Viviano reported, "At every stop, amid the student marches and police charges, I found a Democracy Movement supporter willing to escort me to a clandestine fax machine." Out to America went his stories, and in arrived "an endless stream of subversive documents from a variety of sources."[166]

While exciting and dramatic abroad, shotgun faxing—sending politically subversive information to random fax numbers—did not revive the democratic movement. The Western media coverage actually backfired, alerting the Chinese government to the threat of uncensored information. During the Tiananmen days, security officers physically stood by many fax machines to seize any incoming subversive literature.[167] Unlike the Soviet coup leaders two years later, the Chinese government maintained an effective grip on information and political power. Faxing alone could not a revolution make.

LEGAL ACCEPTANCE

Lawyers adopted faxing quickly to improve access to information such as court decisions, speed internal and external communications, and ease the tedious process of revision. Their most important contribution was establishing the legal validity of faxed signatures and documents, thus giving faxes the same legal and contractual status as an original document or telex.

Especially in rural regions where hours of travel measured the long distances, faxing saved considerable time and resources while improving administrative coordination such as scheduling judges. The "most surprising finding" of a mid-1990 survey of rural lawyers was how rapidly facsimile had become an integral part of their practice, primarily to correspond with clients and serve papers. Ninety percent had fax machines, with two-thirds acquiring their machines only in 1989–90. Most of the attorneys did not file documents by fax, instead walking across the street from their offices to the courthouse.[168]

In November 1988, Minnesota's trial courts became the nation's first to allow fax filing. Other states introduced faxing to speed transmission of criminal histories as well as child support enforcement and protective orders.[169] By 1990, numerous state and federal courts had experimented with filing documents by fax and other uses. Accepting faxing meant changing rules and regulations written for earlier times and technologies. To accelerate this process and promote uniform procedures, the National Center for State Courts published model rules for faxing in 1992.[170]

While experiments introduced faxing, lawsuits created the case law that helped set the legal parameters of faxing at the state level. In New York, civil court judge Richard Lane declared in October 1988 that "faxing patently satisfies the plain intent" of procedural rules for serving legal papers and therefore was sufficiently reliable to serve as an original copy.[171] For warrants, the landmark case occurred in Michigan. A judge had administered an oath to a police officer by telephone and then faxed the affidavit and court order authorizing a blood-alcohol test. Plaintiff Thomas A. Snyder challenged his arrest because the warrant had not been signed in the physical presence of a judge. On December 28, 1989, the Michigan Court of Appeals affirmed the legality of a faxed warrant.[172]

Federal courts were more conservative.[173] The Judicial Conference, the rules and policy making organ of the federal courts, started discussions of faxing in 1989 and issued restrictive guidelines in 1991, allowing fax filing "only in compelling circumstances" or when a local court had already approved it. Only in 1994 did the Conference approve faxing for non-routine, locally sanctioned use, in effect letting local courts decide.[174]

Opposition came from fears of overburdened clerks, potential abuse by *pro so* litigants (who represented themselves), infringing on local court rules, and, of great importance, lawyers unable to file because the fax line was blocked, "since pleadings all tend to be filed at the last possible moment."[175] Another concern was faxing opposing counsel a document in the evening that required a response the next day. This practice became so widespread that in 1990 Texas revised its civil procedures to state that any document received after 5 p.m. legally arrived the next day.[176]

Paralleling the federal courts, government agencies initially rarely accepted faxed documents because they lacked fax machines or because they strictly interpreted administrative rules. As faxing diffused through society, court cases, experience, and revised rules made faxed documents, contracts, and signatures legal.[177] Whether they remained private, however, was another issue.

SECURITY

Because most pre-G3 faxing occurred intra-firm and intra-organization and between known numbers, ensuring authenticity and privacy was not a serious problem, although a few firms and government agencies encrypted their faxes to thwart electronic eavesdropping. The explosion of faxing after G3 made security a serious problem for the first time. Important questions about the legality, authenticity, and security of the faxed message challenged operators, especially the potential lack of confidentiality. The problem was fourfold: ensuring that only the recipient read the fax, missent faxes, intercepted communications, and fraudulent faxes.[178]

Since they arrived without envelopes, faxes could be—and were—easily read by anyone. Such vicarious reading seemed the equivalent of listening on a party line to other telephone conversations, save for the significant difference that the senders did not know their faxes attracted multiple readers. These spectators were semi-active. The faxes appeared without any effort on their part, but they had to pick up the newly arrived pages, if only to know where to deliver the fax.

Something about the ready openness of a fax sitting in a fax machine caused people to disregard normal proscriptions about looking at others' mail. How else could hotel employees report about their guests' "warm personal faxes"? Many a fax with private information—negotiations for a new job or of an intimate nature ("fax flirtations")—inadvertently provided office gossip. Standing by the fax machine to send or receive a fax replaced standing by the water cooler as a place to trade office information and gossip but with the benefits of looking productive as well as reading colleagues' faxes.[179]

Ensuring that no one else would read a transmitted fax meant either having a private fax machine, as Westinghouse Electric provided for its managers handling sensitive information, or standing by the fax machine.[180] Not until the computer-based fax systems of the 1990s restricted access to incoming faxes did privacy automatically increase. Computer-based faxing ended public spectating but not error, state surveillance, and espionage.

Missent faxes stemmed from the ease of faxing. Pushing the wrong button or using the wrong list of recipients could cause serious problems, especially for politicians whose faxed words or actions visibly contradicted their public statements.[181] In 1988, the *Wall Street Journal* received a merger memo meant for a shareholder whose fax number differed by one digit. The paper's story killed the proposed merger. As the FBI warned law enforcement agencies in 1991, prevention was the best way to avoid missent faxes.[182] Strict administrative procedures, such as keeping recipient lists separate, reduced but did not eliminate errors.

Cover sheets offered some deterrence. In 1991, the accidental transmission of a defense's jury-selection strategy to an opposing lawyer required the creation of a new jury, delaying an asbestos liability trial. Had the defense, by faxing the wrong party, waived its claims to confidentiality? Or had the defense, by identifying the transmission as confidential and private in its cover page with instructions on how to return a missent fax, taken all reasonable precautions?[183] The legal world responded quickly. In 1992, the American Bar Association issued its "Inadvertent Disclosure of Confidential Materials" to guide attorneys and judges.[184] Marking every page confidential, directing everyone handling the fax not to read it or leave it unguarded, and asking the recipient to verify its receipt strengthened the claim of attorney-client privilege as well as reducing the risk of inadvertent disclosure.

The rarely mentioned area of security was deliberate interception, a secret world populated by governments with equipment to monitor, analyze, and even, according to some, alter fax transmissions.[185] Mining faxed messages for information actually proved easier than voice calls, because written faxes tended to be more succinct and optical character recognition enabled automated searching. As one vendor warned, "In effect, every transmission is an open-envelope invitation to preying eyes."[186]

Governments actively intercepted faxes, gathering economic, political, and military intelligence on friends, foes, and neutrals. The Canadian Communications Security Establishment intercepted telephone and fax calls in 1991 between the South Korean foreign ministry and its embassy in Ottawa about the proposed sale of a Candu nuclear reactor, and the European Parliament assumed the American National Security Agency routinely collected fax transmissions.[187] What governments did with that data was another question. How, for example, did its monitoring of faxing between Palestine Liberation Organization headquarters in Tunis and the West Bank and Gaza in the 1980s affect Israeli government actions?[188]

Growing economic espionage as well as traditional intelligence gathering meant "a bright future" for interception equipment manufacturers. By 1994, at least six firms produced equipment, but neither they nor their clients publicized their efforts.[189] South Africans only learned that their National Intelligence Agency read their faxes in 1997 when thieves stole a machine capable of simultaneously intercepting thirty calls from its headquarters.[190] The world learned a bit about CIA prisons when a Swiss paper leaked faxed documents intercepted by Swiss intelligence.[191]

Unsurprisingly, a market also emerged to protect against interception with at least fourteen non-U.S. and American firms selling fax encryption technology and training in 1999.[192] Western military, diplomatic, and intelligence communities took fax security very seriously, equipping some internal networks with machines meeting STU-III, Tempest, or NATO STANAG security certification.[193] The British Foreign and Commonwealth Office in 1997 had fifty-one crypto-fax machines overseas as well as regular fax machines.[194] Such security measures, however, cost money and restricted communications to compatible equipment or within a network.

Some businesses took security seriously. To secure data from competitors, the oil industry has, since the 1970s, encrypted faxed drilling data from rigs.[195] Westinghouse Electric, assuming all its foreign communications were monitored, encrypted its internal transmissions, as did Kodak and other firms in the mid-1990s.[196] Faxing also served as a counterintelligence tool. In 1995, the Federal Bureau of Investigation unveiled its Awareness of National Security Issues and

Response Program (ANSIR) fax network to update 25,000 American corporations about economic and technological targets of foreign spying.[197] Most users, however, did not encrypt faxes, a reality reflected in the lack of publicly available cryptographic equipment and lax procedures. Consequently, intercepting faxes remained "a piece of cake for professionals."[198]

Just as a fax proved more effective than a letter in convincing people to pay their bills and respond to surveys, so too did it benefit criminals and fraudsters.[199] The "classic Nigerian 'fax scam,' a form of fraud so ubiquitous and so successful that legitimate trade with African countries has begun to suffer" became the exemplar of using modern communication technology to separate people from their money.[200] A fax informed the lucky recipient he or she had a share of millions of dollars trapped in a Nigerian bank account—but getting that windfall first required a little money and personal bank account information. Another scam, migrating from telex, sent an invoice to a firm for its listing in a nonexistent fax directory. The assumption, often correct, was that the bookkeeping department would simply pay without checking because the amount was under $1000.[201]

Authenticity had not been a problem with dedicated fax lines and intra-firm sending. Inter-company faxing and directories, however, created easy opportunities for fraud. Faxing made forgeries easier since letterheads and signatures could be photocopied and pasted, while watermarks and seals did not transmit. As Securities and Exchange Commission attorney Jeffrey Norris stated, "More and more of our cases have involved defendants who are downright outlaws. They do it with fax machines instead of a gun, but they are outlaws nonetheless."[202] Some forgeries, following a long tradition, were political. Responding to the Tiananmen Square massacre, the French journal *Actuel* organized a "fax-in" of a phony *People's Daily* (faithful to a front-page article attacking American actions in Central America), prepared by exiled journalists and students, to fax into China.[203]

International higher education suffered particularly from fax fraud. So bad were the "disproportionate problems with forged, falsified, or altered documents" from abroad that the American Association of Collegiate Registrars and Admissions Officers in 1996 recommended not accepting any faxed foreign transcripts.[204] For domestic transcripts and other documents, the Association suggested its members establish procedures to validate and verify faxes, including creating lists of trustworthy fax and telephone contact numbers, putting specific information on cover sheets, reviewing received documents, and calling to confirm dubious transmissions. Such social engineering—creating systems of standard operational procedures—worked only if people paid attention to them. If not, a forged fax with misspellings sent from a McDonalds could free a convict or worse.[205]

Verification and confidentiality remained serious, continual challenges. For the truly paranoid or realistic, Albert R. Belisle, chair of the American Bar Association Information Systems Security Committee, warned, "Think of a fax machine as a telephone. If the information is something you wouldn't say over a telephone, don't send it over a fax."[206] Because security measures like calling to verify a faxed order or accepting only known fax numbers detracted from fax's convenience, they were often ignored unless mandated and enforced.

JUNKFAX

Faxing's virtues of easy operation and low costs quickly created a torrent of information, often unwanted or unnecessary and essentially paid by the recipient. In a variation of Gresham's law, bad data crowded out good information. The problem of undesired mail was not new, but the simplicity and low cost of faxing made sending messages much easier. The greatest threat to the common space for faxing quickly became unsolicited advertising. The growth of unsolicited direct mail and telemarketing in the 1980s did not inflict material costs on recipients. Fax advertising, "the public nuisance of the late eighties," did.[207] Not only did junk fax consume recipients' paper, but receiving the ad tied up the telephone line, preventing incoming or outgoing traffic.

Junk fax's initial success came from its novelty. Faxed ads produced a much higher response rate—7 percent in 1988—compared with 1 percent for conventional mailings.[208] By early 1989, Mr. Fax, a paper supplier, was America's largest junk faxer, transmitting over 60,000 advertisements weekly to its database of over 500,000 fax numbers, nearly 10 percent of all machines. Providing those numbers became a small, low-profile industry, with brokers offering low-paid operators in corporate fax centers inducements like a Sony Walkman for every hundred numbers.[209]

The unwanted faxes provoked a counter-reaction. By 1990, nearly half the states were considering or had restricted such ads.[210] As Oregon Representative Ken Jacobsen stated, "You get a message you didn't want from people you don't know on paper they didn't buy."[211] The Direct Marketing Association, which represented telemarketers, responded initially by arresting the usual suspects. Some abuses had indeed occurred; however, the solution was not onerous government regulation but industry self-policing and a voluntary directory of people wishing not to be faxed.[212] Telemarketers, however, proved their own worst enemy. When Mr. Fax asked businesses to fax Connecticut governor William A. O'Neill to veto a bill banning unsolicited faxes, the governor's machines were jammed for a day while he waited for a report on flood damage. O'Neill signed the bill.[213]

Although Congress passed its first junk fax bills in 1990, final passage occurred only in 1991 because of disputes about automated dialing of unsolicited telephone calls. Effective December 20, 1992, the Telephone Consumer Protection Act banned unsolicited fax advertising or recorded telemarketing calls.[214] Although mostly disappearing or metamorphosing into the far more obnoxious but less costly e-mail spam, enough junk fax continued to trigger lawsuits, FCC sanctions, and the 2005 Junk Fax Prevention Act.[215] Nonetheless, junk fax ceased to be a major issue for most fax users.

Favorable economics and the fusion of technologies in Japan and the United States drove the explosion of faxing in the 1980s as market pull finally surpassed technology push. Faxing proved more revolutionary and widespread than even its most vocal advocates had expected. In addition to the anticipated acceleration of business, fax provided ideal for quickly spreading memes, whether jokes, political propaganda, or the latest discovery in science.[216]

Change from above enabled that innovative change from below. Most importantly, the new CCITT standard ensured compatibility among G3 machines, while procedures to encourage proprietary features enabled continual evolution in a period of rapid technical change. Eliminating incompatibility restructured the fax market. Always important, concerns about cost and features now dominated decisions. The number of manufacturers sharply expanded as G3 enabled newcomers to compete on a near-level playing field, turning fax machines into a high-technology commodity.

Technologically, Japanese manufacturers finally achieved the previously contradictory goals of high speed, high resolution, and low cost. Prices dropped sharply, pushed by Japanese manufacturers battling for market share and pulled by consumers willing to buy—if the price was right, which, increasingly, it was. The average cost of a fax machine collapsed from over a million yen in 1980 to 94,000 yen in 1990 and 63,000 yen in 1995.[217] Operating costs shrank as machines faxed faster and telephone charges declined, pushed by deregulatory competition.

Fax's commercial success did not derive solely from corporate competition. Without the Japanese government, the fax machine would not have achieved its phenomenal success. Yet the government did not arbitrarily impose demands. Instead, it fostered domestic competition by deregulation, supporting research by multiple manufacturers, and providing a communications network for faxing. The government's most important contribution was prodding the major Japanese players into agreeing on a single communications standard and then leading the effort to turn it into an international standard. The creation of READ and then of the G3 standard was not preordained but was achieved only through long negotiations, compromises, and perseverance.

Even excluding this government support, faxing in Japan diffused faster and deeper than in the United States. The greater need for image transmission in the "nation of *kanji*," the earlier and more thorough deregulation of telecommunications, and a more competitive market spurred Japanese domination in manufacturing and consumption. Entering the 1990s, fax's trajectory worldwide seemed increasingly upward. Yet even as the pull of market demand finally overtook the push of technology in standalone machines, engineers and entrepreneurs were forging new technological directions.

CHAPTER 6

The Fax and the Computer

Sending a fax is as simple as creating an Internet e-mail message.
— *E-Sync Networks press release, 1999*

Scathingly proclaiming, "The fax machine is a serious blemish on the information landscape, a step backward, whose ramifications will be felt for a long time," Nicholas Negroponte, founder of MIT's Media Lab, seemed oblivious in 1995 to faxing's popularity.[1] Usage had grown enormously, comprising approximately 15 percent of all telephone calls in the United States, perhaps twice as much in Japan, 40 percent of the telephone bills of Fortune 500 firms, and as much of 50–80 percent of international calls.[2] More than 35 billion sheets of paper whirled from fax machines, sixteen times more than in 1986.[3] By 1998, manufacturers had produced over 100 million stand-alone machines and tens of millions of fax-capable modems, faxboards, and other computer-based equipment.[4]

A major reason for the continued growth of faxing was its expansion from transmitting documents and images to transmitting information. The most important of the rapidly proliferating fax services and products was fax-on-demand, which allowed people to request and receive customized information immediately. Fax-on-demand was part of the integration of the computer with faxing, bringing the latter more fully into the growing world of digitalization.

Fax's continued acceptance reflected the weaknesses of digital communications as well as the attributes of faxing itself, defined primarily by convenience, capability, cost, and compatibility. Negroponte and other advocates of digitalization secured resources, including the aura of inevitability, to promote the seamlessly interconnected digital future over the imperfect analog present, a vision that gained over time as digital technology and its economics improved. Faxing began declining in the late 1990s when the digitalization of data—embodied in the Internet, PDFs, texting, and many other technologies—finally fulfilled the expectations of its promoters, sending messages, images, and files and providing information more easily and less expensively than faxing.

This decline followed different paths in the United States and Japan, reflecting the different technical, economic, and cultural environments. Although faxing's relative and absolute importance decreased after 2000, Japanese factories produced another 100 million machines over the next decade. The reason for that seeming paradox lay in the changing nature of what constituted a fax machine, an increasing diffusion of machines to households and small users, and the continuing value of faxing for shrinking but still large pools of users.

Starting with how its problems of success pushed faxing's automation and increasing computerization, this chapter explores the development of computer-based faxing; the explosive popularity of fax-on-demand services; the rise, fall, and recovery of internet faxing; and the decline of faxing as digitalization finally fulfilled its promises. The integration of faxing with the computer proved far more technically challenging than the development of conventional faxing, especially because of interface issues. While the interactions and technical environments proved far more demanding, the challenges were political and cultural as well as technical and financial.

COMPUTER-BASED FAXING

As futurist Paul Saffo delightfully described in 1990, fax machines were rich in communication but poor in processing and manipulation, whereas personal computers (PCs) were rich in processing and manipulation while poor in communication. Joining these Manichean worlds promised major benefits for faxing by automating repetitive procedures and accessing databases and other computer applications but also great challenges that demanded major advancements in signal processing and digitizing fax transmissions.[5]

Even more than stand-alone fax, computer-based facsimile developed in a very competitive market with a rapid evolution of products and firms. The lower barriers to entry and large installed base of personal computers and computer networks attracted scores of startups and other companies in the 1990s to provide services, support, software, and hardware. Producing a product demanded good software skills more than manufacturing capability. Mostly anchored in California's Silicon Valley or Massachusetts' Route 128, these small startup firms created a bedazzling (when working) and bewildering array of products, options, and interfaces.

Equally kaleidoscopic was the business side, as firms entered, exited, merged, and became absorbed by other firms.[6] Overall, these startups were more successful than their fax machine predecessors because the investment to produce a product was lower, and firms only had to sell their product, not the larger concept of faxing. Business models had to evolve too. Entering the computer fax market in late 1990, Delrina quickly became the major PC-fax software firm by convincing computer and modem makers to include its free fax software, Winfax Lite, in their equipment. The Canadian firm earned money if those users upgraded to Winfax Pro.[7]

Manufacturing remained in Asia, albeit increasingly outsourced from Japan to other less-expensive countries. The nexus of computer faxing returned to the United States, reflecting its dominance of software, the slower adoption of personal computers in Japan, and the continuing, competitive growth of the stand-alone Japanese fax machine market.

Driving the computerization of facsimile was not just the technology push of advances in electronics but, significantly, the market pull of corporate demand for more efficient faxing. Paradoxically, faxing's success became its biggest challenge. Simply handling the flow of incoming faxes challenged users. In 1990, the World Bank received 1,500 faxes daily, making manual management impossible. On Mondays, a flood of undulating thermal paper from weekend transmissions covered the floor of its Telex and Facsimile Section. Handling the 50,000 monthly faxes at the American Management Association in 1996 required two people to do nothing but send, sort, and receive.[8]

Such overwhelming volumes delayed sending and receiving faxes. Solutions were organizational and technical. Dedicating machines to receive only, an increasingly affordable option as equipment prices declined, eliminated the problem of incoming faxes preventing outgoing transmission. Technical fixes for handling increased volume included buffering incoming faxes with electronic memory until they could be printed, increasing the size of paper trays, and faster transmission speeds.[9]

For large users, transmuting the stand-alone fax machine into the computer promised significant advantages. The concept was simple: a computer generated, transmitted, and received a fax. Although evolving into different niches,[10] computer-based fax quickly divided into two main paths: external or internal modems in PCs, and more capable equipment on a local area network (LAN) for multiple users. In LAN faxing, a fax proceeded from the user's computer over the LAN, just like a request to print a document, to the fax server, which transmitted the fax. The server routed an incoming fax to its recipient. The fax server could be as simple as a faxboard running from a PC or as sophisticated as a switch on a mainframe. The difference steadily blurred as systems became more capable and costs decreased.

While PC faxing with its potential market of millions of faxboards attracted the most attention, LAN faxing attracted major users seeking to save time and money. The savings could be considerable: American General Finance, a firm with 1,400 branches, installed two fax servers at its headquarters and eliminated sixty nine stand-alone machines and ten staffers.[11]

Computer-based fax, however, did not spread in a rapid triumph. Lacking were the enthusiasm and excitement that greeted fax machines, because computer-based fax appeared as another form of faxing, which was now a standard office technology. Furthermore, employees often viewed computer-based fax—correctly—as a tool to control them. Shifting faxing from decentralized stand-alone machines to networks increased the ability of higher management to observe, audit, and bill transmissions while at the same time enforcing corporate rules.[12] Corporations liked computer-based faxing because it could reduce internal and external transaction costs of routine operations like obtaining truck permits.[13]

Users found computer-based fax far more difficult and frustrating than stand-alone faxing, emblematic of changes in technologies undergoing computerization. New to faxing but frustratingly familiar to the computer world was developing trouble-free interfaces between software and hardware, hardware and hardware, and software and software. Both market-based and cooperative standards failed to keep pace, slowing acceptance and diffusion.

In 1986, the first computer faxboard appeared with more than fifty more offered by 1989, reflecting the low barriers to entry. Early sales, however, reached disappointingly small numbers—only 15,000 in 1987.[14] Not until 1992 did fax-capable PCs outsell stand-alone fax machines. By 1994, preinstalled fax modems in PCs vastly outsold fax machines.[15] Capability, however, did not necessarily imply use.

Among the institutional obstacles to the diffusion of computer-based faxing was a lack of knowledge. Even in 1991, *Business Week* had to inform executives that a faxboard was "a plastic rectangle imprinted with electronic circuitry that fits inside the computer."[16] Like Xerox and Graphic Sciences two decades earlier, the pioneering firms had not only to offer capable technology but to engage in what GammaLink CEO Hank Magnuski labeled "missionary selling."[17] Providing good documentation, easily understood procedures, and technical support were not as technologically exciting as developing new products but could determine their commercial success.

An indication of the caution with which many firms approached PC fax was "the fact that boards are usually purchased in quantities of one."[18] The caution proved to be justified. Early faxboards were underpowered and overpromised. Users faced too much choice and too little information. How did a novice choose among twenty-six fax server software packages in 1992 or the sixty-eight 9600 bps fax modems offered by twenty-three firms in 1993?[19] Buying and trying offered the only sure way to learn whether one's computer worked with a given faxboard or software. This uncertainty demanded a level of commitment and resources that scared many.

Once purchased, learning, installing, and using computer-based fax demanded more effort compared with a regular fax machine. One psychological challenge was proving that the transmission had occurred. Unlike a stand-alone machine, where the document's disappearance into the machine indicated success, early computer-based fax systems lacked features to tell senders their message's status. This "push and pray" approach did not inspire confidence.[20]

A 1990 overview declared faxboards "annoyingly more modem-like than fax-like—quirky, feature-laden boxes calculated to frustrate anyone accustomed to the simplicity of a fax machine."[21] If a product survived long enough, upgraded versions or new hardware with increased capabilities reduced problems. Reviews often hailed an upgrade as a "vast improvement," leaving readers to wonder how bad the earlier version was—or why it was released.[22]

Underlying these problems was incompatibility due to a lack of specificity in the CCITT standards, poor quality control in both hardware and software, and increasingly complex protocols. As companies introduced more proprietary features, software became more prone to problems when communicating with competitors' equipment. In 1993, an estimated 10–30 percent of American computer facsimile transmissions failed because of mismatches between software and hardware and among different software programs.[23]

Designers could and did interpret CCITT documents—once obtained, a challenge before the Web—differently.[24] Faxboard developers had to start from scratch, unlike Japanese engineers, who drew on years of experience. Nor did faxboard manufacturers have the Japanese networks of colleagues, tradition of informal knowledge sharing, and neutral organization like the CIAJ to encourage cooperation.

To ensure compatibility, firms either had to maintain proprietary protocols or be prepared to modify their handshaking protocols. Delrina adopted its WinFax Pro software to work with over 300 modems, each with its "own nuances and its own little biases in relation to the standard."[25] GammaLink and Brooktrout, two of the most successful firms, had to write, as Brooktrout CEO Eric Giler described it, "spongy software" for their faxboards to accommodate hundreds of variations among software and hardware. One reason Brooktrout did well was that a major early customer, PR Newswire, broadcast to a wide range of fax equipment. Brooktrout had to empirically test fax machines and modify its software for Newswire. By the time Brooktrout publicly released its products, many problems had been fixed, so the firm could promise and deliver high reliability and compatibility. Similarly, providing technical support not only kept customers happy but enabled firms to learn quickly about new problems and opportunities.[26]

Compared with fax machines, creating computer faxing standards involved many more actors and more areas of conflict. In 1991, Intel's Communications Application Specification supported ten image processing programs, seven email-fax gateways, six LAN fax systems, six independent application development toolkits, five mini/mainframe packages, and three fax-on-demand systems with the aid of over 150 third-party software programs.[27] Perhaps the wonder was not that standards, both consensual and market-based, fared so poorly but that they achieved so much.

The classic issue of whether a firm should promote a proprietary or open standard remained as uncertain as ever. Fear of competitors doomed some standards because they were closely identified with specific firms, so opposition and support often depended as much on the promoting firm as the actual standard. Although several cooperatively developed standards appeared, market forces proved more important, especially once Microsoft entered the fray.

Like a mirage temptingly close but always remaining on the horizon, acceptable standards for PC facsimile seemed only a year or two away for several years. The first attempt occurred in mid-1988 when computer chip manufacturer Intel and software firm Digital Communications Associates unveiled their Communications Application Specification (CAS). Instead of welcoming this de facto standard, competing faxboard makers and software developers viewed CAS as an attempt by Intel to dominate the market with its Connection CoProcessor board.[28]

This suspicion, coupled with Intel's failure to upgrade CAS and alleged poor treatment of some software firms, led to FaxBios, a private cooperative standard created by a group of hardware and software companies in 1991. Despite support from every major segment in the computer fax industry, the fears of small firms about large firms using the standard to gain market share (fears not eased when FaxBios promoter Everex became the first firm to release a FaxBios-compliant faxboard) and progress in publicly developed standards meant FaxBios disappeared by 1993.[29]

If market-based standards appeared elusive, so did consensus. This approach, so successful with G3, lagged in computer facsimile, where markets moved more quickly than the TR-29 Facsimile Committee of the Telecommunications Industry Association (TIA, formerly the Electronic Industries Association until 1988). Reflecting the American domination of computer facsimile, most of the discussion and development occurred in the United States.[30] Technical challenges of timing, covert and open opposition by firms offering competing products, and the rapid evolution of computer software and hardware meant that the one ITU and two TIA standards that emerged in the early and mid-1990s never gained full industry acceptance.[31]

Market intervention from above ultimately resolved the fight over PC fax standards. As Microsoft Windows' popularity grew, so did supporting facsimile software. By fall 1992, eight firms offered programs like Delrina's WinFax that faxed from within Windows and operated while another application was running. In June 1993, Microsoft entered computer facsimile by announcing Microsoft At Work, an effort with over sixty companies to create a common language for Windows-based communications among fax machines, photocopiers, and other office equipment. Meeting neither the hopes nor schedules of Microsoft and its supporting companies, At Work quietly died, its components redistributed within Microsoft and many reborn in Windows 95.[32]

Windows 95's Microsoft Fax instantly redefined the PC fax market, creating a single compatible base. Despite shortcomings (quickly noted by fax software firms that saw their market vanishing) like slowness, the new reality was apparent: basic faxing from a PC did not require special software.[33] For many users, that was more than adequate. Microsoft Fax's market dominance effectively ended the standards battles for modems by market domination.

Like faxboards, LANfax did not expand nearly as fast as proponents and market research firms predicted because of optimistic misreading of market need and substantial technical challenges. In 1991–92, Ken Camarro predicted 371,000 new installations in 1993 and over 800,000 in 1995, while Bis Strategic Decisions predicted 53,000 and 178,000, respectively. According to Frost & Sullivan, the actual numbers were 9,000 and 60,000.[34] Computer-based fax enabled faxing documents directly from a computer instead of printing and then faxing them, but a 1992 survey found only half of all faxed documents originated in computers. In 1996, almost 90 percent of Fortune 500 firms still sent most of their faxes from stand-alone machines.[35]

LANfax's major technological limitation was the inability to direct an incoming fax quickly, accurately, and confidentially to its recipient. A problem commonly perceived as important in a competitive market normally attracts multiple firms seeking solutions. For inbound routing, eight solutions developed sufficiently to reach the market. Some were as simple as having an administrator read every fax, while others demanded proprietary equipment, extra work by the sender, or more telephone lines.[36]

Although other options persisted, Direct Incoming Dialing (DID) prevailed, aided by its 1992 acceptance as a G3 option. The premise was simple: by enabling many telephone numbers to share the same telephone line for the LAN, all recipients had their own number. Favorable economics helped. Brooktrout licensed its DID patent at a low cost, while deregulation slashed the monthly cost of an extra telephone number from $30 to 30 cents in some cities by 1993.[37]

FAX-ON-DEMAND (FOD)

The development of new and unexpected applications tremendously expanded faxing's scope and attraction. While people and firms initially bought fax machines to send and receive documents and images, manufacturers, vendors, and other firms quickly developed new services and capabilities that greatly enhanced the value of faxing. One of the most powerful and unexpected markets in the 1990s was automated fax-on-demand (FOD), which merged faxing with computer databases and interactive voice technology to provide unprecedented rapid access to the latest information.[38]

FOD had an enormous impact. Just as the stand-alone fax machine accustomed people to instantly send and receive documents and images, FOD accustomed people to instant access to current information that before had not been easily available — if at all. Whether medical research, ski slope status, rodent identification, or Material Safety Data Sheets for hazardous materials shipments, FOD enabled people to directly and quickly obtain specific information, and organizations to automatically disseminate updated information like press releases to specific recipients.[39]

A FOD system sent stored documents or, in more powerful form, extracted information from a computer database and faxed a customized document. FOD allowed firms to accurately provide customers with information without answering the telephone, writing down the caller's address and request, and then mailing it. Savings in staff time alone usually repaid the system's cost within months. Further savings came from FOD's "demand publishing," which eliminated the cost of printing, storing, and mailing materials.[40]

Development followed the normal path of computer-based faxing. Custom software and hardware proved the concept. The ensuing demand spurred development of services and lower cost turnkey systems. Competition increased capabilities and reduced costs in a virtuous circle for customers, if not manufacturers and providers. Services like Xpedite handled the majority of FOD operations as firms outsourced their technical investment and risk.[41]

FOD provided an "astonishingly simple marketing tool with nearly unlimited possibilities" by quickly sending potential customers information while providing businesses rapid feedback.[42] The capability to combine faxing with computerized databases enabled public relations firms to create customized "micromessaged" press releases. In 1992, PR Data Systems faxed 9,600 Ford Motor releases, each beginning with the name of the local dealer, and reported that 35 percent of the receiving newspapers used the story compared with 5 percent for a standard release.[43]

Technical support services became enthusiastic FOD users. By automating responses to common enquiries, FOD helped popularize the concept of FAQ (frequently asked questions) while reducing costs. The 1989 $20,000 SpectraFax system that enabled the technical support desk for Hewlett Packard laser printers to answer 5,000 requests monthly repaid its cost in nine months. Not all applications were so corporate: one of Spectrafax's first uses was faxing football point spreads at $5 per printout.[44]

FAXING OVER THE INTERNET PROTOCOL (FOIP)

In 1993, Internet innovators Marshall Rose and Carl Malamud sent the first fax over the Internet with the newly developed Multipurpose Internet Mail Extensions (MIME) protocol. A network of enthusiasts evolved who received faxes over their computers and then completed the transmission with a local telephone call. Internet faxing initially seemed like ham radio, interesting technically but of minor economic significance.[45] This quickly changed when businesses realized that faxing over the Internet (more accurately, Faxing Over Internet Protocol, or FOIP) promised management major reductions in long-distance costs and greater control over outgoing faxes.[46]

Like computer-based fax, FOIP advanced the larger cause of office automation. In an ironic reversal of faxing's dependency on high-quality telephone lines, Internet telephony advocates viewed FOIP as a stepping stone to voice over Internet (VOIP) since delays in Internet transmission affected faxing less than conversations. Beyond VOIP beckoned the ultimate goal of universal messaging (or computer-telephony integration), seamlessly handling voice, fax, data, and multimedia messages in digital form.[47] Chief information officers were warned, "Digital faxing may be a controlled warm-up to the big jump coming: the day organizations go totally digital."[48] Actual users were not as impressed. Even in 2000, businesses considered scanning speed, memory, and paper capacity more important for faxing than accessing the Internet or minimizing transmission costs.[49]

Internet fax shared several comparisons with the G1–G2 era. With large corporations as the main customers, ensuring compatibility seemed less important than other criteria. Like the telephone system, network quality proved to be a major challenge, and efforts to create standards lagged behind the emergence of new products. Unlike those early years, a lack of effort was not the problem. The FOIP market truly began in 1996 when small firms, mainly startups, introduced hardware, software, and services. Demonstrating how international Internet development had become, several Israeli firms like ArbiNet played important roles, reflecting that country's entrepreneurs' desire to leverage their high-tech capabilities. By 1997, nearly ninety firms had entered FOIP.[50]

FOIP suffered problems of credibility, confusion, and compatibility. Unlike Xerox's instant legitimacy, most firms offering FOIP were small, struggling to stay solvent long enough to bring their products to fruition. Credibility came in 1997–98 when larger firms invested in these startups, like Compaq Computer in NetCentric, and introduced their own products and services, sometimes acquired from small firms and "white-labelled" as their own. The benefits proved mutual. Small suppliers like FaxNet did not have to market their products to millions of customers but only to a few technologically competent firms. Customers like Bell Atlantic did not have to expend resources and assume the risk to develop their own services but instead focused on reselling them to their customers.[51] Indeed, one writer proclaimed that FOIP had "come of age" when UUNet introduced UUFax in 1997 using technology from Open Port Technology. Established in 1987, UUNet was the first commercial Internet Service Provider and grew into the biggest Internet backbone provider, routing over half the world's email messages before owner WorldCom went bankrupt in mid-2002.[52]

The confusion grew because "Internet faxing" covered a wide range of approaches, including email-to-fax, fax-to-email, web-to-fax, and fax-to-fax via an intranet or internet.[53] Underlying these approaches were two types of faxing, message and session, and three types of networks: corporate intranets, virtual private networks, and the public internet. Session was real-time transmission with confirmation of the fax's receipt. Messaging was a send-and-forward approach, or, as a skeptical engineer labeled it, "send and wonder."[54] Corporate intranets ("intra-firm networks") were private digital networks, some of which had carried faxes years before the word "internet" entered public consciousness. Providers like PSINET and RightFax managed virtual private networks with proprietary software over their own networks ("private backbones") or rented capacity on other firms' networks.[55]

Last to develop but first to attract major public attention was the Internet, "a best-effort delivery system, with no one entity able to guarantee service levels."[56] Faxes sent over the Internet traveled over multiple networks, creating challenges of technical compatibility, coordination, and financial compensation. Unlike intranets, senders lacked control over the quality of the networks.

Providing FOIP compatibility proved less challenging than computer fax but still demanding. To create those standards, the ITU and TR-29 in 1996 worked with the Internet Engineering Task Force (IETF), the internet standards organization.[57] Possibly because they had missed the initial boom in computer-based faxing, Japanese fax machine manufacturers became closely involved.[58] A major challenge was ensuring compatibility not just for FOIP but also with VOIP and internet protocols, which meant negotiations with more groups and technologies.[59] Nonetheless, by 1997–98, the IETF and ITU had produced workable, albeit imperfect, sets of standards. Unfortunately, the market moved faster, so incompatible products competed with each other.[60]

Institutional challenges also frustrated FOIP. For computer-based fax users, moving to internet fax meant minor change. Stand-alone users proved more resistant. As Tom Jenkins, a TeleChoice consultant, stated, "Fax has been so successful that people are scared of messing with something that works. Fax is cheap and fax works." Furthermore, FOIP became a visible tool by corporate managers to control faxing and long-distance charges and was promoted and resisted accordingly. Nor did the promise of unified messaging always facilitate adoption. If a corporation's telecommunications group handled voice and faxing services, while another group handled data, coordinating integration could delay FOIP.[61]

Nonetheless, in 1997 market research firm Dataquest forecast a rapid rise in FOIP transmissions from 44 million pages in 1997 to 382 million in 1998 and 5,686 million in 2000, an optimism other analysts shared.[62] By 1998, FOIP had moved from "a novelty to an important service" for international businesses. NetCentric's Faxstorm handled over 2 million minutes of faxing monthly in late 1999, and Mail.com, 12 million minutes in March 2000.[63] Then the FOIP market imploded, victim of drastic drops in long-distance telephone charges, the dot.com collapse, and the new internet business model of free services.

Dataquest also had presciently warned that the biggest threat to FOIP was not email but the World Trade Organization's 1997 Basic Telecommunications Services Agreement, which pressured PTTs to reduce the price of their international calls.[64] Sharply reduced long-distance charges hurt the economic argument for switching from stand-alone faxing. Average international rates, which had hovered around a dollar a minute since 1985, dropped rapidly to 61 cents in 1998, 34 cents in 2001, and 14 cents in 2004.[65] The bursting of the dot.com bubble reduced funding and increased skepticism of internet-based businesses.

The most serious blow came from a new Internet business model, free services.[66] In February 1999, eFax sparked a wave of Schumpeterian creative destruction by offering free incoming faxes, email, and voicemail. Joe Kraus, co-founder of Excite.com, proclaimed, "The general rule is that anything that has a potential for a broad audience will eventually be free." Income came from advertisements on the received faxes and selling additional services such as the ability to send faxes. More skeptical observers wondered if these firms actually could earn a profit.[67]

Competitors had to follow. In July 1999, eFax had nearly a million users compared with 25,000 for Onebox. When Onebox switched to free membership, less than a year later it boasted 2.5 million subscribers—and revenues of less than $1 million. These firms, like other dot.com firms, consumed large amounts of money. The cost of internet faxing was trivial, but the necessary equipment, like servers, was not. Onebox.com consumed $70 million in venture capital in two years. The real profits came from selling the company. Phone.com purchased Onebox.com for $787 million in stock in February 2000 just before the dot.com bubble burst.[68]

FOIP survived in drastically altered form after 2001, as providers switched from emphasizing low costs to services such as secure and enhanced recordkeeping, services ideal for complying with new American laws for medical and corporate information.[69] In 2006, eleven commercial services offered faxing via the Web or email, but one firm dominated. Market leader j2 Global Communications grew from 27,000 subscribers and $3.5 million in revenue in 1998 to 4.0 million subscribers (200,000 paying) and $48 million in 2002, then to 13.1 million subscribers (1.9 million paying) and $255 million in 2010.[70]

TRIUMPH (FINALLY) OF DIGITALIZATION

In the late 1990s in the United States and early 2000s in Japan, fax succumbed to digitalization as the Internet, email, the Web, PDF, and other technologies shifted economics, perceptions, and the environment against faxing. Digitalization's eventual triumph stood on four foundations: effective compatibility standards, development and diffusion of infrastructure (including hardware and software), the increasing capabilities of equipment and systems, and the creation of the world wide web. The result was a massive and continuing expansion in the range, flexibility, creativity, mobility, and means of communication. Internet technologies appeared as the most visual demonstration of the rise of digitalization, but that term also encompassed cellphones, personal digital assistants, notebooks, Blackberries, iPads, and other computer-mediated devices.

Fax had proved to be its own worst enemy, laying the public groundwork for acceptance of email and the Web. Faxing accustomed people to easily send and receive messages in minutes, not days or weeks, and blurred the distinction between inside and outside the office. Computer-based faxing trained people to communicate from a computer and expect a quick reply. Fax-on-demand taught people to obtain information instantly and companies to provide it. Instead of hindering the electronic office of the future as Negroponte had charged, faxing led the way.

It was not supposed to be like that. For digital devotees, facsimile represented "yesterday's technology," due for replacement as soon as the digital office performed according to promise.[71] Faxing's unexpected rise and persistence insulted the technologically superior (at least theoretically) email. Yesterday's technology, however, usually worked better in today's world than tomorrow's technology. In 2001, Microsoft's Bill Gates confessed, "It was amazing how fax had caught on even though it's a very low-level format. You can't edit things, the resolution is very poor, but it was just so easy to set up that it got to critical mass where everybody was expected to have a fax number and be able to work in that way."[72]

As Forbes noted in 1992, "Electronic mail is largely a tool for computer fans, while the fax is for the masses. All this may change someday."[73] That change took decades, bogged down by challenges of compatibility, complexity, reliability, capability, usability, and economics far more daunting than those experienced by fax. Faxing proved the "techno-peasant's answer" to computer-based message systems, easily learned and used compared with the demanding procedures of email.[74]

Complexity and incompatibility among competing systems hindered electronic mail's long gestation from the 1970s. Into the mid-1980s, "electronic mail" encompassed any technology that electronically sent text or graphics, including fax, communicating word processors, and computer-based message systems such as the Arpanet. By the 1990s, only the last was considered electronic mail.[75] Four major approaches to providing a framework for computer-based communications evolved in the late 1970s to early 1980s: the top-down CCITT X.400 protocol, proprietary conversion services like Electronic Data Interchange, teletex and videotex, and an Internet-based protocol. Only the last fulfilled its expectations.

In 1984, the CCITT approved X.400 as a conversion system to enable competing email systems to communicate. Despite a theoretically attractive architecture, shortcomings like limited backward compatibility led to major changes in 1988, changes so great that the two unwieldy versions could not communicate. As *EMMS* acerbically noted in 1991, "The X.400 standard is ideal for people with a master's degree in computer science. . . . Everyone else will use fax machines."[76]

Another high-profile, highly invested, and highly promoted top-down standard was Electronic Data Interchange (EDI). General Motors and other transportation firms began designing EDI in 1968 to exchange inter-firm invoices and other documents electronically. The first EDI standard in 1975 attracted other industries, understandably enticed by eliminating manual data entry and mailing.[77] EDI suffered from proprietary protocols, high equipment cost, and the unexpected competition of facsimile, which provided EDI's network advantages without the cost and complexity. Indeed, computer fax proponents labelled LANfax as "the poor man's EDI," language that resonated positively with company communication managers.[78]

Far, far more than facsimile, teletex and videotex were major government-industry joint endeavors to promote national telecommunication industries while extending the benefits of computer-based communications and information services to their citizens.[79] Videotex and teletex exemplified how predictions and hopes could go wrong and how dedicating tens of millions of dollars and immense resources to developing a new technology did not guarantee success. Videotex was interactive two-way communications with a television set as the user's terminal, and teletex was the less technically demanding one-way version.

To promoters and observers, "The wedding of television to computerized services looks like it could be the marriage of the century," with expectations that "Teletex is likely to become the most important communications standard of the 1980s."[80] Despite national projects that spanned the globe, such as Bildschirmtext in West Germany and CAPTAIN in Japan, videotex and teletex universally failed. The low-resolution, slow-speed television set proved unsuitable for actively involving people who preferred to remain passive watchers.[81]

While embraced by some fields (especially the sciences), email systems in the 1970s–80s absorbed considerable resources but encountered major resistance from many users (who often had no voice in the design or selection of these systems) because of their incompatibility, difficulty in learning and using them, and limited formatting and graphics. Incompatibility became so bad that General Motors needed a special interface, Diamond, to link its eight incompatible email systems in 1987; one international oil firm had eighty-seven different electronic messaging systems in the early 1990s.[82]

Email also suffered from the same confusing over-competition that had hurt fax machines before they reached critical mass. By 1991, American firms could choose among fifty-five private email networks, fifteen electronic messaging services, and eighty email software suppliers. The "Rolodex problem," email's inability to create a universally accepted address, proved to be a particularly visible shortcoming. In January 1994, the *EMMS* newsletter listed six email addresses versus one fax number as points of contact.[83]

Nor did email systems easily (if at all) transmit documents or files. Sending files between computers became an unexpected application of computer-based faxing. While seeming "somewhat like using horse-drawn carriages to ship airplane parts," faxing data in a binary file transfer via a faxboard was "no different from E-mail attachments, BBS uploads/downloads, or direct modem-to-modem transmissions, except that it's simple enough that a novice who knows nothing about modems can use it."[84] Aided by a 1992 ITU standard and inclusion in the 1993 Microsoft at Work and Windows 95, binary file transfer's simplicity and faster transmission extended faxing's value.[85] Not until sending a file by email became as easy as sending a file by fax, a process of gradual improvements that took years, did email match faxing's capabilities.

Communications protocols forged in the 1980–90s slowly extended compatibility throughout incompatible networks. In 1982, the IETF approved the Simplified Mail Transport Protocol (SMTP) for emailing over the Internet. Compared with the complex X.400, the SMTP designers proved "more interested in getting something that would work."[86] If 1982 was the birth of internet email, it matured like a human, not reaching its potential until it became a teenager fortified with the essential standards of MIME, IPP, and PDF. The 1992 Multipurpose Internet Mail Extensions (MIME) enabled email to convey files of different types, content, and format from computer to computer, negating one of fax's advantages, binary file transfer, although email still had difficulty transmitting large files.[87] Similarly, the 1999 IETF Internet Printing Protocol increased the chance of a transmitted file printing correctly.[88]

The increasing ability to send files by email and programs that created more useful files (e.g., word processing software capable of tracking changes as well as documents that could not be changed), reduced the attraction of faxing. Building on its PostScript desktop publishing, in 1993 Adobe introduced its Portable Document Format (PDF), which "could make it practical to transmit electronically all sorts of things normally distributed on paper—from corporate sales reports to full-color magazines."[89] Adobe PDF was not the first file format for document interchange, but its ability to work with otherwise incompatible software (such as Windows and Apple operating systems) gave it credibility and usability. Making the reader free—a controversial decision that lost Adobe money for years—helped establish the PDF as "the gold standard for electronic documents," a standard that improved with upgrades.[90]

Changing office and home environments benefitted digitalization by creating an infrastructure of equipment, connections, software, and trained users. The spread of broadband and wireless transmission enabled faster downloading of data. High-resolution color monitors—initially on computers but increasingly on more portable devices—provided the detail to read digitized material, encouraging the extension of reading and writing from paper to screens.

Better tools, education, training, and familiarity made the growing amount of electronic information easier to manipulate and handle.[91] A generation of users trained in Microsoft Word and other software found new email systems based on the graphical user interface (GUI) principle of Apple and Windows easy to operate. Improved scanners and software enabled easier conversion of documents into electronic form, allowing their digital manipulation, storage, and transmission. Bulletin boards, listservs, and wikis made distributing messages easier than fax broadcasting and proved more effective in collaborating and exchanging ideas and information. As web developer Scott Matthews proclaimed in 1999, "The Internet is the new fax machine."[92]

Numbers and networks increasingly favored email over faxing. From the first IBM personal computer in 1981, PCs grew to 130 million worldwide by 1991 but rapidly expanded to 775 million in 2004. In the 1970s until the late 1980s, different standards battled to link PCs and peripherals. While less heralded than the struggles to forge compatible email systems, the triumph of Ethernet over its many competitors proved essential in enabling the establishment of Local Area Networks (LAN). As LANs increasingly linked PCs together in offices, and modems became standard equipment on PCs, the number of email accounts exploded in the late 1990s, vastly outnumbering fax machines while benefitting from positive network externalities.[93]

The world wide web made fax-on-demand obsolete. FOD had grown partly because of lackluster competition from online services like CompuServe with low-resolution graphics and high fees.[94] The Web changed that. Propelled by a uniform naming scheme made possible by the Hypertext Transport Protocol (http), which first appeared in 1991, and hypertext browsers like Mosaic, websites grew from 19,000 in 1995 to 1 million in 1997, 20 million in 2000, 100 million in 2006, and 634 million in 2012.[95]

Beyond greater access to information, the Web offered what FOD lacked, the hyperlink and the search engine, one of the most powerful tools ever for finding information.[96] Furthermore, the increased ease of surfing the Web meant FOD "is simply as not as much fun as checking out a Web site for the same information."[97] Downloading or sending maps as attachments also proved easier than faxing. Interactive, web-based maps (introduced by Yahoo and Google in Japan in 2005) linked with a GPS-enabled cellphone provided greater information and flexibility than faxed maps, devastating that important niche. When I visited Japan in 1995, a faxed map preceded every appointment. When I returned in 2008, an email with the website for the map preceded appointments.

Providers and customers increasingly found accessing a webpage or sending an email more attractive, easier, and less costly than faxing. Some firms switched completely from faxing information. In the early 1990s, the international banking firm Morgan Stanley had a team in New York City nightly cutting, pasting, and then faxing a 100-page compendium of current financial data to brokers and traders. By late 1995, automatic database extraction software constantly updated the information on a webpage. The *New York Times TimesFax* became TimesDigest in 2001 with delivery via email, faxing, and the Web. On a smaller scale, Tom Colbert's Industrial R&D, founded in 1992 on faxing, switched to the increasingly faster and cheaper email in 1997–98 as its customers and reporters adopted email.[98] Competition also came from specialized services and technology. The growing number of Bloomberg terminals providing economic data—10,000 rentals in 1990; 100,000 in 1998; and 170,000 in 2002—also reduced the financial world's need for faxing.[99]

Companies providing fax equipment and services had to react to these growing threats or watch their market disappear. Blending fax with digitized technologies, increasing innovation, increasing the value of fax-based products, and leaving the fax market dominated responses. For example, in a corporate move that indicated that it thought faxing's best days had passed, Pitney Bowes, which marketed high-value fax equipment for offices, spun off its fax and photocopier divisions into Imagistics International in 2001.[100]

Like other technologies facing growing competition from a related technology, firms introduced hybrids to link the established faxing with the growing world of digital media. Even as new services provided information from the Web by fax without needing a computer, FOD suppliers integrated email and web capabilities into their products.[101] Nor did FOD immediately disappear. Not until 2003 did Recruit replace its extensive FOD offerings in Japan with websites, while Castelle discontinued its InfoPress FOD products only in 2004.[102]

The fax community viewed email as a growing threat, but the mid-1990s perception, as fax consultant Maury Kauffman confidentially stated, was that "E-mail will kill fax the way cable TV killed the networks."[103] In 1998, Fortune 500 firms still considered facsimile "the most reliable, easy to use, fast, and most universal" way to get a response, with most firms faxing more than in 1997.[104]

Innovation and optimism continued as firms offered new products to dictate, send, and receive faxes from cellphones, BlackBerrys, iPhones, and other wireless devices.[105] In 2000, Gartner Dataquest reported that fax "continues to be a staple business product not only in corporate America, but in the home office."[106] FOIP leader j2 Global Communications, which had purchased eFax in 2000, began warning about the possible decline of the fax market only in 2002.[107]

Indeed, in 2004 Castelle, a leading fax server firm, declared, "We believe fax is here to stay, just as the computer, in its quest for the paperless society, has not replaced paper. . . . Fax remains one of the key business communication tools and is one of the essential components of the corporate messaging environment." Three years later, Captaris, which offered RightFax for network fax and electronic document applications, bought Castelle, but in turn found itself swallowed in 2008 by Open Text.[108] As computer-based faxing became integrated into enterprise information management, so too did fax-oriented firms merge with larger corporations offering clients more capabilities.

Yet as writer Michael Kinsley suggested, just as 1989–90 marked the transition from "What is your fax number?" to annoyance if the responder lacked a fax machine, so 1996–97 marked a similar transition for email.[109] By 1997 email had replaced "huge volumes of cover-sheet [short messages] fax traffic" in the United States, and the next year Frost & Sullivan warned, "The increasingly pervasive use of electronic mail and document conferencing for business communication poses a threat to the stand-alone fax machine industry."[110]

The decision to shift from fax was often made individually. When Hiro Hiranandani, one of the cofounders of Graphic Sciences, began a consulting business in 1995, faxing was the main medium of communication with his geographically dispersed clients. By 2000, his fax machine remained mostly silent, replaced by email and the Web. He was not alone.[111]

Tracking fax's quantitative decline was difficult. Replacing thermal with plain paper in the 1990s eliminated a direct measurement of faxing. Computer-based faxing and multifunctional machines further reduced accurate tracking not just of faxing but of numbers of fax-capable equipment: How many modems actually were used for faxing and how many virtual fax machines did a LAN contain? Faxed telephone traffic was even harder to quantify. Because faxing used the same circuits as voice calls, it could not be obviously distinguished.[112] The spread of

FOIP and private data networks made estimating total fax use essentially impossible although analysts tried, if only to distinguish trends.[113]

Corporate and administrative actions provided visible indicators of decline. *Fax Life* stopped publishing at the end of 1999, viewing increasing competition from websites and a slowing, saturated homefax market as signs that the fax age had ended.[114] In 2000 the American TR-29 Facsimile Committee, founded in 1963, became a subcommittee of the TR-30 Data Transmission Systems and Equipment Committee. The IETF fax working group, created in January 1997, ceased operating in February 2005.[115] Shifting titles and phrases similarly reflected the evolution from stand-alone to networked equipment. The switch from the "telecommunication equipment industry" to the "infocommunication network industry" meant the Communications Industry Association of Japan changed its name in 2002 to the Communications and Information Network Association of Japan.[116] An indication of popular perceptions of decline occurred when faxing started appearing in "dead technology" lists.[117]

Yet even as faxing declined, Japanese manufacturers continued to produce tens of millions of machines—more than 106 million between 2000 and 2009, more than the 95 million built from 1973 to 1999 as table 6.1 shows.[118] Sales peaked in 1997 at over seven million units in the United States, dropped below four million in 1999, but surpassed seven million again in 2002. Japanese manufacturing peaked in 2000.[119]

Three factors accounted for the seeming paradox of growing sales amidst decreasing faxing. First, many new machines replaced older equipment. Second, dropping prices helped keep fax relevant and diffuse machines deeper into organizations. In 1995 a basic machine cost $200, but by 2005 it was only $45.[120] Keeping a fax machine as a backup cost very little, and that insurance occasionally proved to be useful. In the United States after 9/11 and the anthrax scare, the resultant x-raying of regular mail and huge volumes of email meant faxing often offered the best way to reach federal and congressional offices. Worldwide, fax machines provided reliable communications if email or websites proved problematic.[121]

Finally, fax machines became increasingly integrated into a multifunctional "hydra" that also served as a printer, copier, and scanner. Starting in the mid-1990s, several generations of increasingly successful efforts appeared, benefitting from improved technology (especially greater reliability) and dropping prices.[122] These multifunctional devices passed the 50 percent level of all personal fax machines produced in Japan in 2006 and of all business fax machines in 2008.[123] In

TABLE 6.1
Japanese fax machine production

Years	Million
1973–79	0.25
1980–89	17.6
1990–99	77.2
2000–09	106.3
Total	201.4

Source: CIAJ, *Tsushin kiki chuki juyo yosoku* [Communications Equipment Demand Forecast] (Tokyo: CIAJ)

America, these inexpensive, low-end machines, whose fax function may never be used, comprised over 90 percent of the machines sold in 2000.[124]

Even as the digital ocean rose, islands, if not archipelagos, of faxing remained. Many users considered fax more dependable and ubiquitous, and, in an echo of telex's advantages, essential for countries where PCs and the Internet had poorly penetrated. Facsimile provided confirmed delivery and formatted documents; it transferred no computer viruses and provided legal signatures, acknowledged receipt, and exact replicas.[125] Stand-alone fax machines did not preserve outgoing faxes, providing privacy for the sender, an advantage for whistleblowers and other people faxing information they did not want to be traced.[126]

The medical world remained a strong fax market partly because of the need for a doctor's signature, long accepted as legitimate if faxed, and partly because of the non-computerized nature of most medical practitioners, who filled charts and folders via pen, not keyboard. Optometrists, ophthalmologists, and pharmacists all had fax machines in the late 1990s; they were far less likely to have email.[127]

Federal laws and regulations in the United States also encouraged faxing. Compliance with the 1996 Health Insurance Portability and Accountability Act (HIPAA) and 2002 Sarbanes-Oxley Public Company Accounting Reform and Investor Protection Act meant retaining, encrypting, and tracking documents—all actions computer faxing easily performed, which created opportunities for firms like Markit, which processed seven million faxes in 2012 as part of its data services to the financial community.[128]

Not until June 2010 did the Drug Enforcement Agency allow electronic signatures for Schedule II drugs, such as opiates and Ritalin—approximately 10 percent of all American prescriptions. Accepting pharmacies had to employ rigorous procedures to accept, authenticate, and store electronic submissions. For many doctors and pharmacies, faxing remained easier and less expensive: 63 percent of doctors in the 2012 National Physicians' Survey reported they faxed, and only 34 percent used email to communicate with each other. The increasing spread of electronic records, however, eroded this fortress of faxing.[129]

Email's uncertain legal status delayed its acceptance in both Japan and the United States. The law favored existing technologies because their strengths and weaknesses were known and embedded in the daily administrative operations. In the United States, the Electronic Signatures Act of 2000 gave electronic contracts the same weight as those executed on paper, though societal acceptance trailed by several years. Despite a 2001 law, not until April 2005 did Japan's "e-documentation laws" provide firm legal standing for electronic documents. In both countries, practical acceptance lagged legal authorization, reflecting institutional inertia, tradition, and very legitimate concerns about security and authenticity.[130]

JAPANESE DIGITALIZATION

Japan's shift to digitalization depended on the development of electronics capable of handling *kanji, hiragana,* and *katakana* and business models that packaged the new technologies into profitable products. The word processor led the way in the 1980s, attracting tens of millions of Japanese, "charmed by the novelty of the comparative ease and efficiency with which they could now type their own language."[131] The key enabling technology was semi-automatic predictive *kana-to-kanji* translation, which allowed the word processor to respond with a list of possible *kanji* as the user typed the word phonetically in *kana.* In September 1978, Toshiba introduced its 180-kilogram JW-10 word processor for 6.3 million yen ($29,000).[132]

This market, like fax machines, expanded rapidly into offices and homes as Japanese manufacturers fought for market share by introducing new equipment with more features and lower prices. Prices dropped to 500,000 yen ($2,100) in 1983 and under 100,000 yen by 1985 ($400). Sales soared to 2.7 million machines in 1989 and totaled over 30 million by 2000. Word processors reached 50 percent of households in 1997 before slowly falling until 2000 (45%) and then dropping sharply (27% in 2003) as PCs replaced them.[133] Word processors, however, remained stand-alone equipment; fax machines transmitted the documents.

To attract Japanese customers, PCs had to be more capable than their Western counterparts, not just in computing capabilities and memory for *kana* to *kanji* conversion but also in high-resolution printing and displays to display the intricate *kanji* characters. The printhead for the JW-10 word processor had 24 dots compared with 9 dots for an English language printhead.[134]

PC sales significantly surpassed word processor sales by value only in 1994.[135] This slow diffusion occurred partly because Japanese PC manufacturers kept prices high to avoid cannibalizing the market for their word processors. More important was the deliberate path dependency created by manufacturers (primarily NEC with its dominating PC-9800 series), which built incompatible "locked in" hardware and software and also owned or financed retail stores that sold only their products.[136] The resultant high costs and incompatible equipment meant there were twenty-eight PCs per one hundred people in the United States compared with nine per hundred in Japan in 1994. Not until 1996 did all 25,000 Toyota desk workers have PCs.[137]

American firms changed this world. In 1990, IBM introduced DOS/V, which shifted the Japanese PC market (save for NEC) to IBM-compatible designs, a decade after the United States. The November 1995 Japanese version of Windows 95 "triggered a PC boom and dramatically expanded—if not created—the consumer PC market in Japan." Word processor users now had a reason to switch from their single-function machine to a multi-purpose, multi-media system.[138]

Increasing diffusion of computers did not automatically increase electronic communications. Societal factors such as an emphasis on face-to-face communication and the unwillingness of managers to relinquish control also slowed Japanese adaptation of PCs and the Internet. Corporations and governments deliberately built incompatible and proprietary networks, increasing the challenge of wider communications beyond their group.[139] Most visibly to users, NTT's high local telephone rates proved a significant economic barrier to increasing internet diffusion. In 1997, Tokyo dial-up internet users paid nearly double their New York counterparts because NTT charged ten yen for every three minutes of a call.[140]

If a person sitting in a Starbucks with a laptop symbolized American mobile computing, the Japanese equivalent was the fashionably attired woman using her internet-capable cellphone (*keitai*). By 1999, Japan had the world's most competitive cellphone market, with features, functions, and uses unequaled elsewhere.[141] The 1999 launch of DoCoMo's i-mode, KDDI's Ezweb, and J-phone's J-Sky extended that competition to the Internet. For far less than the cost of a PC and expensive telephone service, the *keitai* made the Internet accessible. The demand for browsing services turned DoCoMo into the world's largest Internet Service Provider "almost overnight."[142]

Even with the diffusion of PCs and *keitai,* the Internet did not spread universally. Email demanded a greater investment in equipment, learning, and operating costs than faxing. Lacking the compulsion and attraction of business and other external factors, many older people and small firms declined to make those commitments. Age helped determine whether people continued faxing or voluntarily shifted to email. The older the person, the less likely the need or desire to email or access the Web and the greater the desire to remain comfortably entrenched with faxing.[143] Many family-owned shops, restaurants, and other small businesses found a fax machine easier, less expensive, and requiring less space than a PC.

Firms like Nexway seized this opportunity to offer small companies a Web presence without the expense of time, effort, and money to become computer literate. A fax machine enabled a family store to order directly from its wholesaler or a small *ryokan* (inn) to confirm reservations made by a customer on a website.[144] When I ordered two dozen pens from my elderly stationer in Shibuya, he found the pen number in a paper catalog, then filled in a form which he faxed to his distributor. The pens arrived the next day.

Homefax machines represented the comforting familiar, a fact "particularly appreciated by older people who are not accustomed to using e-mail."[145] While conducting my research, I heard many stories of parents or grandparents who faxed instead of using email. Such resistance was not just personal. In early 2009, 11 percent of the 2,200 members of the Foreign Correspondents' Club of Japan received club notices by fax instead of email. Nearly all were over 50.[146]

Comfort with faxing, social reluctance, and a lack of resources or desire to invest in email extended beyond individuals to businesses, especially firms concerned about email reliability. Japanese firms viewed fax as a company-to-company communicator and email as a person-to-person system. Fax numbers of departments did not change, while people moved within firms.[147]

Nonetheless, faxing declined in Japan too, albeit more slowly than in the United States. In 1999–2000, the CIAJ viewed the threat to faxing less from email than from the younger generation's rapid adoption of internet-capable *keitai*. Mobile phones surpassed landlines in 2000, which dropped from 60 to 38 million from 1999 to 2010, reducing the number of possible fax connections by one-third. By 2006, the CIAJ was reduced to reiterating fax's advantages of speed, certainty, signatures, physical copy, and use as a backup if the Internet failed.[148]

The two major fax services slowed differently, possibly reflecting different customer bases. Recruit's Facsimile Network Exchange (FNX) traffic peaked in 1998, and its subscribers slightly later, as large firms switched to email and the Web, especially for internal communications. Recruit responded with email and web services, renaming FNX the more encompassing Flexible Network Exchange in 2002. Two years later, Recruit turned its fax operations into a separate company, Nexway.[149] Subscribers to NTT's Facsimile Communications Network peaked in

2004 at 1.4 million and slowly stabilized at a million by 2006.[150] Last to decline was the freestanding pay-for-fax. Not until 2007 did the number of faxes sent from the thousands of 7-Eleven™ convenience stores decrease.[151]

Even so, innovation continued, as manufacturers introduced fax machines that could send faxes to *keitai* as email attachments (bridging the generation gap) and had batteries to work if a disaster stopped the electricity supply. Such new products emerged because demand continued. In 2011, when NHK's *Tameshite Gatten* [Try and Understand] began offering a service to fax recipes from the televised health show, 491,000 viewers subscribed. Even in 2013, Tamagoya received half of its 62,000 daily bento lunch orders by fax, compared with 45 percent by telephone and 5 percent by the Web.[152]

Stand-alone and computer faxing grew greatly into the late 1990s, driven by the demand of corporate and other users, the push of technology developers, and the expansion from transmitting documents to information. The integration of faxing with computers and communication networks climaxed in the development of faxing over the Internet, riding the dot.com boom. The collapse of that boom coincided with the decline of faxing as the long-heralded digitalization of information, embodied in email and the Web, finally met its promoters' promises. Reflecting different cultural and economic environments, faxing declined sharply in the United States and slowly in Japan. For digitalization, the critical point was when email became easier to use and less expensive than faxing, an accomplishment that took far longer than its advocates expected. Perhaps the best indicator of this transition occurred when spammers and swindlers switched from fax to email in trolling for victims.

Conclusion

It can be done and if I don't do it somebody will.

—*Thomas A. Edison, 1868*

Intelsat executive Joseph Pelton probably did not think of facsimile in 1981 when he wrote, "Communications development has been a process of trying to send an ever more densely packed collection of information over electronic channels with broader bandwidth and at higher power. Put that way, it seems dreadfully boring, but actually it's rather exciting stuff. Nevertheless, little of it can be considered really spanking new."[1] Yet fax's broad technical evolution fit his description quite well, even if commercial success eluded the technology for decades.

How unique was faxing's evolution? Did the perseverance of its promoters over several decades and multiple incarnations distinguish faxing from other technologies? The evolution of steam and electric vehicles in the late nineteenth and early twentieth centuries and the resurgence of electric vehicles in the early twenty-first century illustrate that technologies can persist for long periods with minimal market pull. The slow acceptance of modern faxing—faxing for offices over several generations of equipment—tracks the history of other technologies such as the e-book and iPad, whose numerous commercially unsuccessful predecessors involuntarily littered the landscape for years before finally reaching a critical mass once price and performance finally met expectations—expectations that kept changing over time.

The cultural imperative of image transmission as a desired future proved key to the continuing waves of development of fax technology. Generations of new advocates did not just keep this dream alive; they expanded its potential applications and audiences. Despite its discouraging record, facsimile continued to attract inventors, promoters, and, by the 1920s, corporations and investors as well as PTTs and governments. Still, another half-century passed before faxing began to fulfill its promoters' larger expectations.

Unfortunately for the promoters, potential users never viewed fax in isolation but always in comparison with other ways of communicating. Comparative considerations of economics as well as ease of use and capabilities played major roles in deciding whether to acquire a fax machine. However, economics did not always dominate. Public prestige as well as practicality played a role in faxing's early adoption. As early as 1907, when the *Daily Mirror* proclaimed that its Picture Telegraphy demonstrated the paper's devotion to providing the latest pictures as fast as possible, symbolic benefits played a role in this public adoption of faxing. Whether used by Pope Pius XI (who also possessed an American safety razor) in the early 1930s, or providing critical information to Steve McQueen in the 1968 movie *Bullitt*, fax machines showed audiences that the machine's possessor was modern. Similar public demonstrations, like the Japanese introduction of faxing into Manchukuo in the 1930s, the American State Department displaying a weather fax recorder at the 1960 Tunis International Fair exhibit in Tunisia, or an executive prominently installing a Xerox Telecopier in his office in the late 1960s, visibly symbolized technological competence and leadership.[2]

The unexpectedness of its actual success distinguished faxing after decades of gaps between promise and reality. Even as fax machine sales grew sharply in the 1980s, American manufacturers and market research firms viewed faxing as an unsuccessful technology past its window of opportunity, a window shut by the inevitable rise of the theoretically more efficient and elegant computer-based technologies of digitalization and Office Automation. Instead, faxing remained more popular by providing a far friendlier, less expensive and easier way of transmitting words, images, and data. Ironically, faxing's success aided the ultimate acceptance and diffusion of digitalization by gently introducing users to the potential of immediate transmission and acquisition of information and images.

After 2000, fax seemed an obsolete technology, increasingly replaced as a technological essential by all-digital communications. Fax did not diminish to the role of a residual media, abandoned by all but a few zealots or hobbyists, but its worlds of applications kept shrinking as an older generation of primary users died or converted and newer generations viewed faxing as a quaint, limited twentieth-century technology.

The shifts of technological leadership in facsimile from Europe to America in the early twentieth century and from the United States to Japan in the late twentieth century reflected larger shifts of innovation and manufacturing but also larger factors of geography, military need, and opportunity. The economically and physically large, unified American market provided a market that smaller European countries could not match. World War II brought faxing to the military's attention and patronage, although its primary competition of telegraphy increased its value even more. The shift from the United States to Japan paralleled larger shifts in consumer and business electronics (and American corporate shortsightedness), but faxing also benefitted greatly from Japan's unique system of writing. Faxing proved far more suited for *kanji* than telegraphy.

Indeed, Japanese firms deserved responsibility for the fax boom in the United States as well as Japan. Not only did Japanese firms supply the machines that spread by the millions in American workplaces in the late 1980s, but the intense Japanese battle for market share rapidly drove down prices, continually introduced new and more capable features, pioneered innovative ways of selling machines, and created differentiated markets while maintaining the crucial compatibility. American manufacturers lost, but consumers worldwide benefitted from the competitiveness of Japanese firms.

As the technology improved, fax equipment evolved from the completely visible to the complete black box. The operation of an electromechanical Bain or Caselli machine became comprehensible by looking at it. As machines incorporated photoelectric cells and other technologies spawned by the electronics and computer revolutions, the obviousness of their components weakened. A Belin or AT&T machine had cylinders and lead screws but also several boxes whose functions required schematics to understand. The post-1980 G3 machines were essentially black boxes with completely hidden processes. Paper entered and exited without any indication how the machine worked. Computer-based faxing became even more abstract as software provided the only indication of transmission. The computer-based integration of faxing with other forms of digital communications even raises the question of whether calling a transmission a fax has a specific meaning.

The shift from obvious to mysterious equipment paralleled a shifting of "the smarts" from the operator to the machine, which vastly expanded the audience of users, thus making faxing a more visible technology. Creating this easier use depended on integrating increasingly complex components into packages designed for less-skilled operators in environments less controlled than a central telegraph or newspaper office. The resultant simplicity of use and decreasing costs encouraged innovation and widespread application far beyond a tool for the office. This ease of experimentation and use, combined with its compatibility and near-universal spread, made facsimile a doubly democratic technology, not only accessible to all but also allowing almost anything—words, images, computer graphics—to be transmitted.

In contrast, telegraphy limited messages to specific formats of typed letters.

Telex's rigid format, special preparations and equipment, and inability to send images and documents limited its appeal. Faxing's ability to transmit imagery played a key role in the first unsuccessful and successful markets: replacing telegrams, newspaper photographs, and military weather maps. The image itself—the sender's exact message transmitted without operator error, the photograph for the newspaper—proved central to the rationale for faxing.

By the G3 era, the concept of faxing a visual image such as a photograph or map had expanded significantly to letters, drawings, menus, articles, edited documents, handwritten notes—anything reducible to two dimensions on paper. Instead of local circulation by photocopy, however, the reproduced jokes, messages, press releases, and political manifestos could be sent anywhere. While the sender retained the original (or a copy of the original), recipients could retransmit their fax, making popular messages quickly go viral. Fax-on-demand further increased faxing's interactivity and flexibility. These capabilities sharply distinguished faxing from telegraphy and opened new worlds of possibilities for users.

One of the steadfast attractions of faxing over the decades was its ability to transmit unaltered copies of messages and images. Faxing's supporters proclaimed its advantages of error-free transmission, accuracy, authenticity, and secrecy over telegraphy. While these were important, ease of use and superior economics proved more significant and explain why telegraphy lasted so long.

Accuracy appears to have been more important than authenticity and secrecy for early fax users because they often communicated on an internal network or between a central office and peripheral units. Preventing errors in telegraphy demanded constant vigilance. Telegraph companies and users developed reasonably effective procedures to reduce, if not eliminate, errors. The challenge was to do so in a timely and affordable manner. The long time needed to verify and correct the tape of a telex message before sending it was a major reason the Strategic Air Command introduced faxing to send its strike orders quickly, accurately, and reliably in the 1950s.

While faithfully transmitting the exact rendition of a message, faxing created its own types of errors in documents and visual images, errors that proponents underplayed. Faxing reproduced exactly transmission errors such as blurring and interference that could render a fax as unintelligible as a transposed telegram. Photographs for newspapers in particular needed to be error-free because the received image included any flaws in transmission. Documents proved more forgiving as did new users, who could mentally correct minor imperfections. Producing totally error-free transmissions ultimately combined higher quality circuits and error-correcting algorithms.

The authenticity and authority of a faxed message unexpectedly became un-important as an argument to switch from telegraphy in the nineteenth century. Telegraph codes and ciphers proved as reliable and legitimate as a faxed signature to verify a message while providing more secrecy. As long as telegraph stations vastly outnumbered fax machines, telegraphy provided more convenient service across the world. In contrast, the explosion of fax machines and inter-firm faxing in the 1980s–90s created a fertile environment for fraudulent faxing and fax scams. Indeed, faxed messages were very susceptible to local unauthorized reading and inadvertent transmission as well as state interception. The laxness of many people in following security measures—measures that decreased the ease of faxing—actually made G3 faxing initially less secure.

Like the assumed value of the authenticity of a signature proving to be less important than a telegram, the permanence of a message often lost out in the ongoing dance between the permanence and ephemerality of messages. One supposed advantage of a radio-broadcast faxed newspaper over television—permanence over ephemerality—proved illusionary. While a television or radio broadcast did indeed need to be seen or heard at the time of reception, economics and a more engaging visual and aural spectacle trumped permanence. Consumers of news were more interested in inexpensive news in the form of a newspaper and the latest news in the form of an extra newspaper edition or flash news from a radio or television than from an expensive hybrid.

Office communications, however, were another matter. Providing reproducibility, not permanence, became the challenge for faxing. The poor capability to reproduce a faxed document for normal handling within an office greatly limited document faxing in the 1930s–50s. Not until the Xerox 914 made photocopying easy and easily available did faxed documents become part of normal office operations.

The rapid spread of the very impermanent thermal fax paper in the 1980s reflected not a desire for ephemeral faxes but the low cost of thermal faxing compared with the alternatives. One major pressure pushing the development and diffusion of plain paper faxing was the desire for a permanent record of the fax without the hassle of making a photocopy before the thermal fax paper faded. Yet faxing for office use also strengthened the ongoing shift of written communications from permanence to ephemerality, a trend further accelerated by email and texting.[3]

The advance of facsimile was never uncontested or simple. Failure, conspicuously displayed in harsh accounting terms, was a close companion. Two types of failure accompanied fax from its conception through obsolescence: the failure of new technologies to succeed against existing technologies, and the failure of old technologies to compete successfully against new technologies.

Much of fax's history concerned the first type of failure, as inventors and firms struggled to create products and services sufficiently attractive to lure and keep customers. Major miscalculations and erroneous assumptions about the market for faxing and its competitors often doomed technically competent equipment. Many fax products and services failed financially, victims of the gap between optimistic visions of a fax-filled future and the reality of specific markets.

Reflecting the competitive nature of faxing and the development of new technologies, these failures often occurred in clusters—such as selenium/photocell-based fax transmitters in the 1890s–1900s and fax service companies in the 1960s–70s. Like other inventions, facsimile was developed independently by people with similar ideas at approximately the same time, thus demonstrating the importance of a shared environment of perspectives, expectations, and expertise. Sharing, though not necessarily coordinating, a sense of technical and commercial ripeness, promoters saw similar opportunities for faxing. Indeed, if no one else envisioned a similar profitable opportunity, then there was likely no future.

Commercial success meant closing the gap with rising expectations. Recording paper embodied the challenge of providing sufficient quality with reasonable speed and affordability. Customers determined the acceptable definitions of sufficient and adequate. In an ideal world, fax recordings would quickly emerge on plain paper with crisp resolution for easy copying. In reality, the high cost and technological challenges of plain paper reproduction prompted the development and employment of less expensive approaches. Photographic, carbon, and chemically treated electrolytic papers dominated the first century of faxing, while the odiferous electrosensitive paper defined the G1 and G2 machines of the late 1960s and 70s.

The low cost of thermal printing made the explosion of G3 faxing in the 1980s possible. Like cellphones, thermal paper provided a specific tradeoff—lower quality for convenience and low cost. The 1990s saw the substitution of plain for thermal paper as the transfer of photocopier and printer technology made higher quality affordable. Like cellphones, thermal printing established demand first, and its successors moved upmarket into higher quality reproduction later, a common practice for a disruptive technology.[4]

Faxing suffered from a dangerous type of failure common to telecommunications: incompatibility. Creating and enforcing standards to enable competing equipment to communicate proved to be a recurring political as well as technological challenge. If standards were best established when an innovative technology entered a phase of consolidation and incremental change, the problem was recognizing that narrow window of opportunity.[5] Too early and the standard constrained the technology. Too late and the trajectory of that technology might already have been set suboptimally. The history of G1 and computer-fax standards demonstrated how easy it was to miss that window. Never did "the market" quickly and effortlessly decide on a specific standard. Fax's history implies that partial market failure and uncertainty due to incompatibility were the norm, especially as the rate of evolution increased. G3's success in the 1980s–90s proved to be the exception.

Through the 1980s, the speed of regulation kept reasonable pace with G3 technological developments. In computer facsimile and internet faxing, innovation outran administrative coordination. The swirling action of technology and new product creation literally moved too fast for the collective standard-setting process to work. These standards proved far more technically demanding to develop than their non-computer-based predecessors. The very large number of actors and the huge number of modems, faxboards, software programs, and other equipment meant that even a small firm could hope for sales in the tens of thousands or more, enough to venture out on a standardless sea and possibly even set the standard.

Crucial were the number of actors and the speed of technological evolution. In the G1–G2 battles, only a few well-funded firms competed, creating an incentive to stick to their own technology and use incompatibility as a strategic tool to lock in users. In G3, a larger number of Japanese firms and telephone carriers cooperated to form a common strategy. In computer facsimile, the very low barriers to entry for software and hardware meant literally scores of firms emerged, making the process of agreement difficult. In internet faxing, the small size of firms meant none had the capability to dominate the market the way Microsoft finally did for PC faxing.

Why did the G3 standard work so well compared with the previous and subsequent efforts? Its creation and success were not preordained. Without the crucial intervention of the Ministry of Posts and Telecommunications in providing a neutral arena for negotiation and testing while also pushing for a single national standard, the Japanese fax industry might have splintered at a crucial time. G3 proved unique also because Japanese manufacturers, unlike their American and European counterparts, cooperated closely both informally and formally to ensure true compatibility even while competing against each other. Critically, the G3 standard created a common area yet left room for new and proprietary features, which encouraged continuing innovation.

Fax's history was not a straight line but a twisting, interrupted tale with many promising but unexpected dead ends such as the numerous discussions between firms about mergers, buyouts, joint ventures, and licensing or selling patents. Most of these proved to be only talk, proposals (if developed that far) that were never consummated, or deals that fell through. A standard occurrence in business history, these talks raised the tantalizing question "What if?" many times. The history of facsimile could have been so different in so many ways: What if World War II had not happened? What if Xerox had bought instead of rejected proffered patents and firms? What if Graphic Sciences had made a compatible machine? What if G3 had not been approved or approved without an option for proprietary features?

Nonetheless, it is hard to imagine a viable general-purpose fax industry before the 1970s, so much did faxing depend on the availability, capability, and affordability of postwar electronics. As long as fax machines required artisans to manufacture them and skilled operators to use them, those costs limited their market. The importance of changing from electromechanical to electronic components cannot be overemphasized. Not only did the size decrease and capabilities increase, but production costs and operator skills dropped while reliability increased. Consequently, these machines vastly expanded the universe of users who needed only basic office skills to fax.

These "faster, better, cheaper" machines turned faxing from an expensive, demanding technology into an affordable commodity. As a result, from a few machines in the nineteenth century, a few hundred machines by World War I, several hundred in World War II, and several thousand in the early 1950s, fax machines grew to the tens of thousands in the late 1960s, hundreds of thousands in the 1970s, millions in the 1980s, and tens of millions in the 1990s.

What sounds odd in an age of annual multimillion fax sales was the excitement and publicity in the 1960s–70s over placing a hundred machines a month. Some scaling may help place sales in perspective. Assuming 250 business days a year, selling 1,000 machines a year meant four per business day; 10,000 equaled 40 a day; 100,000 meant 400 a day; and one million sales a year meant 4,000 machines placed every business day.

Quantity became a quality of its own. As the number of machines grew, so did the number and types of users and communications. Instead of being mediated like telegraphy, senders directly faxed recipients. The inherently greater flexibility in faxing a message or image compared with the restricted structure of a telegram or telex contributed to the excitement about faxing. One reason for fax's triumph over telex—and digitalization's triumph over faxing—was this vast increase of users, becoming a technology used by a few to one used by the many, generating massive network externalities.

Yet faxing largely disappeared in the early twenty-first century due to another type of failure, succumbing to new, more able competitors in a long, uneven process. Judging by the digitalization's domination of predictions, resources, and publicity, faxing never should have achieved such widespread success in the 1980s–90s. Pragmatism outweighed predictions, especially if individuals and offices decided what technology to choose, not a central office.

Ultimately, Nicholas Negroponte was right. Digitalization devastated (but did not destroy) analog communications. Faxing, however, lost not to an abstract concept but to a constantly expanding and improving variety of ways to create, communicate, and collaborate. Just like faxing expanded the options of communicating while reducing barriers of cost and expertise as compared with telex, so too did the Internet, world wide web, email, PDFs, cellphones, PDAs, Blackberries, laptops, notebooks, social media, iPods, iPads, iPhones, and countless other innovations make faxing seem rather simplistic, old-fashioned, and increasingly limited.

Yet even in its decline, faxing continued to confound expectations. Between 2000 and 2009, Japanese manufacturers produced over 100 million machines in a flow of innovative new models. The continuing large-scale production of an allegedly obsolete technology by profit-seeking corporations in a capitalist economy should raise some questions about how obsolete fax really was. Indeed, in 2009 the library in my town installed a public fax machine in response to requests, and fax art shows still continue, though curators may have to scramble to find a fax machine.[6] Cisco only stopped selling its fax server in 2011 and supporting it in 2014.[7]

For my mother and tens of millions of people worldwide, the fax machine was a technology so easy to use that it required minimal skills, so sophisticated that it easily communicated worldwide, and so flexible that it enabled, not restricted, their creativity. That's my definition of a good technology.

Notes

Introduction

1. Jules Verne, *Paris in the Twentieth Century*, trans. Richard Howard (New York: Random House, 1996), 53.

2. Michael Brian Schiffer, "Cultural Imperatives and Product Development: The Case of the Shirt-Pocket Radio," *Technology & Culture* 34, 1 (Jan. 1993), 99.

3. Xerox, *1971 Annual Report*, 13.

4. Clayton M. Christensen, *The Innovator's Dilemma: When New Technologies Cause Great Firms to Fail* (Boston: Harvard Business School Press, 1997), xv.

CHAPTER 1: First Patent to First World War, 1843–1918

Epigraph. Alexander Bain, "Letter to the Editor," *Times of London*, Nov. 19, 1850, 3.

1. "Image produced by electric printing telegraph," Nov. 1850. UK0108 SC MSS 086/2 ITEM, Jacob Brett material. Institution of Engineering and Technology Archive.

2. Alexander Bain, "Automatic Telegraphy," *The Telegrapher*, Dec. 14, 1867, 129–30.

3. "Mr. Culley on 'Printing Telegraphs,'" *The Electrician*, April 15, 1864, 305.

4. "New Phototelegraph Equipment," *Electronic Engineering*, Feb. 1947, 46.

5. Richard H. Ranger, "Transmission and Reception of Photoradiograms," *Proceedings of the Institute of Radio Engineers* 14, 2 (April 1926), 161.

6. "Smee and Bain," *Telegraphic Journal and Electrical Review*, Feb. 1, 1877, 26.

7. Iwan Rhys Morus, "The Electric Ariel: Telegraphy and Commercial Culture in Early Victorian England," *Victorian Studies* (Spring 1996), 343–49.

8. "Wright and Bain's Electro-Magnetic Railway Controller," *Mechanics' Magazine*, Feb. 5, 1842, 97–99; "Mr. Bain's Electro Magnetic Inventions," *Mechanics' Magazine*, July 22, 1843, 66; Robert P. Gunn, *Alexander Bain of Watten: Genius of the North* (Wick, UK: Caithness Field Club, 1976), 1–11.

9. "Mr. Bain's Electric Printing Telegraph," *Mechanics' Magazine*, April 13, 1844, 268–70; "Bain's Patent Electro-Chemical Copying Telegraph," *Mechanics' Magazine*, Feb. 9, 1850, 105.

10. "Bain's Electric Clock," *Mechanics' Magazine*, July 18, 1846, 69; "Electric Telegraph and Clock," *Mechanics' Magazine*, Oct. 3, 1846, 335. For a more critical perspective, see Alexander Steuart, "Alexander Bain. His Inventions and their Influence on Modern Time Distribution," Keith Lecture, Royal Scottish Academy of Arts, Jan. 11, 1941, 13–14.

11. *The Petition of Alexander Bain Against, and the Evidence Before the Committee on the Electric Telegraph Company Bill* (London: Chapman and Hall, 1846); Gunn, *Alexander Bain of Watten*, 6; W. D. Hackmann, ed., *Alexander Bain's Short History of the Electric Clock (1852)* (London: Turner & Devereux, 1973), ix; Jeffrey L. Kieve, *The Electric Telegraph: A Social and Economic History* (Newton Abbot, UK: David & Charles, 1973), 44, 83.

12. "Mr. Bain's New System of Electro-Telegraphic Communication," *Mechanics' Magazine*, July 10, 1847, 26–31; June 19, 1847, 590–92; and July 24, 1847, 74–79.

13. Edwin J. Houston, "Delany's System of Fac-Simile Telegraphy," *Transactions of the American Institute of Electrical Engineers* II (1885), 1.

14. L. Turnbull, "The Telegraphic Lines of the World," *Mechanics' Magazine*, Sept. 4, 1852, 195; F. C. Bakewell, *Electric Science; Its History, Phenomena, and Applications* (London: Ingram, Cooke, and Co., 1853), 169; "Bain, the Inventor of the Chemical Telegraph," *Scientific American*, April 30, 1853, 258; David Hochfelder, "Taming the Lightning: American Telegraphy as a Revolutionary Technology, 1832–1860" (Ph.D. diss., Case Western Reserve University, 1999), 49–61.

15. Richard R. John, *Network Nation: Inventing American Telecommunications* (Cambridge, MA: Belknap Press, 2010), 84.

16. George B. Prescott, *History, Theory, and Practice of the Electric Telegraph* (Boston: Ticknor and Fields, 1860), 127–28; Alvin F. Harlow, *Old Wires and New Waves: The History of the Telegraph, Telephone, and Wireless* (New York: Arno Press, 1971), 143–44, 162–66.

17. Taliaferro Preston Shaffner, *The Telegraph Manual: A Complete History and Description of the Semaphoric, Electric and Magnet Telegraphs* (New York: Pudney and Russell, 1859), 269.

18. H. T. W., "Alexander Bain," *Dictionary of National Biography* (Oxford: Oxford University Press, 1917), 1:904–5; R. W. Burns, "Alexander Bain, a most ingenious and meritorious inventor," *IEE Engineering Science and Education Journal* 2, 2 (April 1993), 92.

19. "The Royal Society and Mr. Bain's Electro-Magnetic Discoveries," *Mechanics' Magazine*, Dec. 13, 1845, 411–13.

20. Perhaps the most famous example was Alexander Graham Bell's attorneys filing a patent for the telephone only a few hours before Elisha Gray; see Jürgen Schmidhuber, "The Last Inventor of the Telephone," *Science*, March 28, 2008, 1759.

21. Other disputes, not all initiated by Bain, included using the earth to transmit a current and printing colors by electricity ("Voltaic Electricity," *Mechanics' Magazine*, June 11, 1842, 469–70); "Mr. Bain's Electro Magnetic Inventions," *Mechanics' Magazine*, July 22, 1843, 66).

22. Alexander Bain, "Specification of Alexander Bain. Electric Time Pieces and Telegraphs. A.D. 1843" (London: George E. Eyre and William Spottiswoode, 1856), 10, 12. Bakewell apparently used the phrase "copying telegraph" first.

23. "Bain's Patent Electro-Chemical Copying Telegraph," *Mechanics' Magazine*, Feb. 9, 1850, 102. In 1996–97, Professor Masayuki Miyazawa led a group at the Nagano Prefectural Institute of Technology that built a working replica of Bain's 1843 machine for the 25th anniversary of the Institute of Image Electronic Engineers of Japan (Miyazawa Masayuki, "Bain's FAX no fukgun monogatari—The Restoration of Bain's

FAX," *Gazo denshi gakkaishi* [Journal of the Institution of Image Electronic Engineers of Japan] 26, 5 (May 1997), 546–51).

24. "Mr. Bain's New System of Electro-Telegraphic Communication," *Mechanics' Magazine*, July 10, 1847, 26–31, and July 24, 1847, 74–79; "The Electric Telegraph," *Mechanics' Magazine* Jan. 29, 1848, 116; "New Telegraphic Ideas," *Scientific American*, Oct. 14, 1848. 29.

25. "Specifications of English Patents Enrolled During the Week Ending 7th of June," *Mechanics' Magazine*, June 9, 1849, 544–45.

26. Bakewell, *Electric Science*, 170–72; Shaffner, *Telegraph Manual*, 304–9.

27. "A New Telegraph," *Scientific American* May 27, 1848, 284.

28. "Bain's Patent Electro-Chemical Copying Telegraph," *Mechanics' Magazine*, Feb. 9, 1850, 102–5 [emphasis in original].

29. "The Electric Telegraph—Mr. Bain—Mr. Bakewell," *Mechanics' Magazine*, Feb. 23, 1850, 143–44.

30. E.g., "Correspondence," *Mechanics' Magazine*, March 9, 1850, 187–89; "Electric Telegraphs—Reply of Mr. Bakewell to Mr. Bain," *Mechanics' Magazine*, March 23, 1850, 223–25; Alexander Bain, "To the Editor of the Times," *Times of London*, Nov. 19, 1850, 3; and F. C. Bakewell, "The Copying Electric Telegraph," *Times of London*, Nov. 29, 1850, 8.

31. Bakewell, *Electric Science*, title page and 170; Frederick Collier Bakewell, *Great facts; a popular history and description of the most remarkable inventions during the present century* (New York: Appleton, 1860), 149–57, 172.

32. "Bakewell's Copying Telegraph," *The Electrician*, Dec. 12, 1862, 68; "Copying Electric Telegraph," *Times of London*, Nov. 14, 1850, 6; Robert Hunt, "The Science of the Exhibition," in *The Crystal Palace Exhibition Illustrated Catalogue* (London: *The Art Journal* 1851), IX (Reprint: Dover Publication, New York, 1970); "The Telegraphic Soiree at Manchester," *The Electrician*, Nov. 16, 1861, 22.

33. Gail Mitchell, Information Specialist, British Library, personal communication, Feb. 6, 2012.

34. "The Electro-Chemical Telegraph," *Mechanics' Magazine*, May 4, 1850, 359; "Bain's Electric Telegraph," *Mechanics' Magazine*, Feb. 8, 1851, 109–10.

35. Tom Standage, *The Victorian Internet* (New York: Walker and Company, 1998), 18.

36. Bakewell, *Great facts*, 157; L. V. Lewis, "The Post Office Picture Telegraphy Service," *St. Martin's-le-Grand*, April 1930, 89.

37. W. E. Sawyer, "Fac-simile Telegraphy," *The Telegrapher*, April 6, 1876, 85.

38. Thomas Edison to John Clark Van Duzer, Dec. 6, 1868, in Reese V. Jenkins et al., eds., *Papers of Thomas A. Edison: The Making of an Inventor, February 1847–June 1873* (Baltimore: Johns Hopkins University Press, 1989), 1:90–95.

39. Dr. Lardner, *The Electric Telegraph* (revised & rewritten by Edward B. Bright) (London: James Walton, 1867), 163.

40. Jules Verne, *Paris in the Twentieth Century*, trans. Richard Howard (New York: Random House, 1996), 53.

41. "The Pantelegraph," *The Electrician*, Nov. 9, 1861, 10.

42. "Bonelli's Telegraphic Apparatus at the International Exhibition," *The Electrician*, June 6, 1862, 53–54; "The May-June Number of Annales Telegraphiques," *The Elec-*

trician, July 3, 1863, 98; "Bonelli's Electric-Printing Telegraph," *The Electrician,* Sept. 18, 1863, 232; "The International Exhibition, 1862," *The Electrician,* Oct. 2, 1863, 260; "Mr. Culley on 'Printing Telegraphs,'" 304–5; George B. Prescott, *Electricity and the Electric Telegraph,* 7th ed. (New York: D. Appleton & Co., 1888), 2: 763–67.

43. "Necrologie," *Journal telegraphique* 15 (1891), 242–43; "Giovanni Caselli," *Dizionario Biografico degli Italiani* (Rome: Istituto della Enciclapedia Italiann, 1978) 21, 331–35.

44. E.g., "Bakewell's Copying Telegraph," *The Electrician,* Dec. 12, 1862, 67–68.

45. Emilio Pucci, *L'inventione del FAX* (Milan: Edizioni SEAT, 1994), 32.

46. "The Automatic Telegraph of the Abbe Caselli," *The Electrician,* Jan. 24, 1862, 134–35.

47. Harlow, *Old Wires and New Waves,* 203.

48. "The Pantelegraph," *The Electrician,* Nov. 9, 1861, 10, "Foreign intelligence," *Times,* Feb. 22, 1862, 10; "Necrologie," *Annales Telegraphique,* 1865, 8, 379; Pucci, *L'inventione del FAX,* 32–33.

49. Prescott, *History, Theory, and Practice of the Electric Telegraph,* 137.

50. Kieve, *Electric Telegraph,* 82; George B. Prescott, *Electricity and the Electric Telegraph,* 3rd ed. (New York: D. Appleton & Co., 1879), 745; Pucci, *L'inventione del FAX,* 36.

51. Giovanni Caselli, "Improvement of Telegraphic Apparatus," U.S. patent 37,563, Feb. 3, 1863.

52. From 1856 to 1866, the number of transmitted messages grew from 360,299 to 2,842,554 and bureaux from 167 to 1,209. A decade later those numbers reached 8,080,984 messages and 2,890 bureaux (Andrew J. Butrica, "From *inspecteur* to *ingenieur:* Telegraphy and the Genesis of Electrical Engineering in France, 1845–1881" (Ph.D. diss., Iowa State University, 1986), 55, 78; Major Webber, R.E., "Multiple and Other Telegraphs at the Paris Exhibition," *Journal of the Society of Telegraph Operators* VII, Nov. 27, 1878, 437.

53. "Loi relative a la taxe, 1 des dispatchs privees, dessins, etc., transmis par le telegraphe au moyen de l'appareil autotelegraphique," 27 Mai–3 Juin 1863, *Collection Complete des Lois, Decrets, Ordonnances, Reglements et Avis du Conseil d'Etat* (Paris: Directeur de l'Administration, 1863), 548; "The May-June number of Telegraphiques," *The Electrician,* July 3, 1863, 98; "*Bulletin et chronique,*" *Annales Telegraphique,* 1865, 8, 365–68.

54. This section is based on M. Ganot, *Pantelegraph de M. l'Abbe Caselli (extrait de la 12th ed. de la Physique de M. Ganot)* (Paris: Typographie Wiesener etc., 1865), and Prescott, *Electricity and the Electric Telegraph,* 3rd ed., 745–56.

55. "Mr. Culley on 'Printing Telegraphs,'" 305; Robert Soulard, "Caselli et Lambrigot," *Revue d'histoire des sciences et leurs applicaitons* 22 (1969), 76–78.

56. Prescott, *History, Theory, and Practice of the Electric Telegraph,* 137; "Arrangements of M. Caselli for Obviating the Effects of the Discharge Current in Signaling," *The Electrician,* July 17, 1863, 121–22.

57. F. L. Pope, "Discussion," *Transactions of the American Institute of Electrical Engineers* II (1885), 24.

58. Edouard Ernest Blavier, *Nouveau traite de Telegraphie Electrique* (Paris: Librairie Scientifique, Industrielle et Agricole, 1867), 2:299–302.

59. Thomas A. Edison to John Clark Van Duzer, Dec. 6, 1868, *Papers of Thomas A.*

Edison, 90; Sawyer, "Fac-simile Telegraphy," 85; "Mr. Culley on 'Printing Telegraphs,'" 305.

60. Augustin Privat Deschanel, *Elementary Treatise on Natural Philosophy* (New York: D. Appleton and Co., 1887), 822–24; "Bulletin et chronique," *Annales Telegraphique*, 1865, 366.

61. "Lois, Decrets et Arretes, concernant l'Administration des Lignes Telegraphiques," *Annales Telegraphique*, 1865, 8, 327; Ernest Saint-Edme, "Physique industrielle: Telegraphie electrique: le telegraphe autographique de M. Meyer," *Annales Industrielles* 1 (1869), col. 504.

62. Louis Figuier, *Les Merveilles de la science* (Paris: Furne, Jouvet, 1867), 2:159; Pope, "Discussion," 24; Standage, *Victorian Internet*, 105–26.

63. Butrica, "From *inspecteur* to *ingenieur*," 78.

64. This section is taken from Tomitsugu Kawanobe, "The telegraph in the end of the Bakufu Era," *Teregaraafu Komonjo Koo (Telegraph Old Documents)* (Chiba: private publication, 1987) 413–49.

65. Thomas Edison to Patrick Kenny, May 28, 1884. TAEM D8416BRS; TAEM 72:774, Edison Papers digital edition, http://edison.rutgers.edu/digital.html; www.iieej .org/vfax/data/tech/E-4.htm.

66. Prescott, *Electricity and the Electric Telegraph*, 7th ed., 2:759–62; Charles L. Buckingham, "The Telegraph of To-Day," in *Electricity in Daily Life* (New York: Charles Scribner's Sons, 1890), 159.

67. "Telegraph Prizes at the Paris Exhibition," *The Telegrapher*, Aug. 31, 1867, 8; "Necrologie," *Annales Telegraphique*, 1884, 11, 182.

68. "Bulletin telegraphique," *Journal des telegraphes* 4, 11 (Oct. 15, 1869), 1–7; "Description of Cook's Modification of the Bonelli-Hipp Letter Printing Telegraph," booklet [1869?]. File 1, Henry Cook minute 4203/69, PO 30-195a, British Telephone archive.

69. "Necrologie," *Annales Telegraphique*, 1884, 11, 289–90; Prescott, *Electricity and the Electric Telegraph*, 7th ed., 2:756–58; Butrica, "From *inspecteur* to *ingenieur*," 90–92.

70. "R. S. Culley Report on Paris Inspection of 5 Telegraphy Systems," Sept. 28, 1869. Minute 4203.1869, PO 30-195a, British Telephone Archive.

71. Saint-Edme, "*Physique industrielle*; Butrica, "From *inspecteur* to *ingenieur*," 78–79, 88.

72. Butrica, "From *inspecteur* to *ingenieur*," 84, 77–78, 113–16; J. R., "Appareil autographique Meyer," *Annales Telegraphique* 1874, 3rd ser. VI, 45; Webber, "Multiple and other telegraphs at the Paris Exhibition," 440–41.

73. Alexander Graham Bell to L. T. Stanley, Nov. 12, 1875 (269 10104. Box 288, "The Telegraph," Alexander Graham Bell Papers, Library of Congress); Edwin S. Grosvenor and Morgan Wesson, *Alexander Graham Bell: The Life and Times of the Man Who Invented the Telephone* (New York: Harry N. Abrams, Inc., 1997), 53–55.

74. Thomas Edison to John Clark Van Duzer, Sept. 5, 1868, and Dec. 6, 1868, *Papers of Thomas A. Edison*, 1:83, 95, and 84 n.2.

75. *Papers of Thomas A. Edison*, 1:146–47; Elisha Andrews to Thomas Edison, June 20, 1871, *Papers of Thomas A. Edison*, 1:299; "Detailed biography," http://edison.rutgers .edu/bio-long.htm.

76. Daniel Craig to Thomas Edison, Feb. 18, 1871, *Papers of Thomas A. Edison*, 1:249–50; Elisha Andrews to Thomas Edison, June 20, 1871, *Papers of Thomas A. Edison*, 1:299;

Francis Jehl, "Edison and the Telegraph," *Menlo Park Reminiscences* (Dearborn, MI: Edison Institute, 1937), in George Shiers, ed., *The Electric Telegraph: An Historical Anthology* (New York: Arno Press, 1977), 39. Edison received patent 479,184 on July 19, 1892.

77. Charles Batchelor to Thomas Edison, Aug. 15, 1883. TAEM D8373Q; TAEM 70:1139, Edison Papers digital edition, http://edison.rutgers.edu/digital.html.

78. Patrick Kenny, William J. Mann, Herman Broesel, and Thomas Edison, contract, June 15, 1891 (HM91AAC; TAEM 144:638); Jerome Carty to Thomas Edison, April 17, 1893 (HM93AAE; TAEM 144:728); Patrick Kenny and Thomas Edison to Paul Latzke, April 24, 1896 (TAEM: D9627AAD; TAEM 136:477). See also William Abbott Hardy to Thomas Edison, Oct. 14, 1927 (Marginalia by Edison). Ann A. Hardy Collection, Rindge, NH, TAEM: [X079F1], Edison Papers digital edition, http://edison.rutgers.edu/digital.html.

79. Houston, "Delany's System of Fac-Simile Telegraphy," 1–18.

80. Pope, "Discussion," 24.

81. "The Amstutz Electro-Artograph," *Scientific American*, April 6, 1895, 215. Words also received attention: Jan Szczepanik predicted journalists would transmit articles by his Telectroscope, saving time compared with the telegraph ("The 'Telectroscope' and the Problem of Electrical Vision," *Scientific American*, Supplement July 30, 1898, 18899).

82. Air mail service between Los Angeles and Canton saved twelve days compared with a ship ("Speedier Oriental Pictures," *Editor & Publisher* Aug. 24, 1935, 9).

83. Smith and May were developing a test circuit for undersea cables. In an excellent example of scientific enterprise (including omitting May's role), Smith wrote to the Society of Telegraph Engineers, which published the letter in *Nature* (Willoughby Smith, "Effect of Light on Selenium during the Passage of an Electric Current," *Nature*, Feb. 20, 1873, 303).

84. For these early proposals, experiments, and demonstrations, see W. T. O'Dea, *Handbook of the Collections Illustrating Electrical Engineering*, vol. 2, *Radio Communication* (London: Science Museum, 1934), 77; and Albert Abramson, *The History of Television, 1880 to 1941* (Jefferson, NC: McFarland & Co., 1987), 7–15.

85. Mons. Senlecq, "The Telectroscope," *The Electrician*, Feb. 5, 1881, 141–42; Abramson, *The History of Television*, 8–9, 13. For a debunking of the rumor, see Russell W. Burns, *Television: An International History of the Formative Years* (London: IEE Press, 1998), 44, 54–56.

86. Shelford Bidwell, "The Photophone," *Nature*, Nov. 18, 1880, 58–59.

87. Shelford Bidwell, "Tele-Photography," *Nature*, Feb. 10, 1881, 344–46; "Tele-Photography," *Telegraphic Journal and Electrical Review*, March 1, 1881, 82–84; "Notes," *Nature*, April 14, 1881, 563; Shelford Bidwell, "Telegraphic Photography," *The Electrician*, Oct. 1, 1881, 310.

88. Shelford Bidwell, "Some Experiments with Selenium Cells," *Philosophical Magazine*, March 1891, 250–56, and "The Electrical Properties of Selenium," *Philosophical Magazine*, Sept. 1895, 233–56; George P. Barnard, "The Selenium Cell: Its Properties and Applications," *Journal of the Institution of Electrical Engineers* 69 (1929), 97–120; Arthur Korn, "Phototelegraphy," Stevens Faculty Lecture at the Galois Institute of Mathematics at Long Island University, Brooklyn, NY, 1935, 5, Stevens Institute of Technology archive.

89. Thomas D. Lockwood to John E. Hutchinson, July 20, 1894. Box 1075, AT&T Archives.

90. A.[lfred] Dawson, "Swelled Gelatine," *Penrose's Annual Process Work Year Book* (London: A. W. Penrose & Co., 1895), 24–26; Korn, "Phototelegraphy," 17; Casper M. Bower, "Facsimile. Communications and Industries Stepchild," May 22, 1950, 4. Times Facsimile Corp. (General) 1941–1959 (3). New York Times archive.

91. "Amstutz Electro-Arcograph," *Scientific American*, 215; N. S. Amstutz, "Acrograph," *Electrical World and Engineer*, Feb. 17, 1900, 247–50; Arthur Korn and Bruno Glatzel, *Handbuch der Phototelegraphie und Telautographie* (Leipzig: Otto Nemnich Verlag, 1911), 312, 318–21. Amstutz later claimed he transmitted his first photograph in 1888 ("Traces the History of Pictures by Wire," *New York Times*, July 19, 1925, 9).

92. E.g., Henry Sutton, "Tele-photography," *Electrical Review*, Nov. 7, 1890, 549–51.

93. Alfred T. Story, *The Story of Photography* (New York: McClure, Phillips & Co., 1904), 145–46; Thomas Thorne Baker, *The Telegraphic Transmission of Photographs* (London: Constable & Co., 1910), 9–10; D. W. Isakson, "Developments in Telephotography," *Journal of the American Institute of Electrical Engineers*, 41,11 (Nov. 1922), 813.

94. "Kleinschmidt Facsimile Telegraph," *Western Electrician*, March 14, 1903, 204; Alfred Gradenwitz, "Recent Developments in Picture Telegraphy," *Scientific American*, Oct. 26, 1907, 288–89; Thorne Baker, *Telegraphic Transmission*, 12–14; Korn and Glatzel, *Handbuch*, 368–71.

95. "The Improved Electrograph—A Fac-Simile Telegraph," *Scientific American*, Nov. 15, 1902, 329–30; "The Electrograph—A New Facsimile Telegraph," *Scientific American*, June 15, 1901, 373–74; Korn and Glatzel, *Handbuch*, 116–17; R. C. Ballentine, "Developments in Radio Facsimile" [mid-1929?], 2, Charles Young files.

96. The papers were the *New York Herald, Chicago Times-Herald, St. Louis Republic, Philadelphia Inquirer,* and *Boston Herald.* "Picture Telegraphy," *The Electrician*, May 26, 1899, 162; "The Transmission of Pictures by Telegraphy," *Electrical Review*, June 9, 1899, 926–27; Charles Emerson Cook, "Pictures by Telegraph," *Pearson's Magazine*, April 1900, 405.

97. Korn and Glatzel, *Handbuch*, 359; Terry and Elizabeth P. Korn, *Trailblazer to Television: The Story of Arthur Korn* (New York: Charles Scribner's Sons, 1950), 85–86, 95.

98. "Korn's New Telephotographic System—I," *Scientific American*, Supplement July 4, 1908, 15; Korn, "Phototelegraphy," 6; Arthur Korn, "Some mathematical problems in early telegraphic transmission of pictures," *Scriptica Mathematica*, June 1941, 93–97.

99. "A New Method of Telegraphing Pictures," *Electrical Review*, March 9, 1907, 420; Gradenwitz, "Recent Developments"; Shelford Bidwell, "Practical Telephotography," *Nature*, Aug. 29, 1907, 344; Thorne Baker, *Telegraphic Transmission*, 25, 48–49; Korn, *Trailblazer to Television*, 102–3, 110–16.

100. "The 'Daily Mirror's' Fourth Birthday Celebrated by the Installation of the Korn Machine for Telegraphing Photographs," *Daily Mirror*, Nov. 2, 1907, 1; "'Daily Mirror' Pictures by Wire," *Daily Mirror*, Nov. 2, 1907, 1, 3; "Professor Korn's Phototelegraphy," *Daily Mirror*, Nov. 8, 1907, 8–9.

101. Korn, "Phototelegraphy," 7.

102. "Korn's New Telephotographic System," 14–15; Thorne Baker, *Telegraphic Transmission*, 52–54, 66; "Picture Telegraphy," *Telegraph and Telephone Journal*, 1914–15, 71.

103. "A New Method of Telautography," *Electrical Review*, April 27, 1907, 689; "Method of Transmitting Photographs by Wire," *Modern Electrics* 5 (1912), 235–37; "Telegraphing Pictures," *Scientific American*, Supplement Sept. 12, 1914, 103; Thorne Baker, *Telegraphic Transmission*, 61–72; E. E. Fournier D'Albe, "The Future of Selenium," *Scientia*, Sept. 1916, 167; Korn, *Trailblazer to Television*, 131, 136.

104. "Pictures by Telegraph from Paris," *Daily Mirror*, Nov. 8, 1907, 3; Thorne Baker, *Telegraphic Transmission*, 60; "Photographs by Cable," *Scientific American*, Aug. 4, 1917, 71.

105. Thorne Baker, *Telegraphic Transmission*, 88–94, and *Wireless Pictures and Television: A Practical Description of the Telegraphy of Pictures, Photographs and Visual Images* (London: Constable & Co. Ltd., 1926), 120–21; "Pictures by Wireless," *New York Times*, May 2, 1910, 7; "Pictures by Wire to 'Frisco," *New York Times*, May 23, 1910, 7; Phillippe Bata and Patrice A. Carre, "Presse, Photographie et Télécommunications de 1850 à 1940," *Telecomunications*, Sept. 1985, 320.

106. Bernard Auffray, *Edouard Belin: Le pere de la Television* (Paris: les cles du monde editeurs, 1991), 53, 86–104; Albert Abramson, *Zworykin, Pioneer of Television* (Urbana: University of Illinois Press, 1995), 57, 71–72.

107. Korn and Glatzel, *Handbuch*, 323, 359–60; William J. Hammer, handwritten notes on history of selenium [after 1908]. File 4, Box 17, Selenium. Series 1, William J. Hammer Collection #069, Smithsonian Institution Archives.

108. Germany's Bachner had proposed the idea in 1901. "Photo-Transmission," *Electrical Review*, Dec. 14, 1907, 942; "The Belin System for the Electrical Transmission of Photographs," *Telegraph Age*, Jan. 1, 1908, 7–8; "Belin's Improved Apparatus for the Electrical Transmission of Pictures," *Scientific American*, June 12, 1909, 440, 442; Korn and Glatzel, *Handbuch*, 322–26.

109. Jacques Boyer, "Sending Photographs over a Telephone Wire," *Scientific American*, Dec. 21, 1912, 529, 543; "Rushes Pictures by Wire," *New York Times*, Jan. 19, 1913, C4; "Science and Discovery," *Current History*, Jan. 1921, 158–61; Auffray, *Edouard Belin*, 60–61.

110. Thorne Baker, *Telegraphic Transmission*, 116; "Belin's Improved Apparatus for the Electrical Transmission of Pictures," *Scientific American*, June 12, 1909, 442.

111. "Knudsen's Process of Transmitting Pictures by Wireless Telegraphy," *Scientific American*, June 6, 1908, 412; Thorne Baker, *Telegraphic Transmission*, 14–15, 128–29; Korn and Glatzel, *Handbuch*, 161.

112. "The Telautograph," *Nature*, May 30, 1901, 107–9; James Dixon, "The Telautograph," *Transactions of the A.I.E.E.*, Oct. 28, 1904, 645–57; Edwin James Houston, *Electricity in Every-Day Life* (New York: P. F. Collier & Son, 1905), 3:374–75.

113. "Cowper's Telegraphic Pen," *Telegraphic Journal and Electrical Review*, March 1, 1879, 76.

114. "Navy Type Telautograph" (clipping), Aug. 31, 1920. Box 529A, SRM 133-193A, Clark Radioana, Smithsonian Archives; Telautograph Corporation, *Annual Report 1923*, 1926; Lewis Coe, *The Telegraph: A History of Morse's Invention and Its Predecessors in the United States* (Jefferson, NC: McFarland & Co., 1993), 20–21.

CHAPTER 2:   **First Markets, 1918–1939**

Epigraph. J. W. Horton, "The Electrical Transmission of Pictures and Images," *Proceedings of the Institute of Radio Engineers* 17, 9 (Sept. 1929), 1563.

1. A. Russel Gallaway Jr., to Guy Hamilton, Feb. 6, 1939. Box 4, "Letters File," McClatchy Collection, Sacramento Archives & Museum Collection Center (SAMCC).

2. "Lost Choral Score Facsimile Radioed," *New York Times*, Dec. 29, 1937, 16; "Hail Radioing of Score," *New York Times*, Dec. 31, 1937, 3.

3. Eugene Lyons, *David Sarnoff* (New York: Harper & Row, 1966), 190–203.

4. "Telephotography and Television—Report on Contemporary Systems," Jan. 31, 1928. Jewett 74 09 01, AT&T Archives.

5. In December 1933, 47 Siemens-Karolus and 15 Belin sets operated worldwide, as well as a few Nippon Electric systems and more than a score of AT&T units ("Telephotography Abroad," April 11, 1934, 5–6. Jewett 74 09 01, AT&T Archives).

6. Chester Carlson, "Electric recording and transmission of pictures," US Patent 2277013 Issue date: Mar 17, 1942. A Selenyi article about recording faxed photographs by electrostatic charges formed one of the theoretical foundations of Carlson's xerography concept. The success of the Xerox 914 photocopier in the 1960s helped faxing spread within offices, completing the circle of innovation in copying technologies (David Owen, *Copies in Seconds: How a Lone Inventor and an Unknown Company Created the Biggest Communication Breakthrough since Gutenberg* [New York: Simon & Schuster, 2004], 86, 138).

7. E.g., "Home Radio Photography," *Radio News* 9, 10 (April 1928), 1163; M. R. Mesny, "La phototelegraphie d'amateur," *Bulletin de la Societe francaise des electriciens*, May 1929, 511–24.

8. "Telephotography and Television." AT&T apparently did not know about the Japanese effort.

9. Robert Howard Claxton, *From Parsifal to Peron: Early Radio in Argentina, 1920–1944* (Gainesville: University Press of Florida, 2007), 181.

10. Radio Corporation of America, *Annual Report 1934*, 4.

11. Arthur Korn, *Die Bildtelegraphie im Dienste der Polizei* (Graz: Verlag von Ulr. Mosers Buchhandlung [J. Meyerhoff], 1927); "May Open a New Era in Telegraph Service," *New York Times*, Jan. 26, 1921, 15; "Telephotography—'Facsimile' System," Feb. 25, 1935, 3. OCB 48 05 03, AT&T Archives.

12. Ethel Plummer, "Telephotography: The Latest Marvel of Science," *Vanity Fair*, Feb. 1921, 44.

13. Paul J. Gordon Fischel, "High-Speed Phototelegraphy," *Wireless World*, June 9, 1926, 778; Vladimir K. Zworykin and Earl D. Wilson, *Photocells and Their Application* (New York: John Wiley & Sons, 1930), 1–13; Terry and Elizabeth P. Korn, *Trailblazer to Television: The Story of Arthur Korn* (New York: Charles Scribner's Sons, 1950), 136–38.

14. "Telephotography and Television," 4.

15. J. N. Whitaker, "Modified Page Facsimile Equipment for Ultra High Frequency Service," CM-21-2, Dec. 29, 1938, 4. David Sarnoff Research Laboratory.

16. "Telephotography—'Facsimile' System," 6. Feb. 25, 1935. OCB 48 05 03, AT&T Archives.

17. L. F. Morehouse to Walter Gifford, July 11, 1923. 85 08 02 02, AT&T Archives.

18. "Newspapers Sent to Sea—by Radio," *Popular Mechanics*, Nov. 1930, 744; D. K. Gannett, "Memorandum," Oct. 1, 1935. Jewett 73 08 02, AT&T Archives.

19. Comite Consultatif International Telephonique, *Xth Plenary Meeting, Budapest, Sept. 3–10, 1934*, vol. 3, *Transmission, Definitions, Recommendations, Specifications* (London: International Standard Electric Corporation, 1936), 252.

20. R. C. Ballentine, "Developments in Radio Facsimile" [mid-1929], 9, 20. Neils Young Collection, Boise, Idaho.

21. Directeur General, PTT, Netherlands, to Secretary, General Post Office, London, Nov. 13, 1933. File 7, PO 33 4668, British Telecom archives (hereafter BT archives).

22. "Facsimile Radio-Telegraphy," *Electrical Review*, Feb. 8, 1929, 235. For similar restrictions on the AT&T tuning forks, see C. R. Hommowun, "Memorandum for File," Feb. 5, 1935; B. W. Kendall to A. F. Dixon, March 7, 1935. OCB 48 05 05, AT&T Archives. For Japanese tuning forks, see Yasujiro Niwa, "A Synchronizing System for Electrical Transmission of Pictures," *Far Eastern Economic Review*, July 1933, 328.

23. "Photographs by Cable," *Scientific American*, Aug. 4, 1917, 84; R. V. L. Hartley, "Transmission of Information," *Bell System Technical Journal*, July 1928, 535–63; R. B. Shanck, "Memorandum," Nov. 27, 1940. Case 36694; W. A. Phelps, "Memorandum for file," Dec. 10, 1940; Case 37727, AT&T Archives; Maynard D. McFarlane, "Digital Pictures Fifty Years Ago," *Proceedings of the IEEE* 60, 7 (July 1972), 770.

24. O. E. Glunt to A. F. Dixon, June 16, 1932. Case 19538, 48 07 01 05A, AT&T Archives.

25. Bureau of the Census, "Postal Rates for Domestic Airmail: 1918 to 1957," *Historical Statistics of the United States: Colonial Times to 1957* (Washington, DC: GPO, 1960), 499; "Pieces of Mail Handled, Number of Post Offices, Income, and Expenses, 1789 to 2009," www.usps.com/postalhistory/PiecesofMail1789to2009.htm; www.usps.com/postalhistory/_pdf/DomesticLetterRates1863-2009.pdf; *Bell System Statistical Manual, 1900–1945* (New York: AT&T, 1946), 1203.

26. "Domestic telegraph industry—messages, wire, offices, employees, and finances: 1866–1987," *Historical Statistics of the United States. Millennial Edition Online* http://hsus.cambridge.org/HSUSWeb/search/searchTable.do?id=Dg8-21.

27. Roy Worth Barton, *Telex* (London: Sir Isaac Pitman & Sons, Ltd., 1968); Chris Drewe, "BT's Telex Network: Past, Present—and Future?" *British Telecommunications Engineering* 12, 1 (April 1993), 17–21; Donald E. Kimberlin, "Telex and TWX History," 1986, http://baudot.net/docs/kimberlin—telex-twx-history.pdf.

28. Nevil Maskelyne, "Improvements in or relating to the Reproduction of Pictures, Drawings, & the like, at a Distance," Great Britain Patent 190800658. Jan. 7, 1909.

29. "News Pictures by Wire in Plain Morse Code," *Scientific American*, Dec. 1921, 140; Bernard Auffray, *Edouard Belin: Le Pere de la Television* (Paris: les cles du monde editeurs, 1991), 63–64.

30. J. H. Leishman, "How I Telegraph Pictures," *Electrical Experimenter*, Dec. 1917, 516–18, 572; D. W. Isakson, "Developments in Telephotography," *Journal of the American Institute of Electrical Engineers* 41, 11 (Nov. 1922), 814–15; "Another Method of Telegraphing Pictures," *Scientific American*, Dec. 1925, 403.

31. *First Annual Report of the FCC for 1935* (Washington, DC: GPO, 1936), 8; P. Mertz, "Memorandum," Sept. 29, 1936, 7. Case 36710-1, AT&T Archives; Russell W. Burns, *Television: An International History of the Formative Years* (London: IEE Press, 1998), 33.

32. For example, one reporter described John Hogan's 1934 Radio Pen as "a sort of slowed-up television system" (Laurence M. Cockaday, "If Not 'Television' Why Not 'Facsimile,'" *Radio News* 16, 2 [Aug. 1934], 77).

33. Arthur Korn, *Bildtelegraphie* (Berlin: Walter de Gruyter & Co., 1923), 6; *First Annual Report of the FCC for 1935*, 8; David Sarnoff, "Radio Art and Industry," in *The Radio Industry: The Story of Its Development* (Chicago: A. W. Shaw Co., 1928), 110; James E. Brittain, *Alexanderson: Pioneer in American Electrical Engineering* (Baltimore: Johns Hopkins University Press, 1992), 185–86; Albert Abramson, *Zworykin, Pioneer of Television* (Urbana: University of Illinois Press, 1995), 57, 71–72.

34. An internal 1930 RCA proposal suggested combining the two (R. W. Carlisle, "Can We Combine Facsimile with Television?" RCA Victor TR-51, July 16, 1930. Box 130, Hagley Library Collection #2069). I thank Alexander Magoun for the reference. See also Charles J. Young, "An Experimental Television System," *Proceedings of the Institute of Radio Engineers* 22, 11 (Nov. 1934), 1286–94; E. B. Craft to Frank Jewett, June 21, 1927, Jewett 74 09 01, AT&T Archives; Auffray, *Edouard Belin*, 53, 86–104.

35. Charles J. Young, "Methods for Exploitation of Broadcast Facsimile," Oct. 11, 1932, 4. Neils Young collection.

36. Orrin E. Dunlap Jr., *The Future of Television* (New York: Harper and Brothers, 1942), 118.

37. Kenneth Bilby, *The General: David Sarnoff and the Rise of the Communications Industry* (New York: Harper & Row, 1986), 111–38.

38. "Berlin-Vienna Wire for Pictures Opens," *New York Times*, Dec. 2, 1927, 3.

39. H. E. Young to J. B. Harlow, March 17, 1928, 48 07 01 05A; A. F. Dixon to Frank Jewett, Dec. 11, 1928, and D. C. Tanner to Jewett, Dec. 22, 1928. Jewett 75 74 08 03, AT&T Archives.

40. "Telephotography and Television," 5–6; A. J. Smith, "Pictures by Wireless," *The Electrician*, Jan. 13, 1928, 27–28.

41. A. Korn and E. Nesper, *Bildrundfunk* (Berlin: Julius Springer, 1926), 9; Korn, *Die Bildtelegraphie im Dienste der Polizei*. After Hitler's takeover, Korn was removed and then dismissed from his professorship. More fortunate than most victims of Nazi anti-Semitism, Korn reached the United States in 1939, teaching at the Stevens Institute and working at the Times Telephoto Corporation until his death in December 1945 ("Refugee Is at Stevens," *New York Times*, Jan. 14, 1940, D4; "Dr. Korn, Pioneer in Radiophoto, Dies," *New York Times*. Dec. 23, 1945, 17; Jochen Bruning, Dirk Ferus, and Reinhard Siegmund-Schultze, *Terror and Exile: Persecution and Expulsion of Mathematicians from Berlin between 1933 and 1945: An Exhibition on the Occasion of the International Congress of Mathematicians, 1998* (Berlin: Deutsche Mathematiker-Vereinigung, 1998), 36).

42. "Picture Telegraphy Services," n.d. File 6, PO 33, 3888, BT archives; E. S. Ritter to Col. Shreeve, Jan. 16, 1930. Jewett 75 74 08 03, AT&T Archives.

43. Engineer in Chief to the Secretary, March 2, 1929. File 9, PO 33, 2553; Treasury Chambers to Postmaster General, July 6, 1929. File 1, PO 33 2874A, BT archives; L. V. Lewis, *Picture Telegraphy* (London: The Post Office Green Papers Number 17, 1935), 13.

44. S. C. Hooper to George Lewis, Aug. 30, 1932. Box 14, Hooper Papers, Library of Congress. For RCA, Lee Galvin to F. R. Deaking, March 13, 1936. SRM 133-152A, Western Union Collection, Smithsonian Institution Archives (hereafter, Western Union,

Smithsonian Archives). For AT&T, E. B. Craft to F. B. Jewett, Nov. 28, 1925. Jewett 75 74 08 03; J. D. Madden to C. H. Fuller, Aug. 29, 1927. 48 07 01 05A, AT&T Archives. For Korn, see Korn, *Trailblazer to Television*, 141. For Belin, Auffray, *Edouard Belin*, 78–83.

45. E. B. Craft to F. B. Jewett, Nov. 28, 1925. 75 74 08 03; T. G. Miller to Colpitts, Sept. 2, 1927. Jewett 06 10 03, AT&T Archives.

46. Na Liu and Yanping Li, "Edouard Belin and the Early History of the Fax in China," *Chinese Journal for the History of Science and Technology* 2 (2010) http://english .ihns.cas.cn/institute/OS/journal/cjhst/201011/t20101109_61195.html.

47. Auffray, *Edouard Belin*, 78–83; Morizo Sugi, "The Dawn of the Facsimile" and "The Untold Story of the Early Era of Phototelegraphy," in "The Era of Home-grown Technology Development: Highlights of Phototelegraphy." Manuscript in possession of the author.

48. Sannosuke Inada, "Public Service of Phototelegraphy in Japan," *Electrical Communication*, July 1931, 26–33; Keijiro Kubota et al., *Fakushimiri bakujoho chosa* [Facsimile back information survey], 3rd ed. (Tokyo: NTT, Nov. 2004), 9.

49. William D. Caughlin, personal communication, Sept. 11, 2011.

50. Niwa, "A Synchronizing System for Electrical Transmission of Pictures," 327; Sugi, "Dawn of the Facsimile."

51. "Telephotography Abroad," April 11, 1934, 5–6.

52. "Recent Telecommunication Developments of Interest," *Electrical Communication*, 14, 2 (Oct. 1935), 181; Tatuo Morikawa, "An Experiment on Facsimile Transmission," *Nippon Electrical Communication Engineering* 8 (Nov. 1937), 460; Matsushita Graphics, *Fakushimiru no oyumi* [History of Facsimile Development] (Tokyo: Matsushita Graphics, 1990), 102; Keijiro Kubota, "Waga kuni ni okenu fakushimiri no hatten" [Development of facsimile in our country], *Denki Gakkai*, Feb. 2, 1995, HEE-95-6, 51.

53. M. Leroy, "Telephotographic Service in France," June 18, 1936, 19–20. Case 36710, AT&T Archives.

54. J. McCarthy to the Secretary, Aug. 30, 1930. File 7, PO 33, 3888; "Private Picture Telegraphy Installations in Great Britain," July 24, 1936. File 3, PO 33, 4877, BT archives.

55. Kubota, *Facsimile back information survey*, 9.

56. "Statement of the Traffic, Income, and Expenditure of the Picture and Facsimile Telegraphy Service." Files, 21, 25, 27, 28, 29, PO 33, 3888, BT archives.

57. C. Morsack to C.P. Cooper, "Telephoto Service," Aug. 23, 1927, 48 07 01 05A; E. H. Colpitts to O. B. Blackwell and L. F. Morehouse, April 27, 1927. Jewett 74 09 01, AT&T Archives; "Statement of the Traffic, Income and Expenditure."

58. A. F. Bunker to O.T.B., Telegraph & Telephone Dept., August 17, 1934. File 19, PO 33, 3888; Postmaster Surveyor, Manchester to Public Relations (Sales), July 10, 1935. File 22, PO 33, 340, BT archives.

59. May 9, 1930 memo, File 2, PO 33 3888, BT archives; J. H. Bell, "Memorandum for File," May 20, 1927, 48 07 07 05B; "Information concerning operation of Type 'A' Telephotograph System," 85 08 02 01; AT&T Archives; Directeur General, PTT, Netherlands, to Secretary, General Post Office.

60. "May Open a New Era in Telegraph Service," *New York Times*, Jan. 26, 1921, 15.

61. R.C.A. Communications, Inc., "Photograms Via RCA" and "Across the Atlantic and Pacific," 1933. SRM 54-036 and SRM 54-046f respectively, George H. Clark Radioana Collection, Smithsonian Archives.

62. General Post Office, "Public Picture Telegraphy Services Instructions to Offices of Acceptance: Extension of service to Denmark," Aug. 27, 1930. Minute 2198/1939, File 1, PO 33, 5441, BT archives.

63. John Mills to E. B. Craft, May 29, 1925, 48 07 07 05B, Milton O. Boone, Captain, Q.M.C., Aide-de-Camp, to Frank Jewett, Oct. 19, 1927, Jewett 74 09 01; "History of Commercial Experience with Telephotograph Service" [c. 1936], 7 in Long Lines Division materials for FCC, 1936, main title, "Telephotograph." 85 08 02 01, AT&T archives.

64. E. H. Colpitts to Jewett, April 10, 1928. Jewett 74 09 01, AT&T archives.

65. "Statement of the Traffic, Income, and Expenditure."

66. Inada, "Public Service of Phototelegraphy in Japan," 33.

67. "Statement of Overseas Picture and Facsimile Telegraphy Service, 1936–37." File 27, PO 33 3888, BT archives. For the 1935 Royal wedding, see File 1, PO 33 5211; for the 1936 Olympics, see Files 2–5.

68. Tuyosi Amisima and Masatugu Kobayashi, "Wireless Picture Transmission between Berlin and Tokyo," Nippon Electrical Communication Engineering, Dec. 1937, 499–510; Kokusai Tenbosha, Yakushin nihon taikan [Overview of the progress of Japan] (Tokyo: Kokusai Tenbosha, 1937), 124.

69. "Germany's Wirephoto Service Expanding," Telegraph and Telephone Age, Dec. 1939, 275.

70. Feb. 5, 1929, memo, File 8, and Engineer in Chief to the Secretary, Aug. 16, 1930, File 9, PO 33, 3553; Imperial and International Cables, Ltd. to General Post Office, Nov. 20, 1933, File 7, PO 33, 4668. BT archives; Osborne Mance, International Telecommunications (Oxford: Oxford University Press, 1944), 13.

71. McFarlene, "Bartlane Brown Book," 1–2; "Yacht Race Picture Cabled to London," New York Times, July 23, 1920, 3; Hugh Cudlipp, Publish and Be Damned! The Astonishing Story of the Daily Mirror (London: Andrew Dakers, Ltd., 1953), 51–52.

72. M10078/28. Files 4 and 8, PO 33 2374, BT archives; "Telephotography and Television," cover letter by E. H. Colpitts to F. B. Jewett, Feb. 1, 1928.

73. "Bartlane Brown Book," 12–14; F. E. d'Humy, A Brief Outline of the Technical Progress Made by the Western Union Telegraph Company, 1910–1934, Dec. 31, 1934, 1:101–3. Box 6. Western Union, Smithsonian Archives; Maynard D. McFarlane, "Digital Pictures Fifty Years Ago," Proceedings of the IEEE 60, 7 (July 1972), 768–70.

74. "Transatlantic Cable Facsimile System," 6–10 [1940]. Folder 5, Box 541, Series 15, Western Union, Smithsonian Archives.

75. "Science and Discovery," Current History, Jan. 1921, 158–61; G. G. Blake, History of Radio Telegraphy and Telephony (London: Chapman & Hall, 1928), 221.

76. E. F. W. Alexanderson, "New Fields for Radio Signalling," General Electric Review 29, 4 (April 1925), 268; David Sarnoff, "Radio in Relation to the Problems of National Defense," address before the Army Industrial College, Feb. 20, 1926, 8. Box 7, Hooper Papers, LC. Radio engineer Alfred N. Goldsmith, fully aware of the technological challenges, dryly noted Young's "poetic imagination and bravery which foresaw new physical phenomena and possibilities and was untrammeled by too dogmatic a knowledge of present-day technological limitations" (Alfred N. Goldsmith to Charles Young, Sept. 25, 1962, Neils Young collection).

77. "Radio Corporation to Inaugurate Trans-Atlantic Picture Service," Editor & Publisher, April 10, 1926, 54; E. F. W. Alexanderson, "Radio Photography and Television,"

General Electric Review 30, 2 (Feb. 1927), 79–84; R. C. Ballentine, General Electric, "Developments in Radio Facsimile" [mid-1929?], Neils Young collection; "Richard Howard Ranger," Owen Dunlap, *Radio's 100 Men: Biographical Narratives of Pathfinders in Electronics and Television* (New York: Harper and Brothers, 1944), 239–40.

78. Richard H. Ranger, "Transmission and Reception of Photoradiograms," *Proceedings of the Institute of Radio Engineers* 14, 2 (April 1926), 165, 178; "Speeding Up Photoradio," *Wireless Age*, April 1929, 15; R. H. Ranger, "Mechanical Developments of Facsimile Equipment," *Proceedings of the Institute of Radio Engineers* 17, 9 (Sept. 1929), 1564–75; J. L. Callahan, J. N. Whitaker, and Henry Shore, "Photoradio Apparatus and Operating Technique Improvements," in Alfred N. Goldsmith et al., eds. *Radio Facsimile* (New York: RCA Press, 1938), 82–87; "First Radiophotos Received from Russia," *Editor & Publisher*, July 12, 1941, 9.

79. "Flash Facsimile Transmission Seen," *Editor & Publisher*, Sept. 22, 1934, 7; "Photoradiogram Expedites Ship Repairs, Saving Expense of Some $7,000," *Wireless Age*, March 1929, 35–36; Table Dg90-102 — International telegraph and telephone industry—messages, calls, ocean cable, finances, and employment: 1907–1987, *Historical Statistics of the United States. Millennial Edition Online* hsus.cambridge.org/HSUSWeb/search/searchTable.do?id=Dg90-102.

80. Work Order No. 8933, "Investigation and Development of Apparatus for Sending and Receiving Pictures by Telegraphy," Dec. 1, 1920; R. D. Parker, "Memorandum," Feb. 26, 1936. 85 08 02 02, AT&T Archives.

81. H. V. Hayes and T. Spencer to R. W. Devonshire, April 27, 1893; HN/RM, "Memorandum," Dec. 6, 1921, 85 08 02 01, AT&T Archives; James Gleick, *The Information: A History, a Theory, a Flood* (New York: Pantheon Books, 2011), 198–202.

82. The Telepix Corporation, "Announcing the *New* Telepix," brochure [after Nov. 1925], "Western Union Automatic Telegraphs," 2 [late 1939?], Box 541, Folder 5, Series 15, Western Union, Smithsonian Archives. See also "Telephotograph — History of Commercial Experience w/ Telephotograph Service," 1–2; "Photographs Sent 1,000 Miles by Wire," *New York Times*, Nov. 15, 1920, 17.

83. Frank Jewett to E. S. Bloom, "Memorandum," March 1, 1923, and attached "Notes of Talk with Mr. Benington." Box 64, Jewett, AT&T Archives.

84. A. W. Drake, "Memorandum for File," April 4, 1923. Box 64, Jewett, AT&T Archives.

85. L. F. Morehouse, to Gifford, July 11, 1923, 85 08 02 02; W. R. Newman, to C. H. Fuller, Feb. 6, 1925, 48 07 07 05B; L. F. Morehouse and A. W. Drake to F. A. Stevenson, Dec. 15, 1924. Box 64, Jewett, AT&T Archives.

86. Frank Jewett to Frank Gill, July 2, 1924, Box 75. Jewett, AT&T Archives.

87. L. F. Morehouse to Walter Gifford, July 11, 1923; R. D. Parker, "Memorandum," Feb. 26, 1936, 85 08 02 02, AT&T Archives.

88. E. H. Colpitts to J. J. Carty, July 10, 1924, 83 07 02; E. B. Craft to Colpitts, April 29, 1925, 48 07 07 05B; C. Morsack to C. P. Cooper, "Telephoto Service," Aug. 23, 1927, 48 07 01 05A; E. H. Colpitts to O. B. Blackwell and L. F. Morehouse, April 27, 1927, Jewett 74 09 01; Commercial Relations Department, "Picture Transmission Expense," Feb. 13, 1936, 85 08 02 03, AT&T Archives.

89. A. F. Dixon to H. H. Lowry, March 19, 1925, 48 07 07 05B; E. S. Bloom to A. W. Drake, July 1, 1925. Box 64, Jewett, AT&T Archives.

90. F. A. Stevenson to E. S. Bloom, Dec. 7, 1925, Box 64. Jewett, AT&T Archives.

91. A train ticket from New York to Chicago in 1935 cost approximately $40, depending on the level of comfort ("New York Central" timetable, effective Feb. 3, 1935, http://www.canadasouthern.com/caso/ptt/images/tt-0235.pdf). For AT&T charges, see J. L. Spellman, Illinois Bell Telephone Co., Chicago, "Telephotographic Service for Commercial Operation, Conference—1925," Feb. 27, 1925, Box 65, Jewitt; Long Lines submission to FCC, c. 1936, main title, "Telephotograph," subtitle "Telephotograph Channel Service," 85 08 02 01, AT&T Archives.

92. J. H. Bell, "Memorandum for File," May 20, 1927, 48 07 07 05B; E. H. Colpitts to F. B. Jewett, April 10, 1928, Jewett 74 09 01, AT&T Archives.

93. Frank B. Jewett, "Radio Telephony," The Radio Industry, 135–37.

94. A. F. Dixon, "Memorandum for File," Jan. 24, 1928, Box 75, Jewett, AT&T Archives. The exception was the 1928 sale to the Daily Express.

95. A. F. Dixon to L. F. Morehouse, June 30, 1932, 48 07 01 05A; J. H. Bell, "Memorandum for File," June 28, 1932, 48 07 01 05A; A. F. Dixon to F. B. Jewett, Nov. 28, 1934, Jewett 74 09 01; R. D. Parker, "Memorandum," Feb. 26, 1936, 85 08 02 02, AT&T Archives.

96. E. B. Craft to F. B. Jewett, June 21, 1927, Jewett 74 09 01; A. F. Dixon to L. F. Morehouse, June 30, 1932, 48 07 01 05A; R. D. Parker and P. Mertz, "Memorandum," March 6, 1936, 85 08 02 01, AT&T Archives.

97. IIC to GPO, July 6, 1933, File 21, Minute 9425/1931, PO 33, 3340, BT Archives.

98. Jack Price, "Smaller Dailies Must Consider Installing Photo Departments," Editor & Publisher, April 11, 1936, 30.

99. Hanno Hardt, In the Company of Media: Cultural Constructions of Communication, 1920s–1930s (Boulder: Westview, 2000), 60; Anthony North, "No, But I Saw the Pictures," New Outlook, June 1934, 21; Jack Price, "Stanley Sees Greater Use of News Pictures," Editor & Publisher, April 16, 1938, 24; Frank Luther Mott, American Journalism: A History: 1690–1960 (New York: Macmillan, 1962), 682–84.

100. Manuel Komroff, "The Little Black Box," Atlantic Monthly, Oct. 1938, 470.

101. "Pigeon Carrier Is Latest Word in Short Distance Photo Transmission," Editor & Publisher, July 6, 1935, 9; John W. Perry, "Dailies Strip Pages, Clear All Wires for Rogers-Post Smash Display," Editor & Publisher, Aug. 24, 1935, 1–2.

102. Including the first newspaper use of the telephone in Indiana in 1906 and the introduction of teletypewriters in 1914 (Kent Cooper, Kent Cooper and the Associated Press: An Autobiography [New York: Random House, 1959], 31–34).

103. Joe H. Brewer, "Pictures for the Papers," The Quill 19, 4 (April, 1931), 9; Cooper, Kent Cooper, 139; Mark Monmonier, Maps with the News: The Development of American Journalistic Cartography (Chicago: University of Chicago Press, 1989), 85.

104. "Telephotograph Channel Service"; R. D. Parker, "Memorandum," Feb. 26, 1936, 85 08 02 02, AT&T Archives; Cooper, Kent Cooper, 213–15.

105. "John Neylan, 74, Lawyer on Coast," New York Times, Aug. 20, 1960, 19.

106. Meyer Berger, The Story of The New York Times, 1851–1951 (New York: Simon & Schuster, 1951), 408.

107. "Neylan Organizing War on Telephoto," Editor & Publisher, June 16, 1934, 9, 3; "Sees Big Press Cost in Telephoto Plan," Editor & Publisher, April 21, 1934, 82.

108. H. Jeavons, to W. J. McCambridge, Nov. 3, 1933, 85 08 02 02, ATT Archives.

109. "Howard Hits Again at A.P. Telephoto," *Editor & Publisher*, May 5, 1934, 6; Cooper, *Kent Cooper*, 214; John W. Perry, "A.P. Members Approve Telephoto Service after Heated Debate in Convention," *Editor & Publisher*, April 28, 1934, 3, 110.

110. "Neylan Organizing War on Telephoto"; Stephen A. Bolles, "Where Are We Going With the Radio or Where Is the Radio Going With Us?" in American Society of Newspaper Editors, *Problems of Journalism. Proceedings of Fourteenth Annual Convention*, April 16–18, 1936, 134–41; Cooper, *Kent Cooper*, 212.

111. [AP advertisement], "Every Day a New Scoop by AP Wirephoto Newspapers," *Editor & Publisher*, Sept. 14, 1935, 18–19.

112. "Norris Huse Dead; AP Photo Leader," *Editor & Publisher*, Jan. 16, 1937, 42; Cooper, *Kent Cooper*, 215.

113. A. F. Dixon to F. B. Gleason, Jan. 4, 1934; H. H. Lowry, "Memorandum for File," Jan. 25, 1934; A. F. Dixon to F. B. Gleason, Sept. 25, 1934; F. B. Gleason to W. T. Teague, Nov. 12, 1934; A. F. Dixon to F. B. Jewett, Nov. 28, OCB 19 34 48 05 05; A. F. Dixon and A. B. Clark to E. H. Colpitts, April 8, 1936, 3–4; J. J. Pelliod to H. P. Charlesworth, Nov. 25, 1936, 5. Case 36710, AT&T Archives.

114. R. D. Parker to A. F. Dixon, Sept. 17, 1934; F. B. Gleason to W. T. Teague, Nov. 12, 1934. OCB 48 05 05, AT&T Archives. See also "Installing Telephoto," *Editor & Publisher*, Oct. 6, 1934, 18; Oliver Gramling, *AP: The Story of News* (New York: Farrar and Rinehart, 1940), 392.

115. C. R. Hommowun, "Memorandum for File," Feb. 5, 1935; B. W. Kendall to A. F. Dixon, March 7, 1935. OCB 48 05 05, AT&T Archives; "Wirephoto 'Kinks' Rapidly Disappear," *Editor & Publisher*, April 11, 1936, 3, 15.

116. Bice Clemow, "'Co-Operation' Is A.P. Meet Keynote," *Editor & Publisher*, October 19, 1935, 7, 16.

117. Bice Clemow, "Wide World Telephoto System Ready," *Editor & Publisher*, Feb. 22, 1936, 9; Bice Clemow, "Picture Services Rushing into Field of Telephotograph Transmission," *Editor & Publisher*, Feb. 29, 1936, 3–4; Jack Price, "First Details of NEA Wirephoto Equipment Told by Ferguson," *Editor & Publisher*, April 4, 1936, 32; Campbell Watson, "Howey Pleased with Soundphotos; Describes Three New Processes," *Editor & Publisher*, April 11, 1936, 8, 10; "News Pictures by Wire," *Electronics*, Nov. 1937, 12–17, 82–83.

118. Director of Transmission Development to E. H. Colpitts, Feb. 21, 1935; Frank Jewett to Kent Cooper, March 7, 1935. Jewitt 74 09 01, AT&T Archives.

119. F. E. Meinholtz to Harry Jeavons, March 21, 1934; Harry Jeavons to F. E. Meinholtz, April 19, 1934; F. E. Meinholtz, "Memo for Mr. A. H. Sulzberger," April 28, 1934; Times Facsimile Company, 1941–59 (General) *New York Times* Archives.

120. Sulzberger's letter is worth quoting: "You may recall the conversation which I had with you some months ago, at which time I asked you whether or not it made a difference to the Telephone Company if English or Turkish were spoken over the wires. When you replied in the negative, I posed the same question for Turkish or gibberish, and followed that by asking frankly if the noise we put at one end of a phone line were taken off at another end and translated into a picture, would be any concern of yours other than a welcome addition to the Telephone Company's revenue. You assured me at that time that this was the case, and shortly after that our first experiments were started" (A. H. Sulzberger to Walter Gifford, Feb. 27, 1935. Jewitt 74 09 01, AT&T Archives).

121. Berger, *Story of The New York Times*, 410–11; "First Pictures of the Survivors of the Macon Disaster," *New York Times*, Feb. 14, 1935, 3, and untitled editorial, *New York Times*, Feb. 15, 1935, 18.

122. R. D. Parker, "Conference Memorandum," March 27, 1936, 3. Case 36710, AT&T Archives; W. J. McCambridge, "More Wirephoto Developments Coming, McCambridge Says," *Editor & Publisher*, Jan. 4, 1941, 9.

123. R. B. Shanck, "Memorandum," Nov. 27, 1940. Case 36694, AT&T Archives.

124. "Wirephoto 'Kinks' Rapidly Disappear."

125. Clemow, "Picture Services Rushing into Field of Telephotograph Transmission," 3–4.

126. "Sound-Photo Line Opened," *Editor & Publisher*, Dec. 21, 1935, 8.

127. "AP," *Fortune*, Feb., 1937, 90.

128. F. H. H., "Broadcasting Photographs," *Wireless World*, March 24, 1926, 437; "Picture Broadcasting Soon," *Wireless World*, July 25, 1928, 89–90.

129. Captain Eckersly to Secretary, June 13, 1928; Lee to Secretary, July 3, 1928. File 5, M10060/1928 7970/26 min, PO 33, 2371B, BT archives; R. W. Burns, "Wireless pictures and the Fultograph," *IEE Proceedings* 128, Part A,1 (Jan. 1981), 78–88.

130. "Salon de T.S.F.," *Wireless World*, Nov. 7, 1928, 647; "Portable Radio Set Receives Pictures," *New York Times*, Sept. 21, 1929, 16; "Facsimile—'the home radio printing press,'" *Electronics* 7, 11 (Nov. 1934), 338–40; "Otho Fulton Dies, Radio Inventor, 70," *New York Times*, March 2, 1938, 19.

131. "What do you mean—radio facsimile?" *Radio Today*, Sept. 1935, 22–23; "Facsimile: Radio Threatens to Reach into Country's Mailboxes," *News-week*, Nov. 23, 1935, 41–42; "Chain of Facsimile Papers," *Business Week*, March 11, 1939, 28–30; "The Periscope," *Newsweek*, Oct. 9, 1944, 24.

132. The other technologies were the mechanical cotton picker, television, air conditioning, plastics, photoelectric cells, synthetic fibers, synthetic rubber, prefabricated houses, automobile trailers, gasoline from coal, steep-flight aircraft, and tray agriculture. (National Resources Committee, *Technological Trends and National Policy Including the Social Implications of New Inventions* (Washington, DC: GPO, 1937), 228–33).

133. "Home radio will print newspaper while you sleep," *Electronics* 7, 9 (Sept. 1934), 278.

134. "'Facsimile' to the aid of the broadcasters," *Electronics* 6, 5 (May 1933), 139; Lester Nafzger, "Facsimile Broadcasting at Weld," *FM and Television* 4, 6 (June 1944), 30–37, 58–59.

135. Edward Harrison, "First Effect Is Seen as Spur to Employment; 211 Papers Already Linked with Radio," Feb. 28, 1938 [clipping], SRM 54 99A, Clark Radioana, Smithsonian Archives; Department of Commerce, *Historical Statistics of the United States, Colonial Times to 1970* (Washington, DC: GPO, 1975), 2:796.

136. Bruce Catton to Frank Thone, Jan. 3, 1938. Folder 11, Box 199, Science Service Records (RU 7091), Smithsonian Archives. I am indebted to Marcel Lafollette for the reference.

137. G. C. Hamilton, "Facsimile broadcasting's place in the newspaper field," Intertype Corporation Dinner, April 23, 1940. Box 4, McClatchy Collection, SAMCC.

138. C. Francis Jenkins to John J. Carty, Sept. 23, 1924. Carty 83 07 02, AT&T Archives (emphasis in original); "Charles Francis Jenkins: Put Pictures on the Air," in Dun-

lap, *Radio's 100 Men of Science*, 141–43; Charles F. Jenkins, *Radiomovies, Radiovision, Television* (Washington, DC: Jenkins Laboratories, 1929).

139. Radiovision Corp., "How to Receive Radio Pictures at Home," brochure (New York, 1928), SRM 133 035, Clark Radioana, Smithsonian Archives; John R. Poppele, "Pre-war Facsimile Broadcasting," in Radio Inventions, *The Story of Broadcasters Faxi-mile Analysis* (New York: BFA, April, 1946), IV-1-4; Dermot Cole, "'Father of Facsimile' dead at 93," *Fairbanks Daily News-Miner* Sept. 8, 1993, 3; Richard D. Lyons, "Austin G. Cooley, 93, Inventor Helped Develop the Fax Machine," *New York Times*, Sept. 9, 1993, D21.

140. "John Vincent Lawless Hogan: Invented a Uni-Control Tuner," in Dunlap, *Radio's 100 Men of Science*, 245–47; Federal Communications Commission, "Official Report of Proceedings Before the FCC at Washington, D.C., March 15, 1948, In the matter of: Promulgation of Rules and Transmission Standards Concerning Facsimile Broadcasting." Docket No. 8751, 1:58–59. Box 3532, RG 173. NARA (hereafter, FCC Docket).

141. "Daily Tests Sending Facsimiles by Radio," *Editor & Publisher*, April 14, 1934, 16; "Displays Radio Pen," *New York Sun*, April 14, 1934 [clipping], SRM 54 041, Western Union, Smithsonian Archives; Cockaday, "If Not 'Television,' Why Not 'Facsimile'?" 76–77, 113. This "radio pen" should not be confused with Allen DuMont's contemporaneous "radio pen" or cathautograph (George H. Waltz Jr., "Radio Pen writes in letters of fire on far-away screen," *Popular Science*, Dec. 1933, 16–17).

142. R. Bown, "Memorandum for File," Sept. 28, 1938. Case 36710, AT&T Archives; R. R. Beal, "Hogan Facsimile," July 26, 1940, Neils Young collection.

143. Samuel Ostrolenk, "Home Facsimile Recording," *Electronics* 11, 1 (Jan. 1938), 26–27; Selwyn Pepper, "Printing News in the Home by Radio," *St. Louis Post-Dispatch*, May 1, 1938; Paul H. Goodell '31, "The Illustrious Career of an Electronics Pioneer — Captain William G. H. Finch," *UC Engineer*, Winter 1990, 24–25.

144. Charles J. Young, first draft autobiography notes, n.d. Neils Young collection.

145. C. J. Young, Radio Engineering Department, "Memorandum of Type of Service for Broadcast Facsimile," Sept. 14, 1928; Charles J. Young, "Notes for a Report on Facsimile," RCA Laboratories, March 15, 1944, 17. Neils Young collection.

146. Young, "Methods," 2–3, 5.

147. J. T. McLamore and R. G. Beerbower, "Facsimile Transmitter Locations and Transmission Costs — Eleven Major Cities," TR-296, RCA Victor Company, Inc., Dec. 20, 1935; RCA Manufacturing Company, Inc., "Facsimile radio broadcasting for regular broadcasting stations," and "RCA Radiopress: Brief description of system and equipment," March 6, 1936; RCA Manufacturing Company, Inc., "Synopsis of New York Facsimile Project," Feb. 7, 1936, 2, 6, 9; Young, "Notes," 4. Neils Young collection.

148. "Radio Would Print News in the Home," *New York Times*, Oct. 21, 1937, 19; "Testing Radio Facsimiles," *New York Times*, Dec. 5, 1937, 12; "3 New Radio Facsimile Stations," *New York Times*, Dec. 26, 1937, 12; "Facsimile Broadcasting in California," *Electrical Engineer*, May 1939, 201.

149. Young, "Notes," 16; RCA Manufacturing Company press release, "New Facsimile Printer Presages Home Service," Feb. 1938. SRM 133 010, Clark Radioana, Smithsonian Archives.

150. H. C. Vance, "Sales Possibilities of Broadcast and Duplicator Facsimile Equip-

ment (Preliminary Report)," March 29, 1940, 11, in Young, "Notes." Neils Young collection.

151. Barbara Everitt Bryant, "Facsimile by the Yard," Jan. 20, 1994, memoir in possession of author, 4; Guy C. Hamilton to G. A. Reimer, Feb. 21, 1939. Box 4, McClatchy Collection, SAMCC.

152. [Advertisement], "First at Macy's, Crosley Reado Facsimile," *New York Times*, Feb. 28, 1939, 7; [advertisement], "Build Your Own Facsimile Radio Printer," *Radio-Craft* 10, 5 (May 1939), 693. See also "Crosley 119 Facsimile Kit," *Service*, Sept. 1939, 428–29, 442, 444; Crosley Corp., "Models 118, 119," in John F. Rider, *Perpetual Trouble Shooter's Manual* (New York: John F. Rider, 1939), 1: 10, 10-41-44.

153. R. D. Parker, "Memorandum," March 21, 1939, Jewett 74 09 01, AT&T Archives; "Radio Progress During 1936," *Proceedings of the Institute of Radio Engineers* 25, 2 (Feb. 1937), 199–203; Young, "Notes," 17–19. Neils Young collection.

154. *Third Annual Report of the Federal Radio Commission* (Washington, DC: GPO, 1930), 22; "Facsimile — 'the home radio printing press,'" *Electronics* 7, 11 (Nov. 1934), 337.

155. "Herald Tribune–RCA Facsimile," *Printers' Ink*, May 4, 1939, 85; Daniel M. Costigan, "'Fax' in the home: Looking back and ahead," *Spectrum* 11, 9 (Sept. 1974), 76–77.

156. "Findings of the McClatchy Broadcasting Company on Facsimile Transmission," [1939]. Box 4, McClatchy Collection, SAMCC.

157. Young, "Notes," 17. Neils Young collection; O. E. Buckley to F. B. Jewett, Oct. 7, 1938. Case 36710, AT&T Archives; William G. H. Finch, FCC Docket, 1:115; *Seventh Annual Report of the FCC for 1941* (Washington, DC: GPO, 1942), 37; "Postwar Radio Newsprinter," *Radio-Craft*, March 1945, 37.

158. *Eighth Annual Report of the FCC for 1942* (Washington, DC: GPO, 1943), 29–31, 37; Lester H. Nafzger, "W8XUM and Facsimile," FCC Docket.

159. Weather Bureau, *Report of the Chief 1927–28* (Washington, DC: GPO, 1929), 5; Max Dieckmann, "Funkbild-Ubertragung im Anschluss an Rundfunk-Gerat," *Elektrische Nachrichten Technik* 3, 6 (June 1926), 200, 207–8; "Telephotography and Television," 15; "Weather Chart Sent by Radio by Germans to Flying Plane," *New York Times*, Oct. 3, 1929, 24; "Radio Facsimile Transmission to Ships," *Marine Engineering* 36, 9 (Sept. 1931), 416; A. B. Moulton and Charles J. Young, "Shore to Ship Facsimile; Part III and Summary to Date," TR-146, March 1, 1932, 21–24, Box 116, RCA collection, Hagley Archives; Daqing Yang, *Technology of Empire: Telecommunications and Japanese Expansion in Asia, 1883–1945* (Cambridge, MA: Harvard University Asia Center, 2010), 193.

160. Radiomarine Corporation of America, "Radio Weather Map Service to Ships" (New York: Radiomarine Corporation of America, [c. 1934]) SRM 133-154A, 3–4, Clark Radioana, Smithsonian Archives.

161. Weather Bureau, *Report of the Chief 1928–29* (Washington, DC: GPO, 1930), 4; *Report of the Chief 1933–34* (Washington, DC: GPO, 1936), 2–4, 9; Donald R. Whitnah, *A History of the United States Weather Bureau* (Urbana: University of Illinois Press, 1961), 198.

162. Weather Bureau, *Report of the Chief 1929–30* (Washington, DC: GPO, 1931), 8–9; Whitnah, *History*, 171–85.

163. National Research Council, *Report of the Science Advisory Board, July 31, 1933 to September 1, 1934* (Washington, DC: NRC, 1934), 56; "The New Weather Program," *Science*, May 11, 1934, 434.

164. "Activities of the U.S. Weather Bureau," *Science*, Dec. 1, 1933, 501.

165. F. W. Reichelderfer, "Weather Maps for Radio Broadcast," *Radio News*, April 1944, 21–23.

166. Lloyd Espenschied, "Memorandum," July 13, 1943. Case 36694, AT&T Archives; F. Vinton Long, "AACS Radioteletype Weather Transmission System," *Communications*, Sept. 1946, 16–18, 52–55; H. Jeavons, "Weather Charts by Wire for Air Force," *Bell Telephone Magazine*, 26, 4 (Winter 1947–48), 228–39.

167. "Memo of April 18 meeting with Vice Admiral Mornet, President of the Conseil d'administration des Establissements Edouard Belin, and the Post Office," April 19, 1928. File 1, Minute 2759/1929, PO 33-2553, BT archives; M. G. Ogloblinski, "Derniers Progres de la Transmission Belinographique en France," *L'Onde electrique*, Oct. 1928, 446–55.

168. "Téléautographie. Télephotographie," *Les Derniers Perfectionnements en Télégraphie* 19 (1930), 655.

169. Lewis, *Picture Telegraphy*, 3; "Facsimile Transmission," *Literary Digest*, April 13, 1929, 78.

170. "Memo of April 18 meeting."

171. P. J. Howe, *A Brief Outline of the Technical Progress Made by the Western Union Telegraph Company, 1935–1945*, 2:163–64. Box 6, Western Union, Smithsonian Archives; Charles R. Jones, *Facsimile* (New York: Murray Hill Books, 1949), 41–42.

172. D' Humy, *A Brief Outline*, 1:5.

173. Howe, *Brief Outline*, 2:165–69, 184–89; "Buffalo-New York Facsimile Telegraph Circuit," *Communication and Broadcast Engineering*, Dec. 1935, 21; "Facsimile Wire Service Begun," *New York Times*, Jan. 2, 1936, 50; "Wire 'Fax': Literary Digest First to Use New Western Union Transmission System," *Literary Digest*, Jan. 30, 1937, 38; P. Mertz, "Memorandum," Nov. 7, 1936, and "Memorandum for File," March 8, 1939, 2. Case 36710, AT&T Archives; Table Dg8-21—Domestic telegraph industry, *Historical Statistics*; "Radio Progress During 1945," *Proceedings of the Institute of Radio Engineers*, 34, 4 (April 1946), 174W.

174. "Communications," *Electrical Review*, Sept. 29, 1939, 433.

175. Western Union, "Report on Plans for the Improvement of Telegraph Service," Oct. 7, 1944, 10–11. Box 51, Western Union, Smithsonian Archives.

176. "'Picture' Message Sent," *Wall Street Journal*, Aug. 27, 1938, 3; Gaius W. Merwin to R. D. Parker, Sept. 26, 1938, cover for Roger Fraser, Aug. 30, 1938 memo. Case 37727, AT&T Archives; J. H. Hackenberg, "Facsimile . . . OLD in Principle, NEW in Practice," *Electronics*, March 1943, 106–7, 200; Howe, *Brief Outline*, 2:176.

177. P. Mertz, "Memorandum," Nov. 7, 1936. Case 36710, AT&T Archives; Howe, *Brief Outline*, 2:184–89; G. H. Ridings, "Facsimile Transceiver for Pickup and Delivery of Telegrams," *Electrical Communication* 26, 2 (June 1949), 129.

178. "Short Waves," *Business Week*, Jan. 25, 1933, 11; "Flash Facsimile Transmission Seen," *Editor & Publisher*, Sept. 22, 1934, 7; "Radio Progress During 1936," *Proceedings of the Institute of Radio Engineers* 25, 2 (Feb. 1937), 201-02.

179. H. H. Beverage, "The New York–Philadelphia Ultra High Frequency Facsimile Relay System," *RCA Review* 1, 1 (July 1936), 15–31; J. N. Whitaker, "Modified

Page Facsimile Equipment for Ultra High Frequency Service," CM-21-2, Dec. 29, 1938; J. L. Callahan, "High Speed Facsimile Business Survey," CM-21-1, Dec. 30, 1938. RCA Collection, Hagley; P. Mertz, "Memorandum for File," March 8, 1939. Case 36710, AT&T Archives.

180. Martin Codel, "Pictures by Radio Near," *New York Sun*, Jan. 13, 1934, SRM-54-045A, Clark Radioana, Smithsonian Archives.

181. C. J. Young, M. Artzt, and A. Blain, "Facsimile Duplicator," TR-456, Aug. 31, 1939; M. Artzt, A. Blain, and Charles J. Young, "Facsimile Message Duplicator," TR-477, Feb. 9, 1940. Box 119. RCA Collection. Hagley.

182. Young, "Notes"; H. C. Vance, "Sales Possibilities of Broadcast and Duplicator Facsimile Equipment (Preliminary Report)," March 29, 1940; P. Mertz, "Memorandum. Telephotography—New RCA Machine File 36710," Nov. 4, 1940. Case 37727, AT&T Archives; RCA Laboratories Division, *Research Report 1950* (Princeton, NJ: RCA, 1951), 109–10, 117–19; "Electrofax," *Radio Age* 14, 1 (Jan. 1955), 20–21; Acoustical and Mechanical Laboratory, RCA Laboratories, "Electrofax," Nov. 19, 1957. Harry F. Olson files, RCA Collection. Hagley.

183. W. A. Phelps, "Memorandum for File," Oct. 11, 1940; "Description of Western Union 'Multifax,'" Case 37727, AT&T Archives; "Development in Facsimile," *Electronics*, June 1941, 72, 74; "Cutting Stencils by Facsimile," *Barron's*, June 16, 1941, 5; John Robert Gregg, *Applied Secretarial Practice*, 2nd ed. (New York: Gregg Publication Co., 1941), 134–48.

184. DKG–MLD, "Memorandum," Oct. 24, 1927; "Transmission of Text—Telephotography vs. Telegraphy," 06 10 03; "Telephotography—'Facsimile' System," Feb. 25, 1935. OCB 48 05 03; W. A. Phelps, "Memorandum for File," Dec. 10, 1940. Case 37727, AT&T Archives.

185. R. D. Parker, "Memorandum for File," June 6, 1939. Case 36694-2, AT&T Archives (emphasis added).

186. T. A. McCann, "Memorandum," Nov. 17, 1938. Case 33508; W. A. Phelps, "Memorandum for File," May 9, 1940 and Dec. 10, 1940; E. F. Watson, "Statement on 1938–40 Investigation," Dec. 17, 1948. Case 37727, AT&T Archives.

187. "Autographic Telegraphy," *London Times*, Dec. 9, 1878, 4; Lewis Coe, *The Telegraph: A History of Morse's Invention and Its Predecessors in the United States* (Jefferson, NC: McFarland & Co., 1993), 20–21.

188. "Pictures by Wireless," *New York Times*, May 2, 1910, 7.

189. Dieckmann, "Funkbild-Ubertragung im Anschluss an Rundfunk-Gerat," 200; Korn and Nesper, *Bildrundfunk*, 64; Col. F. A. Iles, "Picture-Telegraphy and Electrical Vision," *Royal Engineers Journal*, Sept. 1926, 396; "Dr. Korn, Pioneer in Radiophoto, Dies"; Korn, *Trailblazer to Television*, 134.

190. George Dewey, President, General Board to Secretary of the Navy, "Transmitting photographs by radio," March 25, 1916. Subject File 419, Records of the General Board, RG 30. NARA; E. Belin, "Experiences Nouvelles sur la transmission des images a distance," *Bulletin de la Societe francaise des electriciens*, Feb. 1924, 185–94; Auffray, *Edouard Belin*, 76–77.

191. Radio Division, Bureau of Engineering, "The Korn Photographic Reproducing Apparatus," *Monthly Radio and Sound Report* [1922]; R. DeL. Hasbrouck to Director of Naval Intelligence, May 11, 1922; E. H. Hansen to Commanding Officer, USN Radio

Station, Bar Harbor, May 15, 1922. SRM 133 017, Clark Radioana, Smithsonian Archives; Lofton to Hooper [c. 1921?], File: "Jenkins," Box 7, Entry 1084, RG19, NARA.

192. "Belin," Nov. 9, 1921, memo for chief of bureau, file "Belin." Entry 1084, Box 2, RG19, NARA. See also "Transmitting Photographs and Drawings by Radio," *Scientific American*, Sept. 3, 1921, 163, 173; "Why Not Long-Distance Stenography?" *Scientific American*, Jan. 1922, 57.

193. H. A. Brown to E. B. Craft, Nov. 18, 1924. Jewett 75 74 08 03, AT&T Archives; Robert Rothschild to Robert G. Vansittart, Oct. 8, 1928. Note on back of letter. File 6, Minute 2759/1929, PO 33, 2553. BT archives.

194. L. S. Howeth, *History of Communications-Electronics in the United States Navy* (Washington: GPO, 1963), 417.

195. Louis A. Gebhard, *Evolution of Naval Radio-Electronics and Contributions of the Naval Research Laboratory* (Washington, DC: Naval Research Laboratory, 1979), 69–70.

196. S. C. Hooper to C. Francis Jenkins, Dec. 17, 1923. SRM 133 017, Clark Radioana, Smithsonian Archives; *Monthly Radio and Sound Report*, Oct. 1–Nov. 1, 1925, 55–56, and March 1–April 1, 1926, 106; Office of Chief Naval Officer, "Data for Director's Relief," July 13, 1928, 2. Box 9; Milton S. Davis to S. C. Hooper, Feb. 9, 1927, 3. Box 8. Hooper Papers, LC.

197. S. C. Hooper to Captain R. W. McNeely, Aug. 10, 1926. Box 7; S. C. Hooper to Captain J. M. Reeves, March 31, 1927. Box 8; author's proof of "Weather Maps" Aug.–Oct. 1927. Box 8; C. W. Horn to Hooper, Aug. 17, 1928. Box 9, Hooper Papers, LC; "New Message Plan to Cut Operator Need," *Christian Science Monitor*, July 27, 1927, 13.

198. *Monthly Radio and Sound Report*, May 1–June 1, 1926, 59–61, and July 1–Aug. 1–Sept. 1, 1926, 24–25.

199. Joseph R. Redman to S. C. Hooper, Feb. 4 and May 23, 1928. Box 9; S. C. Hooper to Commander H. P. LeClair, May 6, 1927. Box 8, Hooper Papers, LC; *Monthly Radio and Sound Report*, Nov.–Dec. 1927– January 1928, 64–65.

200. S. C. Hooper to Rear Adm. W. H. Stanley, March 25, 1931. Box 13, Hooper Papers, LC; *Radio and Sound Bulletin*, July–Dec. 1931, 20.

201. J. N. Wenger, "Military Study of Facsimile," [1938?], 7. U. S. Navy Records Relating to Cryptology, Box 1, SRMN-001, RG 457, NARA.

202. "Army Developing System to Transmit Maps by Radio," *New York Times*, Nov. 23, 1924, 18.

203. Captain George I. Back, "History of the Communication Equipment Development Projects and Certain Miscellaneous Development Projects during the Period 1924 to 1933." Box 4, Fort Monmouth Post Headquarters, Histories and Historical Reports, 1918–41, RG 338, Fort Monmouth; "The Army Test of the Telephotograph, Fort Leavenworth, Kansas," Oct. 2, 1925. Box 1065, Carty, 83 07 02, AT&T Archives; Major John F. Curry, "Aviation," *General Electric Review* 29, 4 (April 1926), 216; Haydn P. Roberts to CSO, "Report on Radio Facsimile Transmission," May 9, 1930. Box 502, File 413.44 Radio Telegraphy, RG 342; Office, Chief of the Air Corps, to CSO, Nov. 8, 1932. Box 885, File 413.4 RG 18, NARA.

204. T. A. M. Craven, "Memorandum for Commander Raguet," May 16, 1930. Box 4, Entry 1084, "Director, Naval Communications, 1929," RG19; C. W. Howard to General

Inspector Naval Aircraft, June 9, 1930. Box 502, File 413.44 Radio Telegraphy, RG 342, NARA.

205. S. C. Hooper to R. S. Culp, Nov. 14, 1931, Box 13; C. N. Ingraham to S. C. Hooper, Sept. 21, 1932. Box 14, Hooper Papers, LC; Gebhard, *Evolution of Naval Radio-Electronics*, 67–68; Captain H. P. Browning, "The Application of Teletypewriters to Military Signaling," *Signal Corps Bulletin*, May-June 1934, 32, 38, 40; Chief of Air Corps to Commanding Officer, Air Corps, June 10, 1938. Box 896, File 413.4, RG 18. NARA; Thomas S. Snyder, ed., *Air Force Communications Command, 1938–1991: An Illustrated History*, 3rd ed. (Scott AFB, IL: AFCC Office of History, 1991), 1–8.

206. Ferdinand d'Humy to R. B. White, Dec. 16, 1937, 10–11, and April 15, 1938, 4–6. Western Union, Smithsonian Archives. I thank David Hochfelder for the references.

CHAPTER 3: **Facsimile, 1939–1965**

Epigraph. "Editor Hails Facsimile in New Era," *Miami Herald*, March 15, 1947.

1. "Herald's Facsimile," Miami *Beach Sun*, March 31, 1947.

2. Eugene J. Smith, "Word-and-Picture Sending System Shows New Signs of Growth," *Wall Street Journal*, Sept. 18, 1950, 1, 6; G. H. Ridings, "Facsimile Communication Past, Present, Future," *Signal*, Nov. 1962, 34.

3. "Facsimile Fit to Print," *Time*, Sept. 3, 1956, 54.

4. "New Wrinkles," *Wall Street Journal*, May 14, 1953, 6; Ian Batterham, *The Office Copying Revolution: History, Identification, and Preservation* (Canberra: National Archives of Australia, 2008), 11, 66–67; "Appraisal of the Company's Future," Aug. 28, 1953, 1–3, and A. G. Cooley, "Memorandum for Directors of TFC," Dec. 4, 1957. Times Facsimile Corp. (General) 1941–1959 (3), New York Times Archives (hereafter, TFC archives.)

5. Alfred B. Zipser Jr., "Communication Industry Infant, Facsimile, Is Growing Steadily," *New York Times*, Jan. 17, 1954, 4.

6. German bombs hit the British Central Telegraph Office on December 29, 1941, and damaged the stored Siemens-Karolus machine ("Statement of Overseas Picture and Facsimile Telegraphy Service, 1939–40," and Engineering Dept. to OTB Telecommunications Dept., Feb. 3, 1941. File 29, PO 33, 3888, BT archives); P. J. Howe, *A Brief Outline of the Technical Progress Made by the Western Union Telegraph Company, 1935–1945*, 2:175–76. Box 6, Western Union Telegraph Company. Collection No. 205, Smithsonian Institution Archives (hereafter, Western Union, Smithsonian Archives).

7. "SCGDL Proj. No. 4-19 Facsimile Systems for Wire Lines," July 6, 1942; "Project 4-19 Facsimile Systems for Wire Lines 1942–1943," Sept. 1942. Box 130, SCEL 1937–1945 Project File 1937–1943, RG 338, NARA (hereafter, Project 4-19).

8. Several independent inventors submitted proposals to the Signal Corps (e.g., Ralph H. Kofski to War Dept., June 9, 1941. Box 1711, Radio-Photo Transmission, 1941–42, RG 342, NARA).

9. L. A. Thompson to Kenneth A. West, Feb. 11, 1942; Elwood K. Morse to L. A. Thompson, Jan. 30, 1943. Project 4-19.

10. War Dept. Technical Manual TM11-375B, *Facsimile Equipment RC-120, RC-120-A and RC-120-B and Facsimile Set AN/TXC-1* April 5, 1944, 353; Roland C. Davies and Peter Lesser, "Facsimile equipment communication units," *Electronic Industries*, Feb. 1945, 96–99, 170.

11. Col. R. M. Osborne, to Chief Signal Officer, June 21, 1944. Box 3205, June 1, 1944, to July 3, 1944, 413.44 Wireless Equipment. 1940–45, AG Decimal File, RG 407, NARA.

12. A. G. Cooley to Director, Aug. 6, 1942. Project 4-19. See also Rebecca Robins Raines, *Getting the Message Through: A Branch History of the U.S. Army Signal Corps* (Washington, DC: Center for Military History, 1996), 274–75.

13. Wilton J. Norris, "Conference at Finch Telecommunications, Inc.," Jan. 2, 1943; Bruce V. Magee, "Facilities for Manufacture of Facsimile Equipments RC-58 and RC-120," March 3, 1943; Elwood K. Morse to Chief Signal Officer, March 5, 1943. Project 4-19.

14. "Postwar Research and Development Program of the Signal Corps Engineering Laboratories," Dec. 1945. Box Postwar R&D Programs, HL R&D, SCEL archives, Ft. Monmouth, NJ.

15. Don Z. Zimmerman to Chief Signal Officer, "Training Facsimile Operators," Sept. 22, 1942. Project 4-19.

16. Vooheis Richeson, "Army Tests New Communications Device," *United States Army Recruiting News* 21, 3 (March 1939), 6–7, 18; Andre La Terza, "Radio Facsimile Aids Defense," *Radio News* 26, 6 (Dec. 1941), 24–25, 60.

17. A. W. Marriner to Chief Signal Officer, Jan. 26, 1943. Box 3212; F. J. Magee to Commanding General, Armoured Force, Ft. Knox, May 22, 1943. Box 3214, RG407, NARA; "Radio Progress During 1945," *Proceedings of the I.R.E.* 34, 4 (April 1946), 175W; George Raynor Thompson and Dixie R. Harris, *The Signal Corps: The Outcome (Mid-1943 through 1945)* (Washington, DC: Office of the Chief of Military History, U.S. Army, 1966), 605.

18. J. D. O'Connell, to Director, SCGDL, July 7, 1942. Project 4-19; "Radio Progress During 1944," *Proceedings of the I.R.E.* 33, 3 (March 1945), 150; Melvin Schlessinger and William Stokes, "These Are the Fax," *Weather Service Bulletin* 3, 9 (Nov.– Dec. 1945), 3–5, 25; "Radio Progress During 1945," 174W; Rita M. Markus et al., *Air Weather Service: Our Heritage, 1937–1987* (Scott Air Force Base, IL: Military Airlift Command, 1987), 7, 116–17.

19. SCL Engineering Report No. 674, "Investigation of Facsimile Equipment," March 3, 1939; Aircraft Radio Laboratory, "Ground to Air Test of Facsimile Equipment," ARL Engineering Report No. 302, July 29, 1939; Major C. E. Duncan to Adjutant General, Jan. 13, 1940. Box 32, 413.44 Wireless Equipment, RG 407, NARA; Ken A. West, "National Defense Research Committee Aircraft Facsimile Model," June 27, 1942; Major John H. Gardner to Chief Signal Officer, Aug. 10, 1940. Project 4-19.

20. J. N. Wenger, "Military Study of Facsimile," [1938?], 7. Box 1, USN Records Relating to Cryptology, SRMN, RG 457, NARA; J. J. Friedman to Procurement Planning Section, Field Laboratory, Jan. 19, 1943. Box 130, Project File 1937–194, SCEL 1937–1945, RG 338, NARA; Communications Section, National Defense Research Committee of the Office of Scientific Research and Development, *Facsimile Privacy Project PDRC-749* (abridged version), Oct. 15, 1943; Elmer E. Bucher, "Radio and David Sarnoff," [1946?], typescript , Part V., Chapter 36, 1294, David Sarnoff Research Library; Kazuo Kobayashi, "IIEEJ wo turi kaette" [Looking Back on the IIEEJ], *Gazo denshi gakkaishi* [*Journal of the Institute of Image Electronics Engineers of Japan*] 36, 3 (May 2007), 304.

21. Col. R. M. Osborne to Chief Signal Officer, Aug. 21, 1944. Box 3204, RG407, NARA; Howe, *Brief Outline*, 2:332.

22. F. A. Resch, "Photo Coverage of the War By the 'Still Picture Pool,'" *Journalism Quarterly* 20, 4 (Dec. 1943), 311–14; Acme Telephoto, "In United Nations Service" [advertisement], *Editor & Publisher*, Feb. 19, 1944, 19.

23. Barbie Zelizer, "Journalism's 'Last' Stand: Wirephoto and the Discourse of Resistance," *Journal of Communications* 45, 2 (Spring 1995), 81.

24. Col. Carl H. Hatch, "Radiophotos," *Radio News*, 31, 2 (Feb. 1944), 218, 316–318; Thompson and Harris, *Signal Corps*, 565, 605–6.

25. Office of Naval Operations, *United States Naval Administration in WWII: Naval Communications* (April 1948), 58–59. Box 92, WWII Command File CNO, Communications History, Naval History and Heritage Command.

26. Thompson and Harris, *Signal Corps*, 92–94, 105–7, 138, 606.

27. Murray E. Tucker to Contracting Officer, Nov. 29, 1940; "Comments on First Hogan Equipment Delivered to Canada," Jan. 29, 1941; A. G. Cooley to E. K. Morse, Jan. 16, 1943. Project 4-19; Chief of Finance, War Department, *Lend-Lease Shipments: World War II* (Washington, DC: War Department, Dec. 31, 1946), V (Signal), 31.

28. Howe, *A Brief Outline*, 2:331–32; "Telegrams in Chinese," *Popular Science*, May 1945, 128.

29. Paul Deichmann, *Spearhead for Blitzkreig: Luftwaffe Operations in Support of the Army, 1939–1945* (London: Greenhill Books, 1996), 34; U. S. Naval Technical Mission in Europe, "A Summary of German Developments in Meteorology during the War," Technical Report N. 230–45, Aug. 1945; Irving P. Krick, *War and Weather: A Report of the AAF Scientific Advisory Group* (Dayton, OH: Air Material Command, May 1946), 13; British Intelligence Objectives Sub-Committee, "Telecommunications and Equipment in Germany during the period 1939–1945," B.I.O.S. Surveys Report No. 29 (London: His Majesty's Stationery Office, 1950), 26.

30. U. S. Navy Technical Mission to Japan, "Electronics Targets, Japanese Radio Equipment," Dec. 22, 1945, 7. Box 101, E-08. Records of Japanese Navy and Related Documents, 1940–1960, Naval History and Heritage Command; Matsushita Graphics, *Fakushimiru no oyumi* [History of Facsimile Development] (Tokyo: Matsushita Graphics, 1990), 102; Keijiro Kubota, "Waga kuni ni okenu fakushimiri no hatten" [Development of facsimile in our country], *Denki Gakkai*, Feb. 2, 1995, HEE-95-6, 51; Kobayashi, "Looking Back on the Institute of Facsimile Engineers," 302–3.

31. F. J. Singer, "Military Teletypewriter Systems of World War II," *Transactions of the American Institute of Electrical Engineers* 67, 2 (Jan. 1948), 1398–1408; Thomas S. Snyder, gen. ed., *Air Force Communications Command, 1938–1991: An Illustrated History*, 3rd ed. (Scott AFB, IL: AFCC Office of History, 1991), 43; War Office, *Signal Communications* (London: War Office, 1950), 138.

32. John J. Mullaney to R. L. Bunch, Dec. 7, 1938. Box 896, 413.4 B Teletype and A (1932–30), RG 18; Col. J. M. Gillespie to Commanding Officer, March 31, 1943. Box 2573, 413.51, RG 342, NARA.

33. Snyder, *Air Force Communications Command*, 93–95; "Telephone and Telegraph," *Public Utilities Fortnightly*, Nov. 24, 1960, 829.

34. "Progress in Facsimile" *Signal Corps Technical Information News Letter* 1949, 4 (Sept.), 23.

35. Alden Electronic & Impulse Recording Equipment Company, *Annual Report Fiscal Year 1963*, 15.

36. Winfred A. Ross, "Signal Corps R&D Program," June 24, 1949, 12–13. Signal Corps Laboratories, 1930–1976, SCEL archives; P. Mertz and R. B. Shanck, "Memorandum for File," Aug. 23, 1954. Case 37727, AT&T Archives.

37. TFC, "Planning Study: AN/TXC-1B Facsimile Equipment," 1948, 1-4-5, TFC (General) 1941–1959 (3), TFC archives; Richard B. Le Vino, "Facsimile Equipment Development in the Signal Corps," Signal, July–Aug. 1954, 32; W. R. Greiling, "Memorandum," Feb. 15, 1955; M. L. Benson, "Memorandum," May 13, 1957. Case 37727, AT&T Archives; Earl D. Anderson, "Facsimile Telegraph Network for Weather Maps," Western Union Technical Review 14, 2 (April 1960), 42.

38. Smith, "Word-and-Picture Sending System," 1, 6; R. B. Shank, "Memorandum," Nov. 5, 1952; M. L. Benson, "Memorandum," June 7, 1955. Case 37727, AT&T Archives.

39. Shank, "Memorandum"; C. W. Smith, "Memorandum for File," Oct. 25, 1954. Case 37727, AT&T Archives.

40. Elliott Crooks, "Facsimile Systems as an Aid to Research," American Documentation 7, 1 (Jan. 1956), 42.

41. Ivan S. Coggeshall and Albert E. Frost, "Modern Telegraphic Communication Systems," Tele-Tech, April 1951, 40.

42. Fred Shunaman, "Telecar Speeds Telegrams," Radio-Electronics, July 1951, 22–24; Warren H. Bliss, "Advancements in the Facsimile Art During 1963," 1964 IEEE International Convention Record, vol. 5, "Wire and Data Communication" (New York: IEEE, 1965), 195.

43. "Transistorized Miniature Facsimile Scanning Head," SCEL Journal, Sept. 23, 1954, 2–3; "Facsimile Set AN/GXC-5," SCEL Journal, June 10, 1958, 3; "Fully Transistorized Facsimile System," Modern Communications, March–May 1962, 10.

44. "Postwar Research and Development Program of the Signal Corps Engineering Laboratories."

45. "Facsimile Scanning and Recording Techniques," SCEL Journal, June 1, 1962, 4; W. D. Buckingham, "A Flat-Bed Facsimile Telegraph Transmitter," Electrical Engineering, April 1956, 356–59.

46. D. A. Huffman, "A method for the construction of minimum redundancy codes," Proceedings of the I.R.E. 40, 9 (Sept. 1952), 1098-111; "Profile: David A. Huffman," Scientific American, Sept. 1991, 54, 58.

47. R. B. Shanck, "Memorandum," April 8, 1955. Case 37727, AT&T Archives; "Variable Scanning Facsimile Techniques," SCEL Journal, March 10, 1958, 6–7; Melpar Research Report 59/1 "On the Efficient Representation of Pictorial Data," March 31, 1959; Ridings, "Facsimile Communication," 35.

48. W. S. Michel, "Analysis of a Coding Scheme for Facsimile," June 1, 1956; E. R. Kretzmer, "Memorandum for Record," July 10, 1956. Case 38763, AT&T Archives; Garvice H. Ridings, "Facsimile Imaging Systems," Western Union Technical Review, April 1965, 19.

49. Auerbach Corporation, "Techniques for Reduction of Data Transmission Requirements in Facsimile Transmission Systems," Jan. 15, 1964. Folder 15, Box 32, Auerbach collection, Charles Babbage Institute.

50. "Telephotography—Testing Methods and Apparatus," Feb. 27, 1957; W. K. MacAdam to F. J. Singer, March 15, 1957. Case 37727, AT&T Archives.

51. W. S. Michel, "Coded Facsimile vs. Conventional Facsimile: A Comparison of Estimated Transmission Rates," Aug. 3, 1956. Case 38763, AT&T Archives.

52. E.g., "Delay Measurement and Equalization of Facsimile Networks," *SCEL Journal*, Oct. 9, 1952, 2; "Photo Facsimile Transmission Techniques," *SCEL Journal*, April 1963, 10.

53. "Investigation of Facsimile Repeater Techniques," *SCEL Journal*, Oct. 22, 1953, 7; W. H. Bliss, R. J. Wagner Jr., and G. S. Wickizer, "Wide-band Facsimile Transmission over a 900-Mile Path Utilizing Meteor Ionization," *IRE Transactions on Communication Systems*, Dec. 1959, 252–56.

54. "Facsimile over 4,000-Mc Relay System," *Electronics*, Oct. 1946, 146–47.

55. "Banks Can Be Sure of Funds Wired Them," *Wall Street Journal*, Oct. 15, 1952, 4; "Federal Reserve to Install Computerized Wire System," *Wall Street Journal*, Feb. 28, 1968, 6.

56. E. Blanton Kimbell, "Wire Systems," *ANPA Research Institute Mechanical Bulletin*, Aug. 11, 1961, 147–50; Ehrhard Rossberg, "Intercontinental Telex Traffic," *1964 IEEE International Convention Record*, vol. 5, "Wire and Data Communication" (New York: IEEE, 1964), 167–68. For France, see Patrice Carre and Martin Monestier, *Le Telex: 40 ans d'innovation* (Paris: Editions Menges, 1987), 76–95.

57. E. M. Mapes to G. W. Gilman, April 9, 1954. Case 37727, AT&T Archives.

58. C. O. Caulton, "RCA's Participation in the Field of Facsimile," Jan. 14, 1960. Neils Young Collection.

59. Table Dg8-21. Domestic telegraph industry—messages, wire, offices, employees, and finances: 1866–1987, *Historical Statistics of the United States, Millennial Edition Online*, hsus.cambridge.org/HSUSWeb/search/searchTable.do?id=Dg8-21. Table DG90-102. International telegraph and telephone industry—messages, calls, ocean cable, finances, and employment: 1907–1987, *Historical Statistics of the United States. Millennial Edition Online* hsus.cambridge.org/HSUSWeb/search/searchTable.do?id =Dg90-102.

60. D. B. Perry, "Memorandum for File," Jan. 31, 1952. Case 37727, AT&T Archives. For an early telex-fax comparison, see William L. Alden, "Cutting Communication Costs with Facsimile," *Data Processing*, Sept. 1964, 14.

61. D. E. S. Isle, "Improving Britain's Telecommunication Service," *Communications and Electronics* 2, 3 (March 1955), 51.

62. "Facsimile Methods for Broadcast Work," *Electronic Industries*, June 1946, 74–75; Frank R. Brick, "Progress in Postwar Facsimile Equipment," *FM and Television* 6, 8 (Aug. 1946), 23–25; "Facsimile Service Soon in 12 Cities," *Printers' Ink*, Aug. 9, 1946, 96; "Spot News Notes," *FM & Television* 7, 7 (July 1947), 28.

63. "The Beginnings of Alden Electronics," n.d. Alden Archives. In 1958, the name changed to Alden Electronic & Impulse Recording Equipment Company to better reflect the firm's orientation.

64. Milton Alden, "Suiting Facsimile Designs to Service Needs," *FM and Television* 4, 9 (Sept. 1944), 32–40; Milton Alden, "A New Facsimile Dispatch and Report System," *FM and Television* 5, 8 (Aug. 1945), 32–36; Milton Alden, "FX Can Help Make FM Stations Pay," *FM and Television* 6, 6 (June 1946), 21–23.

65. "FM Free-for-All," *Business Week*, March 3, 1945, 29–38.

66. U.S. Dept. of Commerce, *Historical Statistics of the United States, Colonial Times to 1970* (Washington, DC: GPO, 1975), 2:796.

67. Hugh M. Beville Jr., "The Challenge of the New Media: TV, FM, and Facsimile," *Journalism Quarterly* 25, 1 (March 1948): 5–7; Federal Communications Commission, Official Report of Proceedings Before the FCC at Washington, D.C., March 15, 1948, In the matter of: Promulgation of Rules and Transmission Standards Concerning Facsimile Broadcasting. Docket No. 8751, v. 1, 58–59, and E. Z. Jones, v. 1, 250. Box 3532, RG 173, NARA (hereafter, FCC Docket).

68. Hugh D. Lavery, "Adman Looks at FM: Sees Real Selling Job," *Editor & Publisher*, Jan. 18, 1947, 28; Jerry Walker, "History of Facsimile Experiment Recorded," *Editor & Publisher*, Oct. 29, 1949, 42; *Historical Statistics*, 2:797–98.

69. Edwin H. Armstrong, "A Method of Reducing Disturbances in Radio Signaling by a System of Frequency Modulation," *Proceedings of the Institute of Radio Engineers* 24, 5 (May 1936), 732–36; Charles J. Young, "Notes for a Report on Facsimile," RCA Laboratories, March 15, 1944, 18–19. Neils Young collection; F. C. Collings and C. J. Young, "RCA Facsimile Equipment," *FM and Television* 4, 7 (July 1944), 21.

70. Theodore C. Streibert, "The Inception of BFA," John V. L. Hogan, "What BFA Has Accomplished," and R.W. Bristol, "The Future of BFA," in Radio Inventions, *The Story of Broadcasters Faximile Analysis* (New York: BFA, April 1946), chaps 6, 7, and 8.

71. Timothy J. Sullivan, "Automatic Facsimile Editions Start Today," April 1, 1948, and "Multiplex Facsimile Broadcasts On," *Miami Herald*, July 15, 1948.

72. John V. L. Hogan, "What WQXQ New York Is Doing," *FM and Television* 6, 10 (Oct. 1946), 36; Jerry Walker, "Miami to Get Facsimile With Some Refinements," *Editor & Publisher*, Feb. 15, 1947, 46; "GE Turning Out 'Fax' Equipment," *Editor & Publisher*, Jan. 10, 1948, 67; Gregory C. Kunkle, "Technology in the Seamless Web: 'Success' and 'Failure' in the History of the Electron Microscope," *Technology and Culture* 36, 1 (Jan. 1995): 100–102.

73. Burton L. Hotaling, "Facsimile Broadcasting: Problems and Possibilities," *Journalism Quarterly* 25, 2 (June 1948), 141; "High Lights and Side Lights," *GE Review* 50, 1 (Jan. 1947), 39, 51; 2 (Feb. 1948), 52; and 51, 9 (Sept. 1948), 47; "Stewart-Warner Shows Radio Facsimile Set It Can Build for $400," *Wall Street Journal*, Sept. 28, 1948, 2; John V. L. Hogan, FCC Docket, v. 1, 64–65.

74. Elliott Crooks, FCC Docket, v. 1, 154–55.

75. Roger W. Clipp, FCC Docket, v. 2, 269, and Lee Hills, 315–24; "U-M to Offer Course in Facsimile," *Miami Herald*, March 30, 1947.

76. Timothy J. Sullivan, "Herald All Set to Start Facsimile Demonstration," *Miami Herald* March 9, 1947; Stephen Trumbull, "Civic and Business Leaders View Newspaper," *Miami Herald*, March 12, 1947; Nixon Smiley, *Knights of the Fourth Estate: The Story of the Miami Herald* (Miami: Banyan Books, 1974), 228–35, 254–56, 296.

77. Edgar H. Felix, "Miami Herald Transmits Facsimile Newspaper," *FM and Television* 7, 4 (April 1947), 36–39.

78. Stephen Trumbull, "Civic Club Award Shown on Facsimile," *Miami Herald*, March 15, 1947; Jerry Walker, "Miami to Get Facsimile with Some Refinements," *Editor & Publisher*, Feb. 15, 1947, 46; Jerry Walker, "Miami Facsimile Test Witnessed by 50,000," *Editor & Publisher*, April 5, 1947, 48.

79. Stephen Trumbull, "'Miracle Mile' to See Broadcasts," *Miami Herald*, March 21, 1947.

80. Clipp, FCC Docket, v. 2, 267; "Phila. Inquirer Gives Fax Show; Service in Fall" and "Facsimile Inquirer 'Astounds' Mayor," *Philadelphia Inquirer*, May 10, 1947.

81. "Phila. Inquirer Starts Regular 'Fax' Edition," *Editor & Publisher*, Jan. 3, 1948, 36; Jerry Walker, "N.Y. Times Starts Fax But Adds to Presses," *Editor & Publisher*, Feb. 21, 1948, 56; Timothy J. Sullivan, "Automatic Facsimile Editions Start Today," *Miami Herald*, April 1, 1948; Meyer Berger, *The Story of the New York Times, 1851–1951* (New York: Simon & Schuster, 1951), 506.

82. Alexander Nyman, FCC Docket, v. 2, 232–38.

83. "Facsimile in Philadelphia," *FM and Television* 7, 5 (May 1947), 34.

84. G. Bennett Larson, FCC Docket, v. 2, 351; Robin B. Compton, v. 2, 392.

85. Hogan, FCC Docket, v. 1, 50; Finch, 119; Crooks, 164–71; "Miami Herald Signs Contract for 'Fax' Ad," *Editor & Publisher*, March 29, 1947, 20, and "7 Firms on Facsimile," *Editor & Publisher*, July 10, 1948, 44.

86. Hogan, FCC Docket, v. 1, 17–19, 33–37.

87. FCC Docket, *Report and Order* FCC 448–1655; *Fifteenth Annual Report of the FCC for 1949* (Washington, DC: GPO, 1950), 45.

88. Milton B. Sleeper, "Facsimile Is Ready for Home Use," *FM and Television* 7, 6 (June 1947), 19–20, 52; "What Happened to FM Facsimile," *FM and Television* 8, 12 (Dec. 1948), 21.

89. "Missouri U. Develops 'Talking Newspaper,'" *Editor & Publisher*, Jan. 22, 1949, 34; "'Fax' to University," *Editor & Publisher*, Nov. 26, 1949, 46.

90. Hogan. FCC Docket, v. 1, 77.

91. [Advertisement], "At the twist of your wrist . . . ," *New York Times*, Aug. 11, 1948, 4; "Emerson's FM receivers," *FM Business* 1, 5 (June 1946), 22.

92. Clipp, FCC Docket, v. 2, 272–4.

93. "How Word Comes from Sandy Hook," *New York Times*, June 13, 1946, 45.

94. RCA Laboratories, "'Ultrafax,' million words a minute communications system, demonstrated by RCA at Library of Congress," press release, Oct. 21, 1948; T. R. Kennedy Jr., "Novel Copied, Sent by Air in 2 Minutes," *New York Times*, Oct. 22, 1948, 1, 20; Donald S. Bond and Vernon J. Duke, "Ultrafax," *RCA Review* 10, 1 (March 1949), 99–115.

95. "Rights to Ultrafax Refused to Russia," *New York Times*, Oct. 24, 1948, 5; "Signal Corps Board Status Reports, 1 Feb. 51–30 Apr. 51," 7. Box Signal Corps Board, 1924–1962, SCEL archives.

96. Warren H. Bliss and Charles J. Young, "Facsimile Scanning by Cathode-Ray Tube," *RCA Review* 15, 3 (Sept. 1954), 289–90.

97. RCA Laboratories Division, *Research Report 1948* (Princeton, NJ: RCA, 1949), 81; Arthur D. Little, Inc., "Report on Electronic Handling Systems and High Speed Printers to Engineering Products Department, RCA Victor Division," C-58059 (Cambridge, MA: Arthur D. Little, Inc., Feb. 12, 1951), Box 145. RCA Collection, Hagley Museum, Z-175; RCA Laboratories Division, *Research Report 1952* (Princeton, NJ: RCA, 1953), 111–13, and *Research Report 1953* (Princeton, NJ: RCA, 1954), 77–78; Bliss and Young, "Facsimile Scanning by Cathode-Ray Tube," 276–90.

98. E. F. Watson, "Memorandum for File," Feb. 11, 1952. Case 37727, AT&T Archives.

99. C. J. Young and H. G. Greig, " 'Electrofax' Direct Electrophotographic Printing on Paper," *RCA Review* 15, 4 (Dec. 1954), 469–84; Charles J. Young, "The RCA Development and Introduction of Electrofax," manuscript, May 18, 1965. Neils Young collection.

100. L. C. Roberts, "Facsimile—Visit to Winchester Industries, Inc.: Memorandum," Nov. 18, 1952; R. B. Shanck, "Memorandum," Jan. 19, 1955; M. L. Benson, "Memorandum," June 3, 1955; M. I. Uchenick, "Memorandum for File," Feb. 13, 1957. Case 37727, AT&T Archives; "Air Associates' New Name Is Electronic Communications," *Wall Street Journal*, Feb. 6, 1957, 13, "Electronic Communications, Inc.," *Wall Street Journal*, Sept. 1, 1960, 2; "National Cash Offer Lures 85% of Common Stock of Florida Concern: Electronic Communications, Inc.," *Wall Street Journal*, Sept. 4, 1968, 4.

101. E.g., Fairchild Graphic Equipment (M. L. Benson, "Memorandum," April 21, 1955) and 3M (Byron F. Murphey to Walter A. MacNair, April 1, 1957; Doren Mitchell to Bryon F. Murphey, May 23, 1957. Case 37727, AT&T Archives).

102. Zipser, "Communication Industry Infant"; WJD, "Memorandum," Feb. 15, 1952. Case 37727, AT&T Archives.

103. Cooley to Orvil E. Dryfoos, Sept. 21, 1946; E. R. E. to Orvil E. Dryfoos, "Times Telephoto Equipment Progress Report," Nov. 6, 1946, and Dec. 4, 1946, TFC 1941–1959 (4); Schedule 3 "Sales summary & backlog as of 1 Sept. 1953." TFC 1941–1959 (3), TFC archives; Smith, "Word-and-Picture Sending System Shows New Signs of Growth," 1, 6.

104. Jack Drummey, "Westboro Systems Clinics Draw Nationwide Attendance," *Industry*, March 1956. Alden files; M. L. Benson, "Memorandum," June 7, 1955. Case 37727, AT&T Archives; A. G. Cooley, "Memorandum for Directors of TFC," Dec. 4, 1957, TFC 1941–1959 (3), TFC archives.

105. A. G. Cooley to Orvil E. Dryfoos, Sept. 21, 1946; "Times Telephoto Equipment Progress Report," Nov. 6, 1946 and Dec. 4, 1946; draft, "Products Manufactured," Aug. 28, 1953, and "Appraisal of the Company's Future," Aug. 28, 1953, 2–3, TFC 1941–1959 (3), TFC archives.

106. Cooley to Board of Directors, Nov. 13, 1957. 1941–1959 (2); Cooley to Phillip M. McCullough, April 5, 1952, TFC Sale 1952–1962 (4); "Appraisal of the Co's Future," Aug. 28, 1953, TFC (General) 1941–1959 (3); C. Raymond Hulsart to George Woods, Sept. 18, 1953, TFC Sale (4), TFC archives.

107. C. Raymond Hulsart to Nelson, Dryfoos & Cox, Aug. 17, 1953, TFC Sale (4); ESG, "Memorandum for File," Dec. 17, 1958, TFC Sale (3), TFC archives.

108. The contenders included Dictaphone, which entered the fax market later via Stewart-Warner, Western Union, and Haloid (Feb. 27, 1959, contract, TFC Sale 1952–1962 (3) and (4), TFC archives).

109. "Other Sales, Mergers," *New York Times*, July 24, 1956, 33; "Stewart-Warner Acquisition," *Wall Street Journal*, July 25, 1956, 13; G. M. Stamps and H. C. Ressler, "A Very High-Speed Facsimile Recorder," *IRE Transactions of the Professional Group on Communications Systems*, Dec. 1959, 257–63; Daniel M. Costigan, *Electronic Delivery of Documents and Graphics* (New York: Van Nostrand Reinhold Company, 1978), 15.

110. "Times Telephoto Equipment Progress Report," Nov. 6, 1946; "Weather Charts by Wire for Air Force," *Bell Telephone Magazine* 26, 4 (Winter 1947–48), 228–39; George M. Stamps, "A Short History of Facsimile," *Business Communications Review*, June–Aug. 1977, 29.

111. Western Union, *Annual Report 1959*, 7–8.

112. "Memorandum as to Closing under Agreement Dated February 27, 1959," TFC Sale (3), TFC archives; "Facsimile Radio to Tell Weather," *New York Times*, Nov. 7, 1946, 37; A. F. Merewether, "The Use of Weather Facsimile in Airline Operations," Royal Meteorological Society, Canadian Branch, Toronto, Oct. 25, 1951, www.cmos.ca/RMS /r020601.pdf; Will Lissner, "News of Weather Gains as Business," *New York Times*, July 11, 1955, 25.

113. Molesworth Associates, "Alden Electronic will equip US weather stations to receive Tiros II cloud cover photos," press release, Oct. 6, 1960. "AEIREC" press releases & clippings, April 1–Oct. 15, 1960. Alden Archive; "Alden Weather Recorder, *Wall Street Journal*, March 1, 1966, 13.

114. May 21, 1953, note on letter from H. R. Huntley to E. F. Watson, March 3, 1952; R. B. Shanck, "Memorandum for File," March 22, 1955; "Facsimile Service—International News Photos: Memorandum," June 24, 1953. Case 37727, AT&T Archives; Kimbell, "Wire Systems," 148.

115. "News Picture Transmission," *A.N.P.A. Research Bulletin* 104 (May 11, 1955), 19–22; "Facsimile Sales Record Upswing," *Electronics*, April 1955, 16; "UP Pictures Put TV Fans 7 Minutes from Anywhere," Feb. 15, 1955 (clipping). Case 37727, AT&T Archives; Warren H. Bliss, "Advancements in the Facsimile Art during 1956," AIEE Winter General Meeting, New York, Jan. 21–25, 1957, 4.

116. "Facsimile Transmission of Pictures to Newspapers," ANPA *Bulletin*, Nov. 18, 1953, 192; Associated Press, "Domestic Radiophoto Reception," [c. 1954]; William J. McCambridge to Cranston Williams, Nov. 1, 1955, and Nov. 3, 1955. ANPA archives.

117. "Job for Facsimile," *Business Week*, Sept. 15, 1945, 84–88; RCA Laboratories Division, *Research Report 1947* (Princeton, NJ: RCA, 1948); W. A. Phelps, "Memorandum," July 13, 1948. Case 37727, AT&T Archives.

118. G. P. Schleicher, "Memorandum," Dec. 26, 1956. Case 37727, AT&T Archives; Times Facsimile Corp., "Report on Facsimile Transmission," June 26, 1957, TFC (General) 1941–1959 (2), TFC archives.

119. "Report on Facsimile Transmission"; Mitsuo Kaji, "Facsimile Transmission in Japan, Fourth Annual Conference of F.I.E.J. in Paris on November 18th, 1965" (Tokyo: Asahi Shimbun, 1966), 1–7; Warren H. Bliss, "Advancements in the Facsimile Art during 1964," *1965 IEEE International Convention Record* (New York: IEEE, 1965), v. 1, 245.

120. R. B. Shanck, "Memorandum," Jan. 19, 1955. Case 37727, AT&T Archives.

121. Ivan S. Coggeshall and Albert E. Frost, "Modern Telegraphic Communication Systems," *Tele-Tech*, April 1951, 40.

122. "Desk Push-Button Sends Telegrams," *New York Times*, Jan. 24, 1952, 51. Western Union documents also used DeskFax and Deskfax.

123. Joseph Rosenberg, "Summit Inventor in the Spotlight," *Newark Sunday News*, Oct. 23, 1960; Garvice Hyte Ridings, "IEEE Fellow Grade Nomination," March 18, 1964; Garvice H. Ridings, "Garvice H. Ridings" [1970?]. Ridings papers; Garvice Ridings interview, June 9, 1993.

124. A. W. Breyfogel, J. H. Hackenberg, and F. G. Hallden, "Multiline Concentrator Apparatus," *Electrical Engineering* May 1950, 406–08.

125. G. H. Ridings, "Facsimile Transceiver for Pickup and Delivery of Telegrams,"

Electrical Communication 26, 2 (June 1949), 132–34, 137; "Facsimile Raises Telegrams' Speed," *New York Times*, July 8, 1948, 44; "Alumnus in the News," *Virginia Tech Review* [1956], Ridings papers; Mel Most, "Record Communication," *Barron's*, March 9, 1959, 17.

126. "Desk Push-Button Sends Telegrams."

127. "Facsimile Raises Telegrams' Speed"; "Gadget to 'Deliver' Telegrams in Offices," *New York Times*, Feb. 16, 1949, 42.

128. George P. Oslin, *One Man's Century: From the Deep South to the Top of the Big Apple* (Macon, GA: Mercer University Press), 89–90; see also, e.g., "WU Promotes Facsimile for Xmas," *Los Angeles Industrial News*, Dec. 21, 1959. Folder 3, Box 615, Newsclipping book, Oct. 1959–Dec. 1959 Wirefax, Series 19 Public Relations Department Records, 1883–1980, Western Union, Smithsonian Archives.

129. Smith, "Word-and-Picture Sending System Shows New Signs of Growth," 1; "Radio Progress during 1952," *Proceedings of the I.R.E.* 41, 4 (April 1953), 472; "Triumph of a Tin Can," *Western Union Telegraph News*, Sept. 1956, 2; Most, "Record Communication," 17; K. H. Fischbeck and D. A. Rose, "Advancements in the Facsimile Art—1970," *IEEE Comtech Symposium on Advances in Data Communications—1970* (New York: IEEE, 1970), 7.

130. W. R. Greiling, "Memorandum," March 9, 1955. Case 37727, AT&T Archives; C. A. M., "Facsimile Aids Business Communications," *British Communications and Electronics* 2, 6 (June 1955), 48–52; "Desk-Fax Comes to Cuba," *Western Union Telegraph News*, Sept. 1957, 11; "Facsimile Research Spreads," *Electronics*, April 8, 1960, 51; Ridings, "Facsimile Communication," 35.

131. Sidney Feldman, "Facsimile Is Coming of Age," *Electronic Industries* 23, 5 (May 1964), 38.

132. Jules S. Tewlow, "Message Facsimile—A Real-Time Communications Tool," ANPA *R.I. Bulletin* 978, Dec. 31, 1968, 461; Stanford Research Institute, *Facsimile Markets* (Menlo Park, Sept. 1971), 1:84; Costigan, *Electronic Delivery*, 37.

133. "Self-Service Telegraph Office Opened," *New York Herald Tribune*, Dec. 13, 1957. Garvice Ridings scrapbooks, vol. 2.

134. "Facsimile System Leased," *New York Times*, May 24, 1952, 55; J. Z. Millar to E. F. Watson, Oct. 7, 1952. Case 37727, AT&T Archives; Western Union press release, Aug. 13, 1959. File 1, Box 620, Series 19, Western Union, Smithsonian Archives; Most, "Record Communication," 17–20.

135. Western Union press release, "First six months of 1959," Aug. 15, 1959. File 1, Box 620; "New Telex Service Due in Denver Soon," *Denver Post*, Sept. 15, 1959. Folder 3, Box 615; "Western Union Products and Services," Aug. 1962. Folder 3, Box 605, Series 19, Western Union, Smithsonian Archives; "Western Union Net Fell in First Half from '61," *Wall Street Journal*, Aug. 6, 1962, 8; "Facsimile Will Be a Versatile and Profitable Communications Service—Only If and After It Succeeds!" *The Knowledge Report* 1, 11 (Oct. 23, 1967), 2.

136. "Facsimile Development," 8–9 [early 1951]. File 9, Box 2, Series 1, Western Union, Smithsonian Archives; Fred Shunaman, "Telecar Speeds Telegrams," *Radio-Electronics*, July 1951, 22–24.

137. Ridings, "Facsimile Communication," 34.

138. Ibid., 1, 4.

139. AT&T declared its technological foci as television transmission, mobile service, automatic answering machines, speaker phones, and television phones (Harold S. Osborne, "Communication Sets Its Sights Ahead," *Bell Telephone Magazine*, Summer 1952, 68–70).

140. W. S. Michel, "Coded Facsimile vs. Conventional Facsimile: A Comparison of Estimated Transmission Rates," Aug. 3, 1956, 1. Case 38763, AT&T Archives.

141. E.g., in 1955, the Air Material Command considered creating its own network (W. R. Greiling, "Memorandum," Feb. 4, 1955). See also Strategic Air Command, "What we are looking for," March 29, 1954, attached to E. M. Mapes to G. W. Gilman, April 9, 1954. Case 37727, AT&T Archives.

142. C. Raymond Hulsart to E. S. Greenbaum, April 29, 1954. TFC: Sale 1952–1962 (4), TFC archives; WJD, "Memorandum," Feb. 15, 1952; Doren Mitchell, "Bell System Facsimile—Notes on Present Status," Feb. 28, 1957; M. L. Benson, "Memorandum," May 13, 1957. Case 37727, AT&T Archives.

143. "Facsimile Proves Successful on Microwave Debut," *Petroleum Engineer*, Dec. 1955, D-52.

144. "Facsimile Helps Army on Weather," *New York Times*, Feb. 8, 1948, 21; "Weather Maps Transmitted by Facsimile in New System," *Electrical Engineering*, Nov. 1953, 1042–43; Crooks, "Facsimile Systems as an Aid to Research," 41.

145. R. Foulkrod to J. J. Pilliod, Dec. 28, 1951; J. J. Pilliod to R. Foulkrod, Feb. 8, 1952; WJD, "Memorandum," Feb. 15, 1952; May 21, 1953, handwritten note on H. R. Huntley to E. F. Watson, March 3, 1952. Case 37727, AT&T Archives.

146. Frank Chapman, "TVA Using Facsimile over Microwave Circuit," *GAS* 31, 10 (Oct. 1955), 219–20.

147. W. R. Greiling, "Memorandum," Feb. 3, 1955, and Feb. 15, 1955; R. B. Shanck, "Memorandum," Feb. 17, 1955. Case 37727, AT&T Archives.

148. RCA Laboratories Division, *Research Report 1950* (Princeton, NJ: RCA, 1951), 109–10; "Facsimile Reproduction for AEC Library Service," *Electronics* 24, 3 (March 1951), 222–26; "Current Topics," *Journal of the Franklin Institute*, March 1951, 384; R. B. Shanck, "Memorandum," Feb. 17, 1955. Case 37727, AT&T Archives.

149. "Appraisal of the Company's Future," Aug. 28, 1953, 4. TFC (General) 1941–1959 (3), TFC archives.

150. H. A. Rhodes, "Memorandum," Nov. 22, 1954; R. B. Shanck, "Memorandum," Feb. 17, 1955; W. R. Greiling, "Memorandum," March 9, 1955. Case 37727, AT&T Archives; "Facsimile System Speeds Interplant Communications," *Iron Age*, May 26, 1955, 100–102; D. G. Baird, "Production Control by Teletype," *Mill & Factory*, April 1946, 115–19.

151. D. B. Perry, "Memorandum for File," Jan. 31, 1952; E. F. Watson, "Memorandum," Feb. 13, 1952; R. B. Shank, "Memorandum," Nov. 5, 1952, and Nov. 7, 1952; "Excerpt from *Industrial Communications*," March 8, 1957. Case 37727, AT&T Archives.

152. H. A. Rhodes, "Memorandum," Nov. 19, 1954, and Nov. 22, 1954; C. A. Barlett to W. K. MacAdam, Jan. 12, 1955; R. L. Farr, "Minutes," Feb. 1, 1955; M. L. Benson, "Memorandum," April 5, 1955. Case 37727, AT&T Archives.

153. A. Tradop to K. F. Morgan, Dec. 28, 1951; G. W. Gilman to F. A. Cowan, Sept. 10, 1954. Case 37727, AT&T Archives.

154. Lt. Col. Charles A. Green to R. G. Osborne, Feb. 3, 1954; Strategic Air Com-

mand, "Projected Graphic Communications System" [Feb. 3, 1954] attached to E. M. Mapes to G. W. Gilman, April 9, 1954; W. R. Greiling, "Memorandum," Jan. 7, 1955. Case 37727, AT&T Archives.

155. R. B. Shanck, "Memorandum for File," April 23, 1954, "Memorandum for File," Feb. 17, 1955, and "Memorandum for File," March 22, 1955. Case 37727, AT&T Archives; C. Raymond Hulsart to E. S. Greenbaum April 29, 1954, TFC: Sale 1952–1962 (4), TFC archives.

156. C. W. Smith to F. J. Singer, June 28, 1955; Frank A. Cowan to R. L. Helmreich, May 2, 1955; M. L. Benson, "Memorandum," May 13, 1957. Case 37727, AT&T Archives; TFC, "Report on Facsimile Transmission," June 26, 1957, TFC (General) 1941–1959 (2), TFC archives.

157. C. W. Smith to F. J. Singer, June 28, 1955. Case 37727, AT&T Archives.

158. R. B. Shanck, "Memorandum," May 4, 1955, revised June 2, 1955. Case 37727, AT&T Archives.

159. R. B. Shanck to V. N. Vaughan et al., April 20, 1955, and "Proposed Facsimile Service (Preliminary Estimate)," April 20, 1955. Case 37727, AT&T Archives.

160. M. I. Uchenick, "Memorandum for File," Dec. 6, 1956; Doren Mitchell, "Bell System Facsimile—Notes on Present Status," Feb. 28, 1957; M. L. Benson, "Memorandum," May 13, 1957. Case 37727, AT&T Archives.

161. M. L. Benson, "Memorandum," April 21, 1955; M. L. Benson, "Proposed Facsimile Service Charges for Service (Preliminary Estimate)," April 26, 1955; R. B. Shanck, "Memorandum," May 4, 1955, revised June 2, 1955; H. L. Benson, "Conference Notes," April 5, 1957. Case 37727, AT&T Archives.

162. "Excerpt from Industrial Communications"; M. L. Benson, "Memorandum," May 13, 1957. Case 37727, AT&T Archives.

163. Peter Temin with Louis Galambos, The Fall of the Bell System (Cambridge: Cambridge University Press, 1987), 15. I thank former AT&T archivist Sheldon Hochheiser for this observation.

164. Richard B. Le Vino, "Facsimile Equipment Development in the Signal Corps," Signal, July–August 1954, 33; "Signal Corps Board Status Reports, 1 July–31 Dec. 1954," 1. Box Signal Corps Board, 1924–1962, SCEL archives.

165. "Experimental Facsimile Service," SCEL Journal, Sept. 16, 1954, 3; "Experimental Facsimile Service for Signal Corps Supply Agency," SCEL Journal, Nov. 26, 1954, 5–6; "Experimental Facsimile Service," SCEL Journal, April 21, 1955, 3.

166. H. A. Rhodes, "Memorandum," Nov. 22, 1954; W. R. Greiling, "Memorandum," Jan. 7, 1955. Case 37727, AT&T Archives. For a broader military perspective, see James M. Wheeler, "The Paperwork Processing Dilemma," Air University Review 18, 2 (Jan.– Feb. 1967), 84–90.

167. Closed-circuit television proved more expensive (F. K. Becker, J. R. Hefele, and W. T. Wintringham, "An Experimental Visual Communication System," Bell System Technical Journal 38, 1 [Jan. 1959], 141–76).

168. "Facsimile system helps eliminate duplicate files at branch offices," Bank Equipment News, April 1967, reprint; "A New Development in Signature Verification: Alden Electronics' 'Signa Fax' Bridges the Gap," Banker & Tradesman, Feb. 8, 1978, S/S. Alden archive; Larry Farrington, interview, March 17, 1993.

169. Alden Electronics, *Annual Report Fiscal Year 1963*, 15.

170. "Reynolds Metals Shows Interstate Network to Transmit Facsimiles," *Wall Street Journal*, March 20, 1957, 5; Joseph Rosenberg, "Summit Inventor in the Spotlight," *Newark Sunday News*, Oct. 23, 1960.

171. "Speed Charlotte News," *Western Union Telegraph News*, Sept. 1957, 11; "News Picture Transmission," A.N.P.A. *Research Bulletin* 104, May 11, 1955, 24.

172. C. A. M., "Facsimile Aids Business Communications," 48–52.

173. Archie S. Hill "Letterfax Equipments for Service Flexibility," *Western Union Technical Review* 12, 1 (Jan. 1958), 14–20; Douglas M. Zabriskie and William F. Moore, "A Message Accumulator for Letterfax Recorders," *Western Union Technical Review* 14, 1 (Jan. 1960), 31–32.

174. "Another Use for 'Intrafax,'" *Wall Street Journal*, May 19, 1954, 5.

175. Stanley W. Penn, "Electronic Courier: Facsimile Snares New Customers with Added Speed and Reliability," *Wall Street Journal*, Nov. 30, 1959, 1, 15.

176. Paul J. C. Friedlander, "No More Queues," *New York Times*, Jan. 30, 1955, 21; "Western Union Introduces Ticketfax," *Public Utilities Fortnightly*, Feb. 3, 1955, 148; "Facsimile for Industrial Ticket Sales," *Railway Age*, Sept. 17, 1956, 25; "Reservations Sold on Sight," *Railway Age*, Sept. 17, 1956, 24–25.

177. R. B. Shanck, "Memorandum," Feb. 17, 1955. Case 37727, AT&T Archives; Ridings, "Facsimile Communication," 34.

178. "Is This Facsimile's Year?" *Electronics business edition* 30, 5A (May 10, 1957), 15–16; "Reynolds Metals Shows Interstate Network to Transmit Facsimiles," *Wall Street Journal*, March 20, 1957, 5.

179. Cooley to Board of Directors, Nov. 13, 1957, 3. TFC (General) 1941–1959 (2), TFC archives.

180. Hugo Gernsback, "Automated Electronic Newspaper," *Radio-Electronics* 34, 3 (March 1963), 31, 74.

181. Penn, "Electronic Courier," 15; "Western Union to Build Weather Map Facsimile Network for Air Force," *Wall Street Journal*, July 24, 1958, 7; Most, "Record Communication," 3, 15–22; "First six months of 1959," press release, Aug. 15, 1959. Folder 1, Box 620, Series 26, Western Union, Smithsonian Archives.

182. Felix Belair Jr., "Post Office Testing Fast Facsimile Mail," *New York Times*, Oct. 28, 1959, A1, 26.

183. Edward Gamarekian, "Mail Delivery at Speed of Light; A Satellite as a 'Mailman,'" *Washington Post*, Nov. 30, 1958, E1.

184. "Minutes of a Meeting between the Members of President Eisenhower's Cabinet, the Post Office Department Advisory Board, and the Departmental Top Staff in the Office of the Postmaster General, Tuesday, November 10 at 10:30 A.M." Nov. 12, 1959. Box 1, Facsimile Mail 1959–6, Management Systems Division. Subject Files 1957–1969, Post Office RG 28, NARA (hereafter, Facsimile Mail, NARA).

185. W. A. Sponsler to Mr. McKibben, Oct. 29, 1959. Facsimile Mail, NARA.

186. TRS, "Tentative Outline for Discussion of Speed Mail," draft, Feb. 17, 1961. Facsimile Mail, NARA.

187. Post Office press release, "United States Post Office Department Continues Study of Electronic Mail," Jan. 26, 1960. 1960. File 3350, Box 16, Facsimile Mail, NARA.

188. "Preliminary Report of the Ad Hoc Cabinet Committee on Facsimile Mail" [Sept. 1959?], 1–3. Facsimile Mail, NARA; Bert Mills, "Sales Orders Bounced off a Satellite," *Sales Management*, Nov. 18, 1960, 52.

189. Mr. McKibben, "Electric Mail," Nov. 6, 1959. Facsimile Mail, NARA.

190. Felix Belair Jr., "Post Office Testing Fast Facsimile Mail," *New York Times*, Oct. 28, 1959, A1, 26; "Telephone and Telegraph," *Public Utilities Fortnightly*, Nov. 24, 1960, 827.

191. "Western Union Sets Facsimiles," *New York Times*, Oct. 27, 1959, 49; "Western Union Wins Letter-by-Wire Race," *Washington Evening Star*, Dec. 1, 1959. Folder 3, Box 615, Newsclipping book, Oct.–Dec. 1959. Wirefax, Series 19, Western Union, Smithsonian Archives.

192. New Products and Services Group, Planning Department, "Western Union Products and Services," Aug. 1962. Folder 3, Box 605. Series 19, Western Union, Smithsonian Archives.

193. Mr. McKibben to Postmaster General, Jan. 8, 1960. Facsimile Mail, NARA.

194. ITT Laboratories, *Final Report. Study Program for Speed Mail System* (Nov. 1959); Mr. McKibben to Postmaster General, Dec. 14, 1959; [W. W.] Sullivan to W. A. Sponsler, "Speed Mail Evaluation," Dec. 17, 1959; Mr. Plummer to Brawley, "Descriptive Summary of the Speed Mail Project," Jan. 26, 1961; Mr. Plummer (Rex Landis) to Mr. Treanor, Feb. 16, 1961. Facsimile Mail, NARA; "Facsimile Research Spreads," *Electronics*, April 8, 1960, 51; Willard Clopton, "Letter to Ike Tests Latest Mail System," *Washington Post*, Nov. 2, 1960, B1; "Speed Mail," *Signal*, Jan. 1961, 16–17.

195. W. A. Sponsler to Mr. McKibbin, Dec. 14, 1960; Arthur E. Summerfield to Harold S. Geneen, Jan. 17, 1961; "Descriptive Summary"; Robert E. Miller to John Trainor [*sic*], "Speed Mail Facsimile System," Feb. 8, 1961; "History of the Speed Mail Project" [May 1961?], 6–7. Facsimile Mail, NARA.

196. Cabinet Ad Hoc Committee on Facsimile Mail, "Minutes of the November 1, 1960 Meeting." Facsimile Mail, NARA.

197. Walter P. Marshall to Arthur E. Summerfield, Nov. 22, 1960; Arthur E. Summerfield to Walter P. Marshall, Dec. 6, 1960; "History of the Speed Mail Project," 8; Walter P. Marshall to J. Edward Day, Dec. 19, 1960. Facsimile Mail, NARA.

198. Jerry Kluttz, "The Federal Diary," *Washington Post*, March 22, 1961, B1; "Electronic Mail Plan Abandoned," *Washington Post*, Aug. 9, 1961, A4.

199. "History of the Speed Mail Project," 7. Facsimile Mail, NARA.

200. Ridings, "Facsimile Communication," 36.

201. "RSMA Hears Facsimile Role," *Railway Age*, April 3, 1961, 33; "High Speed Low Cost Facsimile System," *Modern Communications*, June–Aug. 1962, 21; Ridings, "Facsimile Communication," 37.

202. CCITT, *XVIIth Plenary Assembly, Geneva, 4–12 October 1954*, vol. 6, *Operating & Tariffs* (Geneva: ITU, 1955), 113.

203. CCITT, *Documents of the XVIIIth Plenary Assembly (Geneva, 1956)* vol.1, *Documents* (Geneva: ITU, 1957), 32.

204. Germany, "Subscribers' facsimile service," 14 Jan. 1959, Contribution 44. ITU.

205. Germany, "Conditions for facsimile telegraph transmission other than picture telegraphy," 29 Oct. 1962. Contribution 13. ITU.

206. Feldman, "Facsimile Is Coming of Age," 38–39.

207. Kobayashi, "Looking Back on the Institute of Facsimile Engineers," 302–6, 565–66.

208. Federation of Japan Electric Communication Industrial Associations, *Communication Equipments of Japan 1956* [*sic*] (Tokyo: FJECIA, 1956), 11; The Electronics Association, *Electronics in Japan '60* (Tokyo: EA, 1961), 60, and *Electronics in Japan '61* (Tokyo: EA, 1962), 61; "Fakushimiri ryou daisuu shuukei hyou" [Compilation of all facsimile machines], *Gazo denshi gakkaishi* [Journal of the Institute of Facsimile Engineers of Japan] 14, 3 (Nov. 1967), 78.

209. *Communication Equipments of Japan 1956*, 13; "News Flash," *Japan Telecommunication Review* 1, 1 (Jan. 1959), 58; Electronics Association, *Electronics in Japan '60*, 7, and *Electronics in Japan, 1976–77* (Tokyo: EA, 1977), 18, 59–60; Kubota, "Development of facsimile in our country," 52.

210. CIAJ, *Tsushin kiki chuki juyo yosoku* [Communications Equipment Demand Forecast] (Tokyo: CIAJ, 1984), 118.

211. "Fakushimiri ryou daisuu shuukei hyou" [Compilation of all facsimile machines], 78.

212. Osamu Murakami, *Fakushimiri: Atarashii Tsuushin Medaia* [Facsimile: New Communication Media] (Tokyo: Denkii Tsushin Gyijyutsu Niisusha, 1981), 224.

213. Warren H. Bliss, "Advancements in the Facsimile Art during 1966," *1967 IEEE International Convention Record* (New York: IEEE, 1967), 5:54–56; SRI, I, 107-08, 142–43.

214. V. N. Amarantov and G. B. Davydov, "New Facsimile Equipment," *Telecommunications [Elektrosvyaz']* 1958, 9, 947–56.

CHAPTER 4: **The Sleeping Giant Stirs, 1965–1980**

Epigraph. Frost & Sullivan, *Facsimile Equipment and Systems in the U.S.A.* (New York: Frost & Sullivan, May 1977), 137.

1. Sarah Weddington, *A Question of Choice* (New York: Grosset/Putnam, 1992), 145–50.

2. Stanford Research Institute, *Facsimile Markets* (Stanford: SRI, Dec. 1971), I, 3–4.

3. Robert B. Horwitz, *The Irony of Regulatory Reform. The Deregulation of American Telecommunications* (New York: Oxford University Press, 1989), 230–32.

4. Kevin J. Sroub to J. M. Lewis, Nov. 21, 1968. Robert Wernikoff papers.

5. E.g., ITT and RCA GlobComm objections blocked FCC approval of Graphnet Systems' proposal to link fax and data terminals between Europe and the United States for two years ("New Data Service Cleared by FCC For Four Carriers," *Wall Street Journal*, Jan. 12, 1977, 13; "Graphic Scanning Unit Cleared on Trial Service Between U.S., Europe," *Wall Street Journal*, April 27, 1979, 15).

6. James C. Tanner, "Sales Spurt in Data Transmission Gear Forecast Due to Carterfone Case Ruling," *Wall Street Journal*, Nov. 13, 1968, 8; "ATT Seeks Adoption of Safety-Device Plan on Other Makers' Gear," *Wall Street Journal*, July 22, 1975, 12; Phil Hirsch, "The Long Wait Could Be Longer," *Datamation* 21,12 (Dec. 1975), 102-03; "Standards Relaxed for Gear Connected to Phone Network," *Wall Street Journal*, March 17, 1976, 17.

7. "Link Asked by Western Union International of U.S., Foreign Phones," *Wall Street Journal*, July 17, 1972, 6; "FCC Will Allow Data To Be Sent Overseas Via Telephone Lines," *Wall Street Journal*, Jan. 9, 1976, 6; Jeffrey A. Tannenbaum, "An Era of Cut-Rate

Competition Arrives as Firms Enter International Telex Field," *Wall Street Journal*, Aug. 8, 1979, 40.

8. Facsimile Rapporteurs' Group (SG I), "Extracts from the meeting report (Bern, January 1979)," March 1979, 75. International Telecommunications Union Library, Geneva (hereafter, ITU).

9. The CCITT worked on a four-year cycle. A plenary conference assigned questions to Study Groups whose Working Groups wrote Recommendations, based on contributions submitted by PTTs and other interested parties. The next plenary conference then debated and voted on the new proposals (George A. Codding, "Evolution of the ITU," *Telecommunications Policy* 15 [Aug. 1991], 271–85).

10. F. Bardua, "The Telefax Service in the Federal Republic of Germany," in CCITT, *Proceedings of the First CCITT Symposium on New Telecommunication Services*, Geneva, May 14–16, 1979 (Geneva: ITU, 1979), 216. See also Rank Xerox, "Proposal for amendment to Recommendation T.2," Sept. 1972, 7/XIV, Contribution 28. ITU.

11. TR29.2 Subcommittee on Facsimile Equipment, "Minutes," June 13, 1973 and Oct. 10, 1973. Telecommunications Industry Association archives; Dennis Bodson, Kenneth R. McConnell, and Richard Schaphorst, *FAX: Digital Facsimile Technology and Applications*, 2nd ed. (Boston: Artech House, 1992), 13, 299–302.

12. SG XIV, "Final Report to the VIth Plenary Assembly," May 1976. Contribution 35, 35–42; Correspondent of SG XIV, "Survey of the work of SG XIV," Jan. 1979. Contribution 155. ITU.

13. Special Rapporteur for the study of Group 2 machines (Recommendation T2), "Interim Report," July 1975, Contribution 30. ITU; Jack Needham, "The Improving Image of Facsimile," *Business Equipment Digest*, June 1976, 26; Robert Krallinger, "Evolving Business Fax Standards Mean Widespread Compatibility," *Communication News*, Sept. 1978, 58–59; Susanne K. Schmidt and Raymund Werle, *Coordinating Technology: Studies in the International Standardization of Telecommunications* (Cambridge, MA: MIT Press, 1998), 210–12.

14. SRI, I, 131; Phil Hirsh, "Products and FCC's Wiley at ICA Show," *Datamation* 22 (June 1976), 155.

15. G. H. Ridings, "Facsimile Communication Past, Present, Future," *Signal*, Nov. 1962, 37–39; Sidney Feldman, "Facsimile is coming of age," *Electronic Industries* 23, 5 (May, 1964), 39; Garvice H. Ridings, "Facsimile Imaging Systems," *Western Union Technical Review*, April 1965, 22–23; "Portable Facsimile Device," *Mechanical Engineering* 88, 6 (June 1966), 58–59; SRI, I:18.

16. SRI, I, 34–36; Michael J. Edmands, "Burst of Speed," *Barron's*, Aug. 17, 1970, 11; Roy J. Bruun, "On the Technical Side: Facsimile Systems Update — Is There a Dilemma for Users?" *Infosystems* 21, 7 (July 1974), 3; Ronald Brown, *Telephone facsimile for Business: A Research Report on Systems, Equipment, Costs, Advantages and Markets* (Somerset: Ronald Brown, 1975), 53.

17. E. F. O'Neill, ed., *A History of Engineering and Science in the Bell System. Transmission Technology (1925–1975)* (Murray Hill, NJ: Bell Laboratories, 1985), 708.

18. David H. Axner, "The Facts about Facsimile," *IEEE Data Processing Magazine*, May 1968, 44–49.

19. Kaveh Pahlavan and Jerry L. Holsinger, "Voice-Band Data Communications

Modems—A Historical Review: 1919–1978," *IEEE Communications Magazine* 26, 1 (Jan. 1988), 16.

20. Stuart Zipper, "Data Com, Distributed EDP Push Modems toward Record Year," *Electronic News*, March 14, 1977, 1, 72–76; Robert Gallager, "A Conversation with G. David Forney Jr.," *IEEE Information Theory Society Newsletter*, June 1997, http://golay.uvic.ca/publications/nltr/97_jun/jdvcon.html.

21. E. R. Kretzmer, "Facsimile Communication Via the Dialed Telephone Network," *1965 IEEE International Convention Record*, vol. I, "Wire and Data Communication" (New York: IEEE, 1965), 138.

22. Feldman, "Facsimile is Coming of Age," 41; "Facsimile Will Be a Versatile and Profitable Communications Service—Only If and After It Succeeds!" *The Knowledge Report* 1,11 (Oct. 23, 1967), 3.

23. "Data Communication," *Wall Street Journal*, Jan. 11, 1973, 1.

24. David Spencer, "Facsimile Data Compression," *1971 IEEE International Convention Digest* (New York: IEEE, 1972), 332–33; Bruno J. Vieri, "Data Processing Techniques for Message Facsimile Equipment," *Communication Systems and Technology Conference*, April 30–May 1, 1974 (New York: IEEE, 1974), 112–15.

25. Robert Wernikoff, "Long-Distance Copiers," July 1968. Wernikoff papers.

26. SRI, I, 129; "Qwip Hits Fax Problems Head-on—Connection Ups the Ante," *EMMS*, Feb. 15, 1980, 9.

27. Gideon Haigh, *The Office: A Hardworking History* (Melbourne, Australia: The Miegunyah Press, 2012), 230.

28. Gary Jacobson and John Hillkirk, *Xerox: American Samurai* (New York: Macmillan, 1986), 63; Joseph Mort, "Xerography: A Study in Innovation and Economic Competitiveness," *Physics Today* 47, 4 (April 1994), 32.

29. "Facsimile Mail," *Wall Street Journal*, Sept. 24, 1971, 21; "The Outlook for Facsimile Equipment," Arthur D. Little Industry Comment Letter, Feb. 25, 1974, 5; Harold P. Belcher, "A Survey of the Factors Affecting the Quality of Data Compressed Black/White Facsimile," (Cambridge, MA: Arthur D. Little, Inc., Aug. 1976); "Postal Service's Role in Transmitting Mail by Electronics Studied," *Wall Street Journal*, Dec. 18, 1978, 10.

30. "FCC Asserts Claim to Regulate Certain Electronic Messages," *Wall Street Journal*, March 14, 1979, 22; "U.S. Postal Service Urged Not to Restrict Electronic-Communication Competition," *Wall Street Journal*, March 15, 1979, 7.

31. Table Dg8-21. Domestic telegraph industry-messages, wire, offices, employees, and finances: 1866–1987, *Historical Statistics of the United States: Millennial Edition Online* hsus.cambridge.org/HSUSWeb/search/searchTable.do?id=Dg90-102; Roger W. Benedict, "Western Union Chief Strives to Maintain Wire Service While Building New Systems," *Wall Street Journal*, June 30, 1969, 28; Tannenbaum, "An Era of Cut-Rate Competition Arrives," 40; P. Corbishley and P. G. Hunt, "Telecommunication," *Electronics & Power*, Sept. 1980, 741; Chris Drewe, "BT's Telex Network: Past, Present—and Future?" *British Telecommunications Engineering* 12, 1 (April 1993), 17–18; Anton A. Huurdeman, *The Worldwide History of Telecommunications* (Hoboken, NJ: Wiley-IEEE Press, 2003), 512–13.

32. Axner, "Facts about Facsimile," 44.

33. "RSMA Hears Facsimile Role," *Railway Age*, April 3, 1961, 33; Howard M. Ander-

son, "Fast Fax Coming on Fast," *Infosystems* 22, 3 (March 1975), 44; Sharon L. Shanahan, "Facsimile Joins the Computer Age," *Infosystems* 29, 9 (Sept. 1982), 130–31.

34. Auerbach, *Guide to Facsimile Equipment* (Philadelphia: Auerbach Publishers, 1975), 31.

35. Donald White, "Alden Challenges the Giant," *Boston Globe*, June 1, 1966, 21.

36. Feldman, "Facsimile is Coming of Age," 41; Erik M. Pell, *From Dream to Riches— The Story of Xerography* (Rochester, NY: Erick M. Pell, 1988), 117–18.

37. L. D. Rolland, "Xerox Rents 'LDX,'<N>" *Financial World*, May 20, 1964, 20; Axner, "Facts about Facsimile," 51–52; John H. Dessauer, *My Years with Xerox: The Billions Nobody Wanted* (New York: Doubleday & Co., 1971), 179–80.

38. SRI, I, 45, 70, 72, 83.

39. George M. Stamps, "A Short History of Facsimile," *Business Communications Review*, June–Aug. 1977, 30; George Moreland Stamps, "Pathfinders: Facsimile," Communications Networks '90 Conference, Washington, DC, Feb. 7, 1990, 3.

40. "Portable Facsimile Device," *Mechanical Engineering* 88, 6 (June 1966), 58; "Facsimile Will Be a Versatile and Profitable Communications Service," 2.

41. Tom Alexander, "Lots of Talk, Not Enough Fax," *Fortune*, Feb. 1973, 124.

42. "'Talking Machine'" Telephones Pictures Coast to Coast," *Machine Design*, May 26, 1966, 12; "Portable Facsimile Device," 58–59.

43. "Xerox to Manufacture Telecopier Transceiver with Magnavox License," *Wall Street Journal*, July 21, 1967, 3; William D. Smith, "Xerox, No Copycat, Plans New Devices," *New York Times*, May 20, 1970, 61; F&S, 1977, 11–12.

44. Magnavox Company, *1968 Annual Report*, 16; SRI, I, 114; Paul Plansky, "Magnavox to Drop Fax; Loss $3M after Taxes," *Electronic News*, Nov. 13, 1972, 1, 24; Paul Plansky, "New York Group Renews Magnafax Bid," *Electronic News*, March 19, 1973, 31; "Justice: No Bar to 3M Buying Magnavox Fax Line," *Electronic News*, April 23, 1973, 12.

45. "The Revolution in Office Copying," *Chemical and Engineering News*, July 13–20, 1964, 116; Auerbach, *Guide to Facsimile Equipment*, 77–79; F&S, 1977, 25–28, 91; Frost & Sullivan, *The Facsimile Equipment Market in the United States* (New York: Frost & Sullivan,1982), 62.

46. "Facsimile Will Be a Versatile and Profitable Communications Service," 1–3; "Facsimile Rapidly Enhancing Market Position," *Electronic News*, May 15, 1967, 35.

47. Harris, Upham & Co., "Facsimile—Next Office Machine Boom," *Science and Securities* 8 (April 1968), 6–8; Charles J. Elia, "Heard on the Street," *Wall Street Journal*, April 22, 1968, 33.

48. Robert Wernikoff, speech to Addressograph-Multigraph Annual Stockholders Meeting, Nov. 2, 1967. Wernikoff papers.

49. Alexander, "Lots of Talk," 122; Xerox, *1968 Annual Report*, 3.

50. Harris, Upham & Co., "Facsimile," 6.

51. Edmands, "Burst of Speed"; "Business Letter," *New York Times*, Feb. 7, 1971, 3,3; SRI, I, 5, 103–19; 'New Bell Charge, Carrier Plans Seen Helping Facsimile," *Electronic News*, June 4, 1970, 2, 18; "Business Letter," *New York Times*, Feb. 7, 1971, F3.

52. John Brooks, *The Go-Go Years* (New York: Weybright and Talley, 1973), 279, 183–84.

53. Robert Metz, "Market Place," *New York Times*, Nov. 30, 1971, 66; Dan Dorfman, "Heard on the Street," *Wall Street Journal*, Dec. 13, 1971, 27.

54. W. T. Johns to D. W. Seager, Vice President of Manufacturing, Addressograph-Multigraph, March 5, 1970. Wernikoff Archives.

55. Alexander, "Lots of Talk," 122.

56. Graphic Sciences, Inc., S-1 Registration Statement, Securities and Exchange Commission, June 30, 1967, 5; "Graphic Sciences Registers Initial 401,000 Share Issue," *Wall Street Journal* July 5, 1967, 19; "Graphic Sciences, Inc. Investors Find Out Speculation Can Pay," *Wall Street Journal* ,Sept. 13, 1967, 26; "Graphic Sciences Offer of Debentures Sold Out," *Wall Street Journal*, April 19, 1968, 27; "Graphic Sciences Offer Sells Out," *Wall Street Journal*, March 18, 1970, 37; SRI, I, 114–16; "Over-the-Counter Markets," *Wall Street Journal*, Oct. 10, 1974, 28.

57. E.g., James McCreath, the Xerox Senior Engineer in charge of LDX production engineering became the Director of Manufacturing (Graphic Sciences, S-1 Registration Statement, 5, 9–10). See also "Document Is Sent by Telephone," *New York Times*, July 13, 1968, 31.

58. The firm claimed, "The only components left from the original design are the photodiode and the light bulb." ("Picture This: Point-to-point Paper from the Next Big-time Black Box," *Electronics*, Nov. 24, 1969, 100).

59. Albert L. C. Chu, "Facsimile: Business Gets the Message," *Business Automation* 18, 10 (Aug. 1971), 25; Eric von Hippel interview, Nov. 7, 1996; Hiro Hiranandani, interview, Feb. 14, 2002.

60. F&S, 1977, 13, 154.

61. Graphic Sciences, Inc., *Annual Report for 1973*, 2 and *Annual Report for 1974*, 1–2, 15; Burroughs Corporation, "Prospectus and Proxy Statement," Jan. 21, 1975, 6, 32). See also Graphic Sciences, Inc. v. International Mogul Mines, Ltd. et al. Civil Action No. 74-1188, United States District Court for the District of Columbia, 397 F. Suppl. 112; 1974 U.S. Dist. LEXIS 6212; Fed. Sec. L. Rep. (CCH) P94,834, Oct. 18, 1974.

62. Burroughs, "Prospectus and Proxy Statement," 7–8, 31; Hirandani interview; "Burroughs Reports," *Burroughs Clearing House* 59, 9 (June 1975), 30–31; "Burroughs Creates Office Products Group," *Burroughs Line* 19, 1 (1976), 1.

63. SRI, I, 112.

64. "Companies Working on Facsimile Systems," Febr. 16, 1961. Box 1. Facsimile Mail 1959–61. Management Systems Division. Subject Files 1957–1969. Post Office. Record Group 28. NARA.

65. International Resource Development, Inc., *Facsimile Markets* (Norwalk, CT: IRDI, July 1981), 36, 66–67.

66. "Mail by Phone Picks up New Speed," *Business Week*, Dec. 12, 1969, 120–22; SRI, I, 84; Auerbach, *Guide*, 63–64; Hirsh, "Products and FCC's Wiley at ICA Show," 150–52; F&S, 1977, 47–48; F&S, 1982, 84.

67. White, "Alden Challenges the Giant," 21.

68. John M. Alden letter to stockholders, June 1966. Alden files; "Business Automation Showcase," *Business Automation* 18, 6 (April 1971), 42; SRI, I, 103–4; Auerbach, *Guide*, 23; "Alden 600 order dispatch system" [c. April 1974]; George F. Stafford and John P. Carlson, "Small-Form Facsimile Is More Efficient," *Communication News*, Jan. 1977, Alden archives; F&S, 1977, 54.

69. "Litcom Prepares to Market Japanese Facsimile System," *Electronic News*, Aug. 26, 1968, 35; "Moving Images a Japanese Way," *Business Week*, Aug. 29, 1970, 30;

Bernard M. Rosenheck, "FASTFAX, a Second Generation Facsimile System Employing Redundancy Reduction Techniques," *IEEE Transactions on Communication Technology* Com-18, 6 (Dec. 1970), 772–79; SRI, I, 65, 70; Charles W. Savory, "Tomorrow's Communications—Today," *Signal* 29, 3 (Nov.– Dec. 1974), 22–26.

70. FCC, *37th Annual Report for Fiscal Year 1971* (Washington, DC: GPO, 1972), 72; Western Union Telegraph Company, *1970 Annual Report*, 12; "Western Union Files Proposed Rate Plan For Mailgram Service," *Wall Street Journal*, Dec. 2, 1969, 20; International Resource Development, Inc., *Facsimile Markets* (Norwalk, CT: IRDI, Oct., 1985), 128, 198.

71. Wayne E. Green, "FCC Ponders Setting Standards of Speed for Delivery of Western Union Telegrams," *Wall Street Journal*, Oct. 18, 1968, 36; Roger W. Benedict, "Western Union Chief Strives to Maintain Wire Service While Building New Systems," *Wall Street Journal*, June 30, 1969, 28; "Western Union to Try Message Counters in Supermarkets," *Wall Street Journal*, Sept. 17, 1969, 12; "Telegraph Service Deregulation Urged by Federal Report," *Wall Street Journal*, June 26, 1978, 4.

72. "Facsimile Will Be a Versatile and Profitable Communications Service," 2.

73. Interview with Garvice H. Ridings, June 9, 1993; George P. Oslin, *One Man's Century: From the Deep South to the Top of the Big Apple* (Macon, GA: Mercer University Press), 89–90.

74. Western Union Telegraph Co., *1968 Annual Report*, 12; Jules S. Tewlow, "Message Facsimile—A Real-Time Communications Tool," ANPA *R.I. Bulletin* 978, Dec. 31, 1968, 485–86; F&S, 1977, 117.

75. SRI, I, 103–19.

76. Electronic Industries Association, *Electronic Market Data Book 1971* (Washington, DC: Electronic Industries Association, 1971), 39–40.

77. "Facsimile Systems," *The Office*, July 1970, 55–64; Chu, "Facsimile," 20.

78. H. G. Morehouse, *Telefacsimile Services between Libraries with the Xerox Magnavox Telecopier* (Reno: University of Nevada Library, Dec. 20, 1966), 13.

79. AT&T, "Rate changes through the years," 1999. Box 04 11 01, AT&T Archives; "WU Files Proposed Rate Plan for Mailgram Service," *Wall Street Journal*, Dec. 2, 1969, 20.

80. One manager joked that a technician should have been standard with every six-minute machine (Janet Endrijonas, "What Factors Will Affect Fax Growth?" *The Office*, Aug. 1990, 68).

81. SRI, I, 122; Charles Beaudette personal communication, June 19, 2002.

82. Roger Kenneth Field, "The Goals: A Communications System that Replaces Person-to-person Contact," *Electronics*, Nov. 24, 1969, 86; Chu, "Facsimile," 25.

83. E.g., "Facsimile Machines: They Transmit What Voice Can't Convey," *Administrative Management*, Oct., 1972, 46–49.

84. "Picture This: Point-to-point Paper from the Next Big-time Black Box," *Electronics*, Nov. 24, 1969, 99–100; Chu, "Facsimile," 24; L. Brett Glass, "Fax Facts," *Byte*, Feb. 1991, 302.

85. TR-29 Committee on Facsimile, "Minutes," March 15, 1972; TR29.2 Subcommittee on Facsimile Equipment, "Minutes," Nov. 1–3, 1972. Telecommunications Industry Association archives.

86. Philip Wexler, "Evaluating and Installing a Fax Network at duPont," *The Office*,

Sept. 1984, 136; General Accounting Office, *Reduced Government Facsimile Communication Costs Possible through Better Management* (Washington, DC: GPO, Oct. 22, 1976), 3–4.

87. Interviews with John Morrison, March 31, 1992, and Eric von Hippel.

88. John Brooks, "Xerox Xerox Xerox Xerox," *New Yorker*, April 1, 1967, 53, 55.

89. Hugh G. J. Aitken, *Syntony and Spark: The Origins of Radio* (Princeton, NJ: Princeton University Press, 1985), 20.

90. "Panafax Cranks Up in the Facsimile Biz," *EMMS*, July 3, 1978, 3.

91. E.g., Gary Winkler, Graphic Sciences internal memo, "dex vs. TELETYPE," March 16, 1972; [Panafax sales manual], "Welcome to . . . Panafax" [c. 1978]. Personal files; Tim Moran, "Electronic Mail," *Office Products News* 15, 5 (Oct. 1980), 5, 9, 14–15.

92. "Dedication on Wheels," 3M *Center Span*, Sept. 1980, 5.

93. "Going for Broke at Qwip," *Sales & Marketing Management*, Feb. 5, 1979, 14.

94. Rentals, service fees, and royalties comprised $667 million of Xerox's 1968 $896 million in revenue (Xerox, *1968 Annual Report*, 3). See also Joseph Mort, "Xerography: A Study in Innovation and Economic Competitiveness," *Physics Today* 47, 4 (April 1994), 32–38.

95. F&S, 1977, 14; "Xerox Corp. Unveils More Versatile Model of Facsimile Device," *Wall Street Journal*, Feb. 20, 1975, 26; Auerbach, *Guide*, 71.

96. Howard A. Floyd, "Facsimile Today," *The Office*, July 1970, 48.

97. SRI, II, 295; H. W. Geiger, "Memorandum, The Facsimile Market," Jan. 16, 1976. Teleautograph archives.

98. "Visual Communication Systems," *Business Graphics*, April 1968, 18; IRD, 1981, 53.

99. F&S, 1977, 137. The White House and some agencies also extensively employed secure fax machines (Michael K. Bohn, *Nerve Center: Inside the White House Situation Room* (Washington, DC: Brassey's, 2003), 73).

100. Savory, "Tomorrow's Communications—Today."

101. Auerbach, *Guide*, 71; GAO, *Reduced Government Facsimile Communication Costs*, 8.

102. "Facsimile Raises Telegrams' Speed," *New York Times*, July 8, 1948, 44.

103. Alexander, "Lots of Talk," 122.

104. SRI, I, 45, 72; IRD, 1981, 40.

105. Stanley W. Penn, "Electronic Courier: Facsimile Snares New Customers with Added Speed and Reliability," *Wall Street Journal*, Nov. 30, 1959, 1; "Facsimile Rapidly Enhancing Market Position," *Electronic News*, May 15, 1967, 35; Frances Ridgway, "By Using Facsimile Network, Coordination Becomes Easier," *Product Engineering*, June 5, 1967, 196–97.

106. R. C. Uddenberg, Program Integration Manager, "Boeing-TIE Management Techniques," D2-117099-1, July 29, 1969. Boeing Archives; William Clothier, "Cross-country Conferences," *Boeing Magazine*, Sept. 1968, 10–11; Warren C. Wetmore, "Boeing Network Cuts Apollo Work Costs," *Aviation Week and Space Technology*, Feb. 10, 1969, 43–46; "Voice-Picture Hookups Connect Two Companies," *Machine Design* 42, 2 (Jan. 22, 1970), 43.

107. Ron Schneiderman, "Fax Speeds Up," *Electronics*, Nov. 14, 1974, 76; Howard Anderson, "Facsimile Pulls Out the Stops, and User Is Prime Beneficiary," *Communica-*

tion News, Sept. 1978, 56; "Making Electronic Mail Pay Off in the 1980s," *EMMS*, Oct. 15, 1979, 2–3; "The A to Z of Electronic Messaging—B," *EMMS*, Dec. 17, 1979, 8.

108. SRI, I, 87–93, 99; II, 292; Richard E. Hanson, "Higher Speeds Reduce the Cost of Fax," *Administrative Management* 35, 8 (Aug. 1974), 86; Auerbach, *Guide*.

109. "RSMA Hears Facsimile Role," *Railway Age*, April 3, 1961, 33.

110. Ridings, "Facsimile Communication," 38; SRI, I, 83.

111. One unexpected market was funeral homes, which usually wrote an obituary with the surviving family and sent it to the local newspaper. To show that it had sent the entire obituary, a funeral home employee faxed it to the paper in front of the family, thus escaping blame for any changes (Morrison interview).

112. "Burroughs Reports," *Burroughs Clearing House* 59, 9 (June 1975), 30–31; F&S, 1977, 148–49.

113. "Facsimile Machines: They Transmit What Voice Can't Convey," 47; GAO, *Reduced Government Facsimile Communication Costs*, 8–11; F&S, 1977, 140; R. B. Shanck, "Memorandum," Jan. 11, 1955. Case 37727, AT&T Archives.

114. This reluctance to experiment with new technology might be considered symbolic of Ford's English overall management (Brown, *Telephone facsimile for business*, 155–56).

115. Winkler, "dex vs. TELETYPE"; "Allied Chemical's Fax Motto: 'A Unit Where There's a Need,'" *Communication News*, Sept. 1978, 63.

116. Bureau of Labor Statistics, Department of Labor, *Communications, 1968* (Washington, DC: GPO, 1970), 12.

117. "Transmits Data to Test Product," *Administrative Management* 32, 6 (June 1971), 98.

118. "Facsimile Speeds Distributor Orders," *Data Management* 16, 10 (Oct. 1978), 52–53.

119. Ronald J. Lauland, "Facsimile Network Cuts Bank Crime," *Banking*, Sept. 1975, 73–74; Morrison interview.

120. "Facsimile Transceivers Pave Way for Revolution in Electronic Mail," *Graphic Arts Monthly*, July 1978, 51.

121. Robert Wernikoff to William Schreiber, Oct. 12, 1964. Wernikoff Archives.

122. Jann S. Wenner and Corey Seymour, *Gonzo: The Life of Hunter S. Thompson* (New York: Little, Brown and Company, 2007), 166.

123. "Litcom Prepares to Market Japanese Facsimile System," *Electronic News*, Aug. 26, 1968, 35; Auerbach, *Guide*, 27–29.

124. E. Blanton Kimbell, "Wire Systems," ANPA *Mechanical Bulletin* 746 (Aug. 11, 1961), 149–50; Buren H. McCormack, "Facsimile at the Wall Street Journal," ANPA *Mechanical Bulletin* 749 (Sept. 8, 1961), 195–96; Buren H. McCormack, "Facsimile Newspaper Production," ANPA *R.I. Bulletin* 793 (July 19, 1963), 124–26.

125. "Satellite Fax Newspaper Publication Grows," *EMMS*, Feb. 15, 1981, 10–11; IRD, *Facsimile Markets*, 40, 67; Shirley Biagi, *Media/Impact: An Introduction to Mass Media* (Belmont, CA: Wadsworth Publishing, 2011), 369; "Our History," *New York Times*, http://www.nytco.com/who-we-are/culture/our-history.

126. Frank A. Pietropaoli to Mr. Shank and Mrs. Huffer, June 12, 1970, National Museum of American History Library, Smithsonian Institution; David W. Heron and J. Richard Blanchard, "Seven League Boots for the Scholar? Problems and Prospects of

Library Telefacsimile," *Library Journal*, Aug. 1966, 3601–4; Sharon Schatz, "Facsimile Transmission in Libraries: A State of the Art Survey," (Washington, DC: Library of Congress, June 1967), 9–10.

127. "New York State Pilot Facsimile Transmission Project," *The Bookmark*, Feb. 1967, 139–40; "N.Y. Facsimile Project Judged a Failure," *Library Journal*, April 15, 1968, 1564–65; Nelson Associates, *The New York State Library's Pilot Program in the Facsimile Transmission of Library Material: A Summary Report* (June 1968); James R. Hunt, "The System in Hawaii," *ALA Bulletin*, Dec. 1966, 1145–46.

128. SRI, I, 85; F&S, 1977, 117; Joel A. C. Blum, Helaine J. Korn, and Suresh Kotha, "Dominant Designs and Population Dynamics in Telecommunications Services: Founding and Failure of Facsimile Transmission Service Organizations, 1965–1992," *Social Science Research* 24 (1995), 97–135.

129. "A Public Facsimile Service," *The Office*, July 1970, 50; "Facsimile Network Opens National Line," *New York Times*, Dec. 9, 1970, 107.

130. "Competition Heats Up," *Wall Street Journal*, Oct. 3, 1968, 1; SRI, I, 85; Chu, "Facsimile," 24.

131. "Nationwide Facsimile Transceiver Network Keeps Truckers Rolling," *Communication News*, Sept. 1978, 72.

132. IRD, 1985, 58.

133. Peter E. Heller to Elliot Sivowitch, National Museum of American History, Oct. 10, 1991.

134. "Graphic Facsimile Transmission," *EGG ink*, July–Aug. 1967, 2–4.

135. Robert E. Wernikoff, "Digital Facsimile: Problems, Solutions, and Guidelines for Evaluation," *Signal*, Dec. 1966, 19–22.

136. J. Malster and M. J. Bowden, "Facsimile—a review," *Radio and Electronic Engineer* 46, 2 (Feb. 1976), 59.

137. Alexander, "Lots of Talk," 134.

138. E.g., EIS: E. H. Woodhull to Wernikoff, June 28, 1963. Wernikoff papers; Visual Sciences: F&S, 1977, 88, and Frank DiSanto and Denis Krusos, interview, June 7, 1993. See also Bruun, "On the Technical Side," 30.

139. "The Recession and the Fax Industry," *EMMS* July 1, 1982, 3; International Resource Development, Inc., *Facsimile Markets* (Norwalk, CT: IRDI, May 1983), 15–16.

140. Graphic Sciences, Inc., *Annual Report for the year ended June 30, 1973*, 7.

141. Dorfman, "Heard on the Street"; SRI, II, 308–13; "The Antenna," *Electronic News*, Sept. 18, 1972, 12; Robert Butler and Paul Plansky, "Firming in Fax Due," *Electronic News*, Sept. 18, 1972, 24; "GS Shows Sub-1-Min. Fax Transceiver," *Electronic News*, Nov. 13, 1972, 40; Robert Butler, "Struggling Fax Market Seeks New Technology Boost," *Electronic News*, Dec. 11, 1972, 26.

142. Melpar Research Report 59/1, "On the Efficient Representation of Pictorial Data," March 31, 1959; Robert Wernikoff to Tom Long, Obituaries Editor, *Boston Globe*, June 15, 1997. Wernikoff papers.

143. J. E. Hawes to Paul Epstein, "Summary of our TELIKON Coding Discussions," April 18, 1967; Robert Wernikoff, "State of the Art in Facsimile Communication," March 1968 IEEE Chicago meeting. Wernikoff papers.

144. Robert Wernikoff to William Schreiber, Aug. 24, 1964; EIS, "Design and Development of a Prototype Electronic Photofacsimile Recorder for Satellite Operations,

Phase I. Final Report Design," Sept. 12, 1964; EIS, "A proposal for the development and fabrication of a prototype Delta/Del Video Delta Modulation System," NASA RFP No. BG751-9-5-463P, Feb. 25, 1965. Wernikoff papers.

145. E. H. Woodhull to Wernikoff, June 28, 1963; Benjamin Parran to Wernikoff, July 20, 1965. Wernikoff papers. Ironically, Xerox purchased Wernikoff's patents from Addressograph-Multigraph in 1976, a decade too late to help EIS (Sharon L. Shanahan, "Facsimile Joins the Computer Age," *Infosystems* 29, 9 [Sept. 1982], 127).

146. Wernikoff to Schreiber, Aug. 24, 1964, and Oct. 12, 1964. Wernikoff papers.

147. Wernikoff to Schreiber, Oct. 12, 1964. Wernikoff papers.

148. Robert Wernikoff to William Schreiber, April 7, 1965; Robert Wernikoff, "To the stockholders of Electronic Image Systems Corporation," Sept. 15, 1966; "Facsimile Will Be a Versatile and Profitable Communications Service," 2.

149. "EIS—a strong new teammate," *AM News*, Nov. 1967, 6.

150. Advanced Systems Research, Addressograph-Multigraph, "Telikon Markets—Initial Survey and Projection," Aug. 30, 1967, 2; J. M. Lewis to A. E. Mignone, Aug. 26, 1968; W. T. Johns to D. W. Seager, March 5, 1970. Wernikoff papers.

151. Addressograph-Multigraph, "Project Plan High Speed Printer," Feb. 1968; EIS, "A Proposal for the Development of an Addressograph-Multigraph Non-Impact Printer," TM 70/2, Jan. 22, 1970. Wernikoff papers.

152. EIS, "Telikon II, May 1–5, 1967, Marriott Motor Hotel, Key Bridge," "1968 Telikon Demonstrations Summary," "Telikon and Facsimile" booklet [1967?] and Wernikoff, "Long-Distance Copiers," July 1968. Wernikoff papers; "Message Facsimile," 482–83.

153. Edmands, "Burst of Speed"; Herbert M. Dwight Jr. to EIS, July 14, 1970. Wernikoff papers.

154. "Cambridge Firm Is Realigned," *Boston Evening Globe*, Oct. 23, 1970, 37; "Electronic Image Is Merged into Addressograph," *Electronic News*, Oct. 26, 1970, I, 2; Albert Redman Jr. to Joseph Van Horn, Sept. 14, 1970; Wernikoff to Long. Wernikoff papers.

155. "Addressograph Slashes Quarterly to 15 Cents," *Wall Street Journal*, Nov. 6, 1970, 23.

156. F&S, 1977, 81, 109.

157. "Comfax Introduces System," *New York Times*, April 19, 1973, 69; Bruun, "On the Technical Side," 38: Ron Schneiderman, "Fax Speeds Up," *Electronics*, Nov. 14, 1974, 75–76; Auerbach, *Guide*, 31–33; Rick Minicucci, "Facsimile Begins to Exploit Its Potential," *Administrative Management* 37, 1 (Jan. 1976), 45–51; "Comfax Shows New Fax," April 23, 1973. Box 7, Comfax Communications Industry Folder, Computer Product Literature Collection. Charles Babbage Institute.

158. SRI, v. I, 122: F&S, 1977, 39–41; F&S, 1982, 79.

159. Bruun, "On the Technical Side," 34; Auerbach, *Guide*, 27–29.

160. "CBS-Savin Joint Purchase," *Wall Street Journal*, Jan. 4, 1971, 4; SRI, I, 122: F&S, 1977, 2, 42–46.

161. "Rapifax 100" (advertisement), *Administrative Management* 35, 8 (Aug. 1974), 31; Howard M. Anderson, "Fast Fax Coming on Fast," *Infosystems* 22, 3 (March 1975), 44; John F. Jansson, "Rapifax: The Chicago Tribune's Experience," *ANPA R.I. Bulletin* 1226 (May 7, 1976), 162–6; Minicucci, "Facsimile Begins to Exploit Its Potential," 46; Gary Winkler, "Consider the Total Usage Costs When Selecting Your Fax System," *Communi-*

cation News, Sept. 1978, 64; "The Rapifax 100 Electronic Mail System" (advertisement), *Communication News*, Sept. 1978, 57.

162. Auerbach, *Guide*, 39, 59, 72; "Marketing Strategies for High Speed Facsimile," *EMMS*, Nov. 15, 1978, 1–3; F&S, 1982, 4–5, 81, 91, 106.

163. John Shonnard interview, Aug. 21, 1992; DiSanto and Krusos interview.

164. "Litcom Prepares to Market Japanese Facsimile System," *Electronic News*, Aug. 26, 1968, 35; "Moving Images a Japanese Way," *Business Week*, Aug. 29, 1970, 30; F&S, 1982, 72.

165. DiSanto and Krusos interview.

166. F&S, 1977, 86–88; Mark Harrington, "CopyTele Takes U-turn," *Newsday*, Aug. 28, 2000, C5.

167. F&S, 1977, 88; "Panafax Cranks Up in the Facsimile Biz," *EMMS*, July 3, 1978, 1–4; F&S, 1982, 24, 154; DiSanto and Krusos interview.

168. Steven Titch, "Matsushita Barred on Fax Distribution," *Electronic News*, Dec. 21, 1981, 11; F&S, 1982, 73–74.

169. Dan Dorfman, "Heard on the Street," *Wall Street Journal*, March 19, 1973, 29; "Facsimile Systems Start Moving—Fast," *Business Week*, Industrial Edition, Feb. 13, 1978, 84I; Richard A. Shaffer, "Xerox Faces Problems in Trying to Duplicate Its Own Past Success," *Wall Street Journal*, Feb. 16, 1978, 1, 29; Jeffrey A. Tannenbaum, "Xerox Maps 'Aggressive' Growth for '80," *Wall Street Journal*, Jan. 15, 1980, 12.

170. "Xerox Corp. Unveils More Versatile Model of Facsimile Device," *Wall Street Journal*, Feb. 20, 1975, 26; Phil Hirsh, "Products and FCC's Wiley at ICA Show," *Datamation* 22, 6 (June 1976), 155.

171. F&S, 1977, 23.

172. "Graphic Sciences' New Fax Machines—The Beginning of a New Era or the End of a Passing Phase?" *EMMS*, Feb. 1, 1978, 1–2; Anderson, "Fast Fax Coming on Fast," 42.

173. F&S, 1977, 81–82.

174. Bob Johnstone, *We Were Burning: Japanese Entrepreneurs and the Forging of the Electronic Age* (New York: Basic Books, 1999), 189.

175. Anderson, "Fast Fax Coming on Fast," 42; F&S, 1977, 21, 90–91.

176. Exxon, *1979 Annual Report*, 22.

177. Rick Minicucci, "Facsimile Begins to Exploit Its Potential," *Administrative Management* 37, 1 (Jan. 1976), 50; "Exxon Enterprises vs. IBM: Striking Oil in EMMS/Automated Offices," *EMMS*, Dec. 22, 1977, 5–7; IRD, 1981, 53–54, 160.

178. F&S 1982, 44–45; Auerbach, *Guide*, 61–62.

179. Ron Schneiderman, "New Face in Fax," *Electronic News*, July 29, 1974, 14; Anderson, "Facsimile Pulls Out the Stops," 56.

180. "Qwip Stalks Bigger Game," *Sales and Marketing Management*, March 17, 1980, 20–21; IRD, 1981, 160–61; IRD, 1983, 79; F&S, 1982, 52–54.

181. "United Tel Jumps Into Fax Business—Bell Next?" *EMMS*, Feb. 15, 1980, 10.

182. "New Technology Spurring Greater Use of Facsimile," *Communication News*, Sept. 1978, 54.

183. "Facsimile Systems Start Moving—Fast," 84B.

184. F&S, 1977, 6.

185. Anderson, "Facsimile Pulls Out the Stops," 56; AT&T, "Rate Changes through the Years."

186. "An Electronic Mail and Message Systems Primer," *EMMS*, April 3, 1978, 4; IRD, 1981, 58.

187. "Graphic Sciences' New Fax Machines," 1–2.

188. "The Japanese Facsimile Onslaught," *EMMS*, Sept. 4, 1979, 4; IRD, 1983, 43.

189. F&S, 1977, 7–8.

190. "Dissimilar Gear Could Swap Data under ITT Plan," *Wall Street Journal*, Dec. 8, 1975, 5; Ronald A. Frank, "Faxpak Plays Catch-up," *Datamation* 27, 6 (June 1976), 46–50; Pamela Hamilton, "Fax, Teletypewriters Talk to One Another," *Electronics*, March 27, 1980, 50.

191. "Southern Pacific's Proposed Facsimile Service," *EMMS*, Jan. 16, 1978, 6; "Facsimile Markets Eyed in ITT, MCI Offerings," *Datamation* 22, 1 (Jan. 1976), 101; F&S 1982, 116–21; International Resource Development, Inc., *Facsimile and PC-Based Image Transmission Markets* (New Canaan, CT: IRDI, Feb. 1989), 182–86.

192. F&S, 1977, 7.

193. Thomas Haigh, "Remembering the Office of the Future: The Origins of Word Processing and Office Automation," *IEEE Annals of the History of Computing* 28, 4 (Oct.–Dec. 2006), 19.

194. "The Office of the Future: Executive Briefing," *Business Week*, June 30, 1975, 48–84; Thomas M. Lodahl and N. Dean Meyer, "Six Pathways to Office Automation," *Administrative Management* 41, 3 (March 1980), 32–33, 74, 78, 80, 90; J. Christopher Burns, "The Automated Office," and A. E. Cawkell, "Forces Controlling the Paperless Revolution," in *The Microelectronics Revolution: The Complete Guide to the New Technology and Its Impact on Society*, ed. Tom Forester (Cambridge, MA: MIT Press, 1983), 220–31, 244–57.

195. "Facsimile Market to Surprise Critics," *EMMS*, June 15, 1983, 3; "Japan takes over in high-speed fax," *Business Week*, Nov. 2, 1981, 104–5.

196. "Facsimile Systems Start Moving—Fast," 84I.

197. CIAJ, *Tsushin kiki chuki juyo yosoku* [Communications Equipment Demand Forecast] (Tokyo: CIAJ) 1980, 122; "Japan takes over in high-speed fax," 104–5.

198. The four different *kanji* for 'kisha' in "Kisha no kisha wa kisha de kisha shimashita" mean "Your company's reporter returned to his company by steam train." My appreciation to Steve Myers for this delightful play on words.

199. Tsutomu Kawada, Shin-ya Amano, Ken-ichi Mori, and Koji Kodama, "Japanese Word Processor JW-10," *Proceedings of Compcom* (Sept. 1979), 238; Nanette Gottlieb, *Word-Processing Technology in Japan: Kanji and the Keyboard* (London: Curzon, 2000), 1–26.

200. *Fakushimiri choosa kenkyuu kai* [Facsimile Survey Research Group], *Fakushimiri: Sono genjyoo to tenboo* [Facsimile: Its present state and future] (Tokyo: MPT, 1980), 12. Hereafter, "MPT Fax Survey Group"; CIAJ, 1986, 27.

201. J. S. Courtney-Platt and A. C. Schmidt, "Visit to Nippon Telephone and Telegraph (NTT)," Sep.7, 1977. 97 05 396, Western Electric Company. AT&T Archives.

202. Neil Gross and John W. Verity, "Tokyo's Love Affair with High Tech Stops at the Office," *Business Week*, Oct. 10, 1988, 112; Detmar W. Straub, "The Effect of Culture on IT Diffusion: E-Mail and FAX in Japan and the U.S.," *Information Systems Research* 5, 1 (March 1994), 26; Gavin Whitehead interview, May 27, 2009.

203. MPT, *Report on the Present State of Communications in Japan, Fiscal 1980*,

12–14, and *Communications in Japan, Fiscal 1986* (Tokyo: MPT, 1987), 38; Tessa Morris-Suzuki, *Beyond Computopia: Information, Automation, and Democracy in Japan* (London: Kegan Paul, 1988), 33; Koji Kobayashi, *Computers and Communication: A Vision of C&C* (Cambridge, MA: MIT Press, 1986); "Koji Kobayashi," *Economist*, Dec. 7, 1996, 83.

204. MPT, *An Outline of Posts and Telecommunications Activities of Japan for 1969* (Tokyo: MPT, 1970), 20; K. Nishimura, "Facsimile Communication in Japan: Present Status and Outlook," *Proceedings of the 2nd CCITT Interdisciplinary Collegium on Teleinformatics, Montreal, 9–12 June 1980* (Geneva: ITU, 1981), 265.

205. CIAJ, 1984, 118.

206. Robert Barrett, "Development and Use of Facsimile Systems in Japan," *Electronics and Power*, Dec. 12, 1974, 1119–20; interview with Toyomichi Yamada, Nov. 29, 1991. Between 1971 and mid-1977, 197 papers on digital compression appeared in Japan (Yasuhiko Yasuda and Mikio Takagi, "Bibliography on Digital Facsimile Data Compression in Japan," [Multidimensional Image Processing Center, Institute of Industrial Science, University of Tokyo, August 1977]).

207. MPT Fax Survey Group, 43.

208. J. S. Courtney-Platt and A. C. Schmidt, "Visit to MGCS, Inc.," Sept. 5, 1977; "NEC," Sept. 6, 1977; 97 05 396, Western Electric. AT&T Archives.

209. F&S, 1977, 56–57.

210. Ken Kusunoki, "Incapability of Technological Capability: A Case Study on Product Innovation in the Japanese Facsimile Machine Industry," *Journal of Product Innovation Management* 14 (1997), 376.

211. MITI Notification 601, "Elevation Plan of the Facsimile Equipment Manufacturing Industry," Dec. 21, 1978.

212. NTT, "Fakushimiri fukyuu sukushin no genjyoo" [The Present State of Facsimile Promotion] [1994?], 3.

213. Interview with Toshikazu Tanida and Shigeo Tanaka, June 16, 1995.

214. MPT Fax Survey Group, 35, 123; Courtney-Platt and Schmidt, "Visit to Nippon Telephone and Telegraph"; CIAJ, 1981, 115.

215. Courtney-Pratt and Schmidt, "Visit to MGCS" and "Visit to Nippon Telephone and Telegraph."

216. Masayoshi Orii, Hiroya Inagaki, and Harumitsu Shimizu, "'Mini Fax'—a Trial Small Size Facsimile," *Japan Telecommunication Review* 19, 3 (July 1977), 209–16; Courtney-Pratt and Schmidt, "Visit to NEC"; Hirohito Nakajima, "Recent Trends in the Development of Facsimile and Visual Communication System in NTT," *Japan Telecommunications Review* 25, 3 (July 1983), 157; CIAJ, 1983, 117.

217. MPT, *Report on Present State of Communications in Japan, Fiscal 1981* (Tokyo: Japan Times, 1982), 3; CIAJ, 1986, 31; I. Shirotani interview, June 22, 1995.

218. CIAJ, 1984, 123.

219. MPT Fax Survey Group, 35, 48, 79; Osamu Murakami, *Fakushimiri: Atarashii Tsuushin Media* [Facsimile: New Communication Media] (Tokyo: Denkii Tsushin Gyijutsu Niisusha, 1981), 234.

220. "Facsimile Systems Start Moving—Fast," 84I.

221. CIAJ, 1981, 115; 1983, 117.

222. MPT Fax Survey Group, 63, 306–27; Murakami, *Facsimile*, 234.

223. F&S, 1982, 1, 138, 146.

224. Brown, *Telephone facsimile for business*, 14–16, 35, 38, 46–48, 70–75; F&S, 1982, 146; R. J. Raggett, "New Product Bonanza for British Phone Users," *Telephony*, April 2, 1979, 98–102.

225. Including simply acquiring a PTT's specifications (Mutsuo Ogawa interview, June 7, 1995). See also "Non-EEC Equipment Ban," *Wall Street Journal*, Nov. 16, 1969, 33; SRI, I, 144; Hank G. Magnuski, interview, July 6, 1992.

226. Dieter v. Sanden, "Some Remarks on Man's Telecommunication Needs," *Siemens Review*, Nov. 1975, 462–66; "Japan takes over in high-speed fax," 105.

227. Simon Nora and Alain Minc, *The Computerization of Society: A Report to the President of France* (Cambridge, MA: MIT Press, 1980), vii, 1.

228. J. Grenier and G. Nahon, "Minitel: A Videotex Success Story," *Telephony*, July 27, 1987, 46–48, 53; Henry C. Lucas Jr. et al., "France's Grass-roots Data Net," *IEEE Spectrum*, Nov. 1995, 71–77; Andrew Feenberg, *Between Reason and Experience: Essays in Technology and Modernity* (Cambridge, MA: MIT Press, 2010), 83–101.

229. ". . . and in France," *Communication Technology Impact* 1, 8 (Nov. 1971), 8; Lad Kuzela, "French Firms Eye U.S. Facsimile Connection," *Industry Week*, Dec. 8, 1980, 99–100; "French Recipe for Teletex: Add Some Facsimile," *Data Communications*, Oct. 1982, 50–51; F&S, 1982, 146–47; "Telematique: A French Export That Doesn't Travel Well," *Business Week*, July 4, 1983, 79.

230. "British Try Again with Fax Mail," *EMMS*, Jan. 15, 1981, 12–13; Rapporteur's Group on Facsimile Services, "Report of the first meeting of the Group (22 June 1977)," June 1977, SGI, Contribution 29. ITU; K. Freiburghaus, "Facsimile Services as Viewed by the Swiss PTT Administration," in CCITT, *Proceedings of the Second CCITT Interdisciplinary Colloquium on Teleinformatics, Montreal 9–12 June 1980* (Geneva: ITU, 1980), 247–50.

231. F. Bardua, "The Telefax Service in the FRG," in *Proceedings of the First CCITT Symposium on New Telecommunication Services*, 189, 218–23; Sweden, "Facsimile Service in Sweden," 25 Jan. 1979, SG I Contribution 157. ITU; Walter L. Vignault, *Worldwide Telecommunications Guide for the Business Manager* (New York: John Wiley & Sons, 1987), 297.

232. F&S, 1977, 147–48.

233. "Rapifax Corp. claims record with 20-second fax service," *Infosystems* 23, 5 (May 1976), 19.

234. United States, "High-speed Digital Facsimile Service," Dec. 1978, SGI, Contribution 135. ITU; Kouhei Kobayashi, KDD, "The Techniques and Applications of High-speed Digital Facsimile," in *Proceedings of the First CCITT Symposium on New Telecommunication Services*, 189, 225–37; James C. Hepburn, "New Facsimile Services," in *Proceedings of the Second CCITT Interdisciplinary Colloquium on Teleinformatics*, 234–36.

CHAPTER 5: **The Giant Awakes, 1980–1995**

Epigraph. D. G. Elliman, "Document Recognition for Facsimile Transmission," *IEE Electronics Division Colloquium on Message Handling—Past, Present and Future*, Nov. 11, 1991 (London: Institute of Electrical Engineers, 1991), 3/1.

1. Fraida Dubin, "Checking E-mail and the Fax," *English Today* 7, 1 (1991), 48; in-

terview, Southwest Rocky Mountain Japan Association meeting, April 1995; Jim Mac-Kintosh interview, July 1, 1993.

2. CIAJ, *Tsushin kiki chuki juyo yosoku* [Communications Equipment Demand Forecast] (Tokyo: CIAJ, annual); "Facsimile Machine Shipments (U.S. Market Consumption)," *Information Technology Industry Data Book, 1960–2004* (Washington, DC: Computer and Business Equipment Manufacturers Association, 1994), 78.

3. "Five Commandments," *IEEE Spectrum*, Dec. 2003, 34–35; "Beyond the Ether," *Economist Technology Quarterly*, Dec. 12, 2009, 23–24.

4. David Blum, "Fax Mania: Read It and Weep," *New York*, Nov. 21, 1988, 40.

5. Study Group XIV, "Final Report on the Work of Study Group XIV," Dec. 5, 197, Contribution 33, 19–20; and "Report of the meeting in Geneva (24–26 April 1974)," May 1974, Contribution 18, 8. ITU documents are stored at the ITU Library in Geneva (hereafter, ITU).

6. Stanford Research Institute, *Facsimile Markets* (Stanford: SRI, May 1972), v. 2, 5, 15.

7. Other criteria included coding efficiency, error detection and correction, and control signals (Study Group XIV, "Report of the meeting in Geneva (24–26 April 1974)," Contribution 18, 17. ITU; Jim Jordan, "The Future for Facsimile: Compression and Compatibility," *Telecommunications* 13, 11 [Nov. 1979], 67–68).

8. J. S. Courtney-Platt and A. C. Schmidt, "Visit to Fujitsu," Sept. 2, 1977. 97 05 396 Western Electric Company, AT&T Archives; Yasuhiko Yasuda, "Overview of Digital Facsimile Coding Techniques in Japan," *Proceedings of the IEEE* 68, 7 (July 1980), 830–45; "Profile: David A. Huffman," *Scientific American*, Sept. 1991, 54, 58.

9. Special Rapporteur for the Study of Group 3 Machines, "Report on the meetings of the special rapporteur group during the period 1973–1976," Dec. 1975. Contribution 34. ITU.

10. MPT offered Admix, NEC DBP, Hitachi DEC, and Mitsubishi CP (Akio Matsui, personal communication, June 22, 1995).

11. Interviews with Toyomichi Yamada, Nov. 29, 1991, and Noboru Kobayashi, June 21, 1995.

12. Japan, "Proposal for draft recommendation of 2.d coding scheme," 28 Aug. 1978. Contribution 42; Study Group XIV, "Report of the meeting held in Geneva, 11–15 Dec. 1978," Jan. 1979. Contribution 70, 20–30. ITU; "Group III Facsimile Standards Look Set—Modified READ Code," *EMMS*, Dec. 17, 1979, 6–7.

13. Karen Grassmuck, "Saying It Holds Patent on an Essential Part of Fax Machines, Iowa State U. Demands That Manufacturers Pay It Royalties," *Chronicle of Higher Education*, February 20, 1991, A37–38; Kathy A. Bolten, "Five fax makers to pay royalties to ISU," *Des Moines Register*, March 8, 1991, 7S; "David C. Nicholas," https://www.ece.iastate.edu/profiles/david-c-nicholas/.

14. Graphic Sciences, 3-M, Muirhead, Ltd, Plessey Co., Ltd, Rank Xerox, Rapifax Corp, Xerox Corp., "Contribution to Question 2/XIV, Point B1," March 1977; National Communications System, USA, "Development of a computer program for measuring the compression and error sensitivity of facsimile coding technique," Aug. 1979. Contribution 101. ITU. Study Group XIV, "Report of the meeting held in Kyoto, 7–15 November 1979," Dec. 1979, 54; Study Group XIV, "Final report of Study Group XIV on the work during the study period 1977–1980," July 1980. ITU; Roy Hunter and A. Harry Robinson, "International Digital Facsimile Coding Standards," *Proceedings of the IEEE* 68, 7 (July 1980), 854–67.

15. Stanley M. Besen and Joseph Farrell, "The Role of the ITU in Standardization," *Telecommunications Policy*, Aug. 1991, 311–21; "Several Changes Made to International Standards Process," *Human Communications Digest* 1, 5 (July 1993), 12.

16. "Encoding Proposal for Fax Non-Standard Facilities Unveiled," *Human Communications Digest* 1, 3 (Jan. 1993), 7.

17. "New Features Added to Group 3 Fax," *Human Communications Digest* 3, 4 (April 1995), 3–4.

18. Hiroshi Okazaki interview, Feb. 19, 2009.

19. Dennis M. Roney, "Outlook for Facsimile Is a Healthy One," *The Office*, May 1986, 71.

20. Stuart Zipper, "Data Com, Distributed EDP Push Modems toward Record Year," *Electronic News*, March 14, 1977, 1, 72–76; Masahiro Kawai, "Impact of Digital Technology on Data Modems," in Pacific Telecommunications Conference, *Telecommunications for Pacific Development: Toward a Digital World* (Honolulu: University of Hawaii Press, 1985), 299–302; Toshiro Suzuki and Kenjiro Yasunari, "Family of Telecommunication LSIs," *Hitachi Review* 36, 5 (Oct. 1987), 289–96; Maury Wright, "Modem Chip Sets," *EDN*, April 12, 1990, 100; Glen Griffith, "Modems for Facsimile Application," BISCAP International, *Eurofax '90*, Amsterdam, Nov. 14–16, 1990.

21. International Resource Development, Inc., *Facsimile and PC-Based Image Transmission Markets* (New Canaan, CT: IRDI, Feb. 1989), 49–50; "Consensus Evaporating on New Fax Modem Standard," *EMMS*, Jan.15, 1991, 3; "An Ex-swordsman Ploughs into the Peace Business," *The Economist*, Sept. 23, 1995, 59.

22. "The Japanese Facsimile Onslaught," *EMMS*, Sept. 4, 1979, 5; Kenneth McConnell, Dennis Bodson, and Stephen Urban, *Fax: Facsimile Technology and Systems*, 3rd ed. (Norwood, CT: Artech House, 1999), 146–50.

23. Kouhei Kobayashi, "The Techniques and Applications of High-speed Digital Facsimile," in CCITT, *Proceedings of the First CCITT Symposium on New Telecommunication Services, Geneva, 14–16 May 1979* (Geneva: ITU, 1979), 1:189, 237.

24. A three-minute international daytime call in 1978 cost 3240 yen, compared with 1680 yen in mid-1984 and 680 yen in 1990 (Ministry of Posts and Telecommunications, *Communications in Japan:1988 White Paper* [Tokyo: MPT, 1989], 30; MPT, *Communications in Japan: 1990 White Paper* [Tokyo: MPT, 1991], 14). AT&T long-distance rates decreased from 69 cents for one minute in 1985 to 25 cents in 1990 (AT&T, "Rate changes through the years," 1999. Box 04 11 01, AT&T Archives).

25. Ulf Rothgordt, "Document Printing," *Acta Electronica* 21, 1 (1978), 71–82; Fujio Oda, "Thermal Printing Head for Facsimile," *National Technical Report (Matsushita Electric Industry Company)* 24, 4 (Aug. 1978), 668–73; "Facsimile Paper—the Tail that Wags the Dog," *EMMS*, April 1, 1982, 12–13; Tatsuo Yoda, "Thermal Printing Heads Improving in Performance," *Journal of Electronic Engineering*, May 1982, 76–78.

26. "Thermal Paper Takes Over Fax," *EMMS* Feb. 15, 1983, 5; James D. Scott, Taiji Higaki, Joseph N. Lyons, "Thermal Papers and Thermal Paper Technology," *Wescon/85 November 19–22, 1985 San Francisco, California* (Los Angeles: Electronic Conventions Management, 1985), 11/4, 8.

27. "Difficult-to-Read Faxes Are Getting Phased Out," *Wall Street Journal*, April 25, 1991, B1; "66th NOMDA: Amid the Fast Scan, Is There a Future?" *EMMS*, Aug. 1, 1991, 5; "*EMMS* Goes Shopping for a New Fax Machine," *EMMS*, Sept. 1, 1992, 10–14; "Fax Machine Price Falls below $200 with Store's Cut," *EMMS*, Aug. 7, 1995, 5–6.

28. "Fax: Rough and Tumble at Retail," *Dealerscope Merchandising*, Nov. 1989, 29.

29. Bob Johnstone, *We Were Burning: Japanese Entrepreneurs and the Forging of the Electronic Age* (New York: Basic Books, 1999), xv, 110.

30. CIAJ, 1982, 110; 1988, 52.

31. Fakushimiri choosa kenkyuu kai [Facsimile Survey Research Group], *Fakushimiri. Sono genjyoo to tenboo* [Facsimile: Its present state and future] (Tokyo: MPT, 1980), 36.

32. CIAJ, 1996, 27.

33. Shigeo Sawada, "Telecommunication Services Open Up," *Business Japan*, June 1986, 159–71; MPT, *Report on Present State of Communications in Japan, Fiscal 1985* (Tokyo: Japan Times, 1986), 1; Tasuku Yuki, "An Account of the History of Facsimile," Japan Electrical Culture Center, June 1, 1987; Harumasa Sato and Rodney Stevenson, "Telecommunications in Japan: After Privatization and Liberalization," *Columbia Journal of World Business* 24, 1 (Spring 1989), 31–41.

34. MPT, *1987 White Paper: Communications in Japan* (Tokyo: MPT, 1988), 7. While the data should not be considered exact, the much faster increase of faxing mirrored the larger reality.

35. Fumio Kodama, *Analyzing Japanese High Technology: The Techno-Paradigm Shift* (London: Pinter Publishers, 1991), 143–46.

36. "Word Workout," *Daily Yomiuri*, Aug. 13, 1992; "Survey: 81% of offices have fax," *Daily Yomiuri*, Sept. 22, 1994, https://database.yomiuri.co.jp/rekishikan (hereafter, *Yomiuri* database).

37. Noboru Kobayashi and Takahiko Kamae, "A New Nation-wide Network for Public Facsimile Communication," *Japan Telecommunication Review* 20, 4 (Oct. 1978), 276–81; Masayoshi Ejiri, "Advanced Facsimile Communication Network," *Japan Telecommunications Review* 25, 3 (July 1983), 176–83; Yutaka Nakatani and Masami Yamada, "A Decade of Facsimile Communications Network Services," *NTT Review* 3, 5 (Sept. 1991), 41–44; MPT, *Communications in Japan. Fiscal 1986* (Tokyo: MPT, 1987), 45; NTT, SEC Form 20-F for 2001, 21.

38. *Recruit 1994 Company Brochure* (Tokyo: Recruit, 1994), 25.

39. "Murata Layoffs, Thermal Sales, and Home Market Research," *EMMS*, July 1, 1991, 14–15; Fred R. Bleakley, "Huge and Diverse Home-Office Market Is Hard to Crack," *Wall Street Journal*, May 9, 1994, B1; Shuzo Sasaki, "Facsimile's Role in the Creation of a Multimedia Era," *Japan 21st* 41, 11 (Nov. 1996), 32; David Haueter, "Staying Power? The U.S. Fax Market, 2000–2005," *Gartner Dataquest Forecast Analysis*, July 18, 2001, 2; Ministry of Public Management, Home Affairs, Posts, and Telecommunications, *Information and Communications in Japan: Building a Ubiquitous Network Society That Spreads Throughout the World, White Paper 2004* (Tokyo: MPMHAPT, 2004), 49.

40. Kerstin Cuhls, "Foresight with Delphi Surveys in Japan," *Technology Analysis and Strategic Management* 13,4 (2001), 561–62; Osamu Murakami, *Fakushimiri. Atarashii Tsuushin Media* [Facsimile. New Communication Media] (Tokyo: Denkii Tsushin Gyijutsu Niisusha, 1981), 280–82; Matsushita Graphics, *Fakushimiru no ayumi* [History of Facsimile Development]) (Tokyo: Matsushita Graphics, 1990), 60–61.

41. MPT, *Communications in Japan, White Paper 1998* (Tokyo: MPT, 1998), 68; Mariko A. de Couto, "Fax Services Seep into Home Market," Sept. 26, 1990, and "Salaryman Stranded," *Daily Yomiuri*, July 22, 1994. *Yomiuri* database.

42. Matsushita Graphics, *History of Facsimile Development*, 60–61.

43. Sadayuki Ito and Shinitsu Narazaki interview, April 21, 2009; "FAX johoshi to kyoodoo hakko" [Joint publishing of FAX information journal], *Nikkei shinbum*, Nov. 13, 1992, 5.

44. "Fax Machines Still Going Strong," *Daily Yomiuri*, Oct. 20, 2003, and "Business Update," *Daily Yomiuri*, Nov. 27, 2007. *Yomiuri* database; "network speax," Sept. 4, 2007, http://www.diginfo.tv/2007/09/04/07-0259-r.php.

45. "Facsimile Transceivers Pave Way for Revolution in Electronic Mail," *Graphic Arts Monthly*, July 1978, 50; "SBS Shows Results of Intra-company Mail Study," *EMMS*, Jan. 3, 1978, 1–2; J. Christopher Burns, "The Automated Office," in *The Microelectronics Revolution: The Complete Guide to the New Technology and Its Impact on Society*, ed. Tom Forester (Cambridge, MA: MIT Press, 1983), 222.

46. Megaera Harris, personal communication, July 13, 1992; John Flina, "P.O.'d: Mail moves slower than it did 20 years ago," *San Francisco Examiner*, Feb. 10, 1991, A2; Richard B. McKenzie, "The Economy of Faxing: A Technological Threat to the Mail Monopoly," St. Louis: Washington University: Center for the Study of American Business Occasional Paper 120 (Jan. 1993), 2.

47. B. I. Edelson, H. Raag, and R. Smith, "INTELPOST—An Experimental International Electronic Message System," *Proceedings of the Fifth International Conference on Computer Communication: Atlanta, Georgia, 27–30 October 1980* (Piscataway, NJ: IEEE, 1980), 23–28; "Move Over, National Debt: Here Comes Intelpost," *EMMS*, Feb. 1, 1984, 8–9; U.S. Postal Service, *Annual Report of the Postmaster General 1987* (Washington, DC: USPS, 1988), 28; Burton I. Edson to Stephen Day, June 26, 1990, letter in possession of author.

48. "'Fax' Machines Experiment Set by U.S. Postal Service," *Wall Street Journal*, March 29, 1989, C16; "Postal Service to Test Use of Self-Service Facsimile Machines," *Wall Street Journal*, Nov. 7, 1989, A14; "U.S. Postal Service Notifies Hotelecopy of Northeast Contract Termination," *PR Newswire*, Jan. 18, 1991; "Lawsuit Filed against U.S. Postal Service Alleges Default on Joint Venture Faxmail (sm) Program," *PR Newswire*, March 18, 1991.

49. International Resource Development Inc., *Facsimile Markets* (New Canaan, CT: IRDI, Oct. 1985), 140–41.

50. "Interconnection Debate Continues," *EMMS*, Nov. 1, 1988, 13; Frederick H. Katayama, "Who's Fueling the Fax Frenzy," *Fortune*, Oct. 23, 1989, 156; ITU and TeleGeography, Inc., *Direction of Traffic: International Telephone Traffic, 1983–1992* (Geneva: ITU, 1994), 3–4; Graham Lynch, "The Slow, Slow Death of AMPS and Telex," *Telecom Asia* 9, 12 (Dec. 1998), S16–18.

51. Tony Borg interview, Sept. 9, 2010.

52. IRD, *Facsimile Markets*, 1985, 110, 114; Adam Greenberg and Pamela A. Toussaint, "The Fax Market: Survival of the Fittest," *HFD—The Weekly Home Furnishings Newspaper*, Dec. 26, 1988, 1, 4; "Fax: Rough and Tumble at Retail," 26; Kate Evanas-Correla, "Fax Channels Shift as Market Shrinks," *Purchasing*, Feb. 18, 1993, 73.

53. "Advertising," *New York Times*, Feb. 3, 1978, D9; "Japanese Try Fax Mail; May Market Consumer Fax in U.S.," *EMMS*, Aug. 17, 1981, 8; "Facsimile," *Industry Week*, Sept. 1, 1986, BC12; IRD, *Facsimile Markets*, 1985, 119.

54. "Facsimile 1976: A Selected Guide for Managers," *Administrative Management*

37, 1 (Jan. 1976), 50–51; Mark H. Brooks, "Telex/Fax Worksheet: A Model for Comparisons," *The Office*, April 1985, 73–76.

55. E.g., Jeffrey S. Prince, "Facsimile: Getting the Message across Faster," *Administrative Management* 42, 2 (Feb. 1981), 40; Erik Mortenson, "Fax: a Simple Technology in a Complex World," *The Office*, Sept. 1987, 79–80.

56. CIAJ, 1988, 53; "Fax: Rough and Tumble at Retail," 26–29; NTT, "The Present State of Facsimile Promotion" [1994], 6.

57. CIAJ data; Katayama, "Who's Fueling the Fax Frenzy?"

58. International Resource Development, Inc., *Facsimile Markets* (Norwalk, CT: IRDI, May 1983), 159, 168; "Fax Vendors Break New Barriers," *EMMS*, Oct. 17, 1983, 1; IRD, *Facsimile Markets*, 1985, 146; "Fax Market Hyperventilation?" *EMMS*, Jan. 15, 1988, 1; "*EMMS* Goes Shopping for a New Fax Machine," *EMMS*, Sept. 1, 1992, 10–14.

59. E.g., Canon's Faxphone B-170 listed for $1,695, but large retailers offered it for $900, while Lanier/Harris offered its identical version of Okifax's $2,999 2100 machine for $3,995 (M. David Stone, "Canon Faxphone B-170," *PC Magazine*, June 13, 1995, 184; "Encryption and Network Fax Issues Surface at TCA," *EMMS*, Oct. 15, 1991, 7).

60. Matsushita Graphics, *History of Facsimile Development*, 60; Hiroshi Okazaki interview, Feb. 19, 2009.

61. "New Stability for Facsimile's Leaders," *EMMS*, Dec. 1, 1982, 3–6; Shelley Bakst, "Need for Fast Information Spurs New Interest in Fax," *The Office*, Sept. 1987, 117; David Thurber, "Japanese Fax Firms Vie over Global Market," *Washington Post*, July 5, 1989, B1.

62. IRD, *Facsimile Markets*, 1985, 88.

63. Carol E. Curtis, "Qyx, Qwip . . . and quit?" *Forbes*, Feb. 16, 1981, 33–34; "What's Wrong at Exxon Enterprises," *Business Week*, Aug. 24, 1981, 87–90; Exxon, *1984 Annual Report*, 21; Robert M. Donnelly, "Exxon's 'Office of the Future' Fiasco," *Planning Review*, 15, 4 (July/Aug. 1987), 12–15; Hollister B. Sykes, "Lessons from a New Ventures Program," *Harvard Business Review* 64, 3 (May/June 1986), 69–74.

64. "IBM Goes Public with Its (thoughts on) High Speed Facsimile," *EMMS*, Sept. 4, 1979, 1–2; "IBM Announces Facsimile/PC Connection," *EMMS*, March 1, 1984, 3; "PB Lands IBM Fax Contract," *EMMS*, Oct. 1, 1986, 12–14; IRD, *Facsimile and PC-Based Image Transmission Markets*, 16.

65. Peter Drucker, *Managing for the Future: The 1990s and Beyond* (New York: Truman Talley Books/Plume, 1993), 252–53.

66. Also needed were "a new set of organizational capabilities embedded in structures, communications channels, and info-processing procedures" (Ken Kusunoki, "Incapability of Technological Capability: A Case Study on Product Innovation in the Japanese Facsimile Machine Industry," *Journal of Product Innovation Management* 14 [1997], 369).

67. "Fax Scorecard for the U.S., 1985," *EMMS*, Sept. 3, 1985, 6–7.

68. E.g., "Like the collator, facsimile will have become mostly a feature or function of other machines" ("The Simple, Human, Problems with Fax," *EMMS*, Feb. 15, 1980, 5); Joseph N. Pelton, *Global Talk: The Marriage of the Computer, World Communications, and Man* (Brighton: Harvester Press, 1981), 134.

69. Jens Rasch, "Facsimile: Effective Mode of Business Communication," *The Office*, Nov. 1983, 144; "Facsimile Is the Hit at NOMDA Show," Aug. 1, 1986, 2. Skeptics'

acronyms of "It Still Does Nothing" and "I Still Don't Need" captured the slow spread of ISDN.

70. Jim Jordan, "The Future for Facsimile: Compression and Compatibility," *Telecommunications* 13, 11 (Nov. 1979), 67; F&S, *Facsimile Equipment Market*, 13–14, 41; IRD, *Facsimile Markets*, 1985, 45–46, 87, 118, 170–79, 145; R. Michael Franz, "Group 4 Fax Technology: Its Effect Will Be Felt," *The Office*, Aug. 1985, 64–66; Hiroshi Tanaka, "The Direction of the Japanese Market," *Eurofax '90*; CIAJ, 1987, 41; 1991, 47; Richard Stollery, "The Future of the Group IV fax," *Communications Technology International 1992* (London, 1992), 128–29.

71. Chairman of Study Group VIII, "Report of the first meeting of Study Group VIII in the study period 1989–1992 (Part 1) (Geneva, 12–20 April 1989)," 6–7. ITU; Mary Jander and Paul R. Strauss, "Affordable Fax/Data Gear on Way," *Data Communications*, March 1990, 43–45; Dana Blankenhorn, "Fax Standard Dispute Threatens Growth," *Networking Management*, March 1991, 38; UK, "G3 facsimile—64 kbps option," Jan. 1992, Study Group I, Contribution 117. ITU. The proposals had several names, including G3fax/64k/ISDN, G3C/64k, T.30 Annex ZB, T.30 Annex C, G3/64, and G4 Class Zero ("CCITT Mulls Competing Digital Fax Standards," *EMMS*, Sept. 15, 1992, 6–8).

72. Study Group VIII, "Report of the meeting of WP 1 of Study Group VIII, Part I, Geneva, 5–14 September 1990," Sept. 1990, 20; Susanne K. Schmidt and Raymund Werle, *Coordinating Technology: Studies in the International Standardization of Telecommunications* (Cambridge, MA: MIT Press, 1998), 227.

73. SGVIII, "Report of the final meeting of Study Group VIII for the period 1989–1992, of the four Working Parties of Study Group VIII and of the Rapporteurs Group on Q17/VIII—meeting held in Geneva from 22–30 April 1992," May 1992. ITU; "Group 3 Fax Standards for Digital Networks Approved," *Human Communications Digest* 1, 4 (April 1993), 1–2.

74. International Resource Development, Inc., *Facsimile Markets* (Norwalk, CT: IRDI, July 1981), 142; "T1 Networks Expected to Promote Usage of High-End Facsimile," *EMMS*, Jan. 2, 1987, 15–17.

75. IRD, *Facsimile and PC-Based Image Transmission Markets*, 125; Laura O'Brien, "Will Fax Boom Go Bust?" *Telephony*, Sept. 25, 1989, 46.

76. "GammaLink Shifts Presidents, Looks at Fax 400," *EMMS*, March 15, 1991, 18.

77. IRD, *Facsimile Markets* 1985, 87, and *Facsimile and PC-Based Image Transmission Markets*, 132–33; FIND/SVP, *The Facsimile Market* (New York: FIND/SVP, Sept. 1987), III.

78. Mark H. Brooks, "Telex/Fax Worksheet: A Model for Comparisons," *The Office*, April 1985, 73–76.

79. "Nippon Electric Makes Its Move in Facsimile," *EMMS*, May 17, 1982, 2; Sharon L. Shanahan, "Facsimile Joins the Computer Age," *Infosystems* 29, 9 (Sept. 1982), 130; F&S, *Facsimile Equipment Market*, 12; IRD, *Facsimile Markets*, 1983, 78, and 1985, 77, 85–86.

80. "Fax Vendors Break New Barriers," *EMMS*, Oct. 17, 1983, 3; IRD, *Facsimile Markets* 1985, 148.

81. "Fax, Not Rumors," *EMMS*, Jan. 2, 1985, 15; "Facsimile Market Surged in 1985, But Will It Last?" *EMMS*, March 3, 1986, 9.

82. "Buyers' Guide to Facsimile Systems," *The Office*, Nov. 1986, 151–54, and Nov. 1988, 73–78.

83. IRD, *Facsimile Markets* 1985, 96; J. D. Hildebrand, "Signed, Sealed, and Transmitted," *CFO*, Feb. 1988, 16–22; Larry Johnson, "The Joy of Fax," *ABA Journal*, July 1989, 102–4.

84. Philip Wexler, "Evaluating and Installing a Fax Network at DuPont," *The Office*, Sept. 1984, 136.

85. "78 Local Officials Warned for Misuse of Fax Machines," *Daily Yomiuri*, June 22, 1992, https://database.yomiuri.co.jp/rekishikan. A not-so-apocryphal example of definitely prohibited activity occurred on the *Murphy Brown* show when one character declared she became so drunk at an office party that, "I faxed my chest to the West Coast" (Ralph Keyes, "Do You Have Fax Appeal?" *Cosmopolitan*, Sept. 1989, 139).

86. E.g., John Caldwell, *Fax This Book* (New York: Workman Publishing Co.), 1990.

87. Ann Elise Rubin and Lamont Wood, "Conventional and Electronic Mail," in *The Professional Secretary's Handbook* (Boston: Houghton Mifflin, 1984), 299–301; Guy Kawasaki, "The Macintosh guide to fax etiquette," *Macworld* November 1993, 306.

88. Letitia Baldrige, *Letitia Baldrige's New Complete Guide to Executive Manners* (New York: Rawson Associates, 1993), 146; Judith Martin, *Miss Manners' Guide for the Turn-of-the Millennium* (New York: Pharos Books, 1989), 26.

89. Mike Zwerin, "Jane Bunnett on Sax: The Spirit of Havana," *International Herald-Tribune*, Jan. 23–24, 1993, 24.

90. Joan Challinor personal communications, Nov.10 and Dec. 14, 1993.

91. "When Filmmaking Is Child's Play," *Daily Yomiuri*, March 27, 1993, and "Return of 'stupid little Boston band' the Pixies," *Daily Yomiuri*, Dec. 1, 2005. *Yomiuri* database; Mary B. W. Tabor, "Faxes Steal Some Fun at Frankfurt Book Fair," *New York Times*, Oct. 14, 1995, B9; Gary Levin, "Fax Fixation," *Advertising Age*, Aug. 15, 1988, 3, 60.

92. Melvin Kranzberg, "Technology and History: 'Kranzberg's Laws,'" *Technology & Culture* 27 (1986): 545–48.

93. Michael J. McCarthy, "No Place to Hide," *Wall Street Journal*, Nov. 9, 1990, R21–23.

94. Jeffrey Young, "Breaking Away: Knowing the Way to Santa Fe," *Forbes ASAP*, March 29, 1993, 112–16.

95. Ronald Brown, *Telephone Facsimile for Business: A Research Report on Systems, Equipment, Costs, Advantages and Markets* (Somerset: Ronald Brown, 1975), 12–13; "Facsimile Used for Source Document Transmission Service," *EMMS*, March 1, 1982, 6–7.

96. Blum, "Fax Mania," 43.

97. Jill Andresky Fraser, "The Sound of the Ocean, the Hum of the Fax," *New York Times*, Aug. 1, 1993, 3–23.

98. Mark Bisnow, "Dear Wife," *Washington Post*, May 30, 1993, E1, 5; Maggie Jackson, "Using Technology to Add New Dimensions to the Nightly Call Home," *New York Times*, Oct. 22, 2002, C9.

99. Wm. Penn, "Televiews," *Daily Yomiuri*, July 9, 1994. *Yomiuri* database.

100. Brenda Fine, "The Fax Are In: Many Hotels Aim to Bring Travelers Closer to Offices," *National Law Journal*, Jan. 26, 1987, 35; "Pampering Faxophiles," in "Business Bulletin," *Wall Street Journal*, Nov. 12, 1992, A1; Damon Darlin, "You Are Where You Stay," *Forbes*, Dec. 18, 1995, 340–41.

101. IRD, *Facsimile Markets*, 1985, 53; Brian Dumaine, "Turbulence Hits the Air Couriers," *Fortune*, July 21, 1986, 101–2.

102. Plog Research, Inc., "Toward the Future Growth and Development of Federal Express Corp: A Proposal for Research," No. 77–107, Feb. 1977. Box 4, subseries B, Series 1. Federal Express Advertising History Collection. Smithsonian Archives.

103. J. Vincent Fagen, "Federal Express and Telecommunications: An Overview of a Potential Growth Strategy," March 1, 1981, 51–52. Box 5. Federal Express Collection. Smithsonian Archives; "Federal Express Edging Towards Electronic Message Market?" *EMMS*, April 1, 1981, 7–8.

104. "Items of Interest," *EMMS*, Sept. 17, 1984, 16–17; Jim Montgomery, "Federal Express Has $1.2 Billion Plan to Expand Its Electronic Mail Service," *Wall Street Journal*, July 30, 1984, 7.

105. IRD, *Facsimile Markets* 1985, 17, 50–51; John Merwin, "Anticipating the Evolution," *Forbes*, Nov. 4, 1985, 163, 166.

106. "The Federal Express Zapmailer," *EMMS*, March 1, 1985, 4–5; John J. Keller and John W. Wilson, "Why Zapmail Finally Got Zapped," *Business Week*, Oct. 13, 1986, 48–49.

107. "Items of Interest," *EMMS*, Sept. 2, 1986, 18; "Beam Me Up, Scotty," *EMMS*, Sept. 15, 1986, 2–6.

108. "Federal Express Still Backs ZapMail Despite Huge Losses," *EMMS*, Oct. 1, 1985, 5–7; "Beam Me Up, Scotty," 1; George Moreland Stamps, "The Next Generation of Facsimile and the Electronic Communicating Copier," IGC Conference on Intelligent Copiers/Printers, Monterey, California, Oct. 19–21, 1986; Calvin Sims, "Coast-to-Coast in 20 Seconds: Fax Machines Alter Business," *New York Times*, May 6, 1988, 1.

109. Joel A. C. Blum, Helaine J. Korn, and Suresh Kotha, "Dominant Designs and Population Dynamics in Telecommunications Services: Founding and Failure of Facsimile Transmission Service Organizations, 1965–1992," *Social Science Research* 24 (1995), 112, 117.

110. Tara Trower, "For Your Convenience," *Dallas Morning News*, July 26, 1995, D1–2.

111. George L. Beiswinger, "Fax Manufacturers Seek to Offer a Better Mousetrap," *The Office*, Sept. 1988, 118; Martha M. Hamilton, "Setting Up Shop in a 600 Mph Office," *Washington Post*, Nov. 26, 1992, D1–2.

112. "Utility Firm Merges Fax and Cellular Telephones," *The Office*, Sept. 1988, 52; Gene Bylinsky, "Saving Time with New Technology," *Fortune*, Dec. 30, 1991, 100; "Can Wireless Fax Overcome Its Shortcomings?" *EMMS*, April 18, 1994, 11–14.

113. "Photos in Color Sent by Phone," *New York Times*, Oct. 29, 1939, D7; George Raynor Thompson and Dixie R. Harris, *The Signal Corps: The Outcome (Mid-1943 Through 1945)* (Washington, DC: Office of the Chief of Military History, U.S. Army, 1966), 566. Recording by crayons failed because the colors were not bright enough ("Radio Facsimile Gets Color," *Business Week* Aug. 16, 1947, 40–41). See also Federal Republic of Germany, "Facsimile Transmission of Colour Documents," CCITT 8/XIV, Sept. 1970, Contribution 15. ITU; "Thermal Transfer Printing and Color Fax," *EMMS*, Dec. 15, 1983, 9–10.

114. "Color Fax Working Group Studies JPEG Compression," *Human Communications Digest* 1, 4 (April 1993), 2, 4; CIAJ, 2000, 25–26; 2004, 50; Communications and Information Network Association of Japan, *Yearbook 2003* (Tokyo: CINAJ, 2003), 19; "Fast-printing Color Fax," *Daily Yomiuri*, Aug. 3, 2000. *Yomiuri* database.

115. Andrew Radolf, "The Fax Revolution," *Editor & Publisher*, April 22, 1989, 22–26; Patrick M. Reilly, "Newspapers Warily Study Fax Editions," *Wall Street Journal*, Dec. 8, 1989, B1; E. R. Sander, "Newspapers Tiptoe into Electronic Age Via Fax Services," *Investor's Business Daily*, March 25, 1992, 10.

116. Mark Fitzgerald, "Chicago Tribune Folds Fax Paper," *Editor & Publisher*, Aug. 18, 1990, 13.

117. "On the Fridge," *Washington Post*, Jan. 6, 1993, F1; "Travel Info by Fax or Mail," *Washington Post*, April 25, 1993, E4; "Post Updates Its Markets, Tables, Charts," *Washington Post*, Nov. 23, 1993, C1, 3.

118. Dave Rampe interview, June 16, 1992.

119. Philip C. W. Sih, *Fax Power: High Leverage Business Communications* (New York: Van Nostrand Reinhold, 1993), 10–11.

120. M. L. Stein, "*Missoulian* Goes Fishing," *Editor & Publisher*, Aug. 17, 1991, 46; Paul Farhi, "Folding the Faxed Front Page," *Washington Post*, July 26, 1994, D1, 6.

121. Eduardo Kac, "Call for participation—International Fax Art Project Elastic Fax 2," in *Fine Art Forum Art + Technology Netnews* 8, 9/10 (Sept/Oct 15, 1994), msstate.edu /Fineart_Online/Backissues/Vol_8/faf_v8_n10/faf810.html; Claudia Giannetti, "All Symboises Are Temporary Resonances" [1996 or later], http://www.marisa-gonzalez.com/to dalassim_im.htm.

122. "Fax for Artists and Lovers," *EMMS*, March 2, 1987, 15–17; Blum, "Fax Mania," 44; Arturo Silva, "Barney's reproduced authentic," *Daily Yomiuri*, Nov. 8, 1990. *Yomiuri* database; "Telecommunications Art," http://ekac.org/telecom.html.

123. to.or.at/~radrian/BIO/telecom.html; http://residence.aec.at/rax/24_HOURS/in dex.html.

124. Disembodied Art Gallery http://www.dismbody.demon.co.uk/home.html.

125. Heather Norville-Day, "The Conservation of Faxes and Colour Photocopies with Special Reference to David Hockney's *Home Made Prints*," in *Modern Works, Modern Problems? Conference Papers*, ed. Alison Richmond (London: Institute of Paper Conservation, 1994), 66–72.

126. Guy Bleus, "Telecopying in the Eternal Netland," G L O S S O L A L I A, *Electronic Journal for Experimental Arts* 2 (Sept. 1995), http://thing.net/~grist/cyano/gloss2 /glos2.htm#tel; "The World in 24 Hours," http://residence.aec.at/rax/24_HOURS/index .html; Michael Sousis, "Chicago Exhibit Touts the Fax of Transient Art," *Miami Herald*, July 22, 1990, B5.

127. "Shanghaifax," http://alcor.concordia.ca/~kaustin/cecdiscuss/1996/0477.html.

128. "Can Anyone Curb Africa's Dogs of War?" *The Economist*, Jan. 16, 1999, 41.

129. Keyes, "Do You Have Fax Appeal?" 138.

130. "Baseball Funnels Facts to the Media Via Fax," *The Office*, July 1987, 88–89.

131. http://www.joke-archives.com/oddsends/safefax.html; Michael J. Preston, "Traditional Humor from the Fax Machine: 'All of a Kind,'" *Western Folklore* 53, 2 (April 1994), 147–69. *New York* proclaimed that the five worst (or best, depending on one's perspective) puns created by a "new and enormous cottage industry" of faxers were: "Just the fax, ma'am. Crazy like a fax. Outfaxed. Faxually speaking. The joy of fax" (Blum, "Fax Mania," 41, 43).

132. "Message to Santa," New Jersey *Star-Ledger*, Dec. 11, 1991, 33; Arieh O'Sullivan, "Heavenly Hot Line Getting Second Number," www.subgenius.com/subg-digest

/v4/0197.html; "The Wailing Wall 2.0," *Der Spiegel*, Oct. 5, 2006, http://www.spiegel
.de/international/spiegel/0,1518,441010,00.html. See also Joyce Shira Starr, *Faxes to God:
Real-Life Prayers Transmitted to the Heavens* (San Francisco: HarperSanFrancisco, 1995).

133. Mariko A. de Couto, "Fax Services Seep into Home Market," *Daily Yomiuri*,
Sept. 26, 1990. *Yomuiri* database.

134. Udayan Gupta, "Fax Machine Craze Sends a Message of Opportunity," *Wall
Street Journal*, Aug. 29, 1989, B2; Betsy Wiesendanger, "Electronic Delivery and Feedback Systems Come of Age," *Public Relations Journal*, Jan. 1993, 10–14.

135. "Will the FCC Junk-Fax Ban Impact the Marketplace?" *EMMS*, March 1, 1993,
8; "Will 1993 Be the Year of the Fax-Cost Audit?" *EMMS*, May 15, 1993, 12; James Brooke,
"Tackling a Crucial Highway to Keep Skiers Happy," *New York Times*, Dec. 23, 1995, 7;
Suzanne Rene Possehl, "Russian roulette at 31,000 feet," *Aerospace America*, Jan. 1996,
26–29.

136. Akio Kokubu, "Facsimile Information Services in Japan," *CIAJ Journal* 34, 9
(1994), 16–20; "Fax Information Services Flooding the Market," *Daily Yomiuri*, Jan. 26,
1995, and "Executives' View," *Daily Yomiuri*, April 18, 1995. *Yomuiri* database.

137. "Launch Exposed Problems in Relaying Info," *Daily Yomiuri*, April 7, 2009,
"TEPCO Narrowly Avoided Worst-case Scenario," *Daily Yomiuri*, Aug. 24, 2007, "Police
Stations Shaping up with New Gadgets, Services," *Daily Yomiuri*, June 15, 1994. *Yomuiri*
database.

138. Sadayuki Ito and Shinitsu Narazaki interview, April 21, 2009.

139. Kevin Sullivan and Mary Jordan, "When in Tokyo, Don't Leave Home without
a Map," *Washington Post*, September 21, 1995, A18; Jeff Neff interview, June 21, 1995;
T. R. Reid interview, June 22, 1995.

140. Mike Freeman, "Tabloid Tip Sheet Has the Fax," *Broadcasting & Cable*, April
12, 1993, 40; Eugene Carlson, "News Service Sends Radio Stations Nothing but the Fax,"
Wall Street Journal, Dec. 4, 1991, B2.

141. "Fax in the Foyer?" *EMMS*, July 15, 1988, 9–10.

142. Yolanda Villarreal, "How Fax Can Cut Drug Order and Delivery Time," *Hospital Pharmacist Report*, April 1988, 13.

143. Candace Talmadge, "The Fax of Life," *Texas Lawyer*, Nov. 19, 1990, 4.

144. Ivars Peterson, "The Electronic Grapevine," *Science News*, Aug. 11, 1990, 91;
Bruce V. Lewenstein, "From Fax to Facts: Communication in the Cold Fusion Saga,"
Social Studies of Science 25 (1995), 403–36.

145. "Overnight Mail via Facsimile Serves International Company," *The Office*, Nov.
1982, 62, 73; "Exporting Pays Off," *Business America*, June 3, 1991, 23; Kenneth Wolf,
"Engine Parts Firm Faces Fax, Phases Out Overseas Agents," *Journal of Commerce*, April
23, 1993, 1.

146. Fortune, *Changing Trends in the Market for Facsimile Equipment* (New York:
Fortune, Oct. 1990), 15.

147. MPT, *Communications in Japan: White Paper 1986* (Tokyo: MPT, 1986), 62, and
1988 White Paper: Communications in Japan (Tokyo: MPT, 1989), 33; *Communications
in Japan: White Paper 1998* (Tokyo: MPT, 1998), 66.

148. Joe Saltzman, "Fax for Freedom," *USA Today Magazine*, Sept. 1989, 67.

149. Eric Barnouw, "Historical Survey of Communications Breakthroughs," in *The*

Communications Revolution in Politics, ed. Benjamin Gerald (New York: Academy of Political Science, 1982), 13.

150. "Hata Solicits Opinions with New Fax Service," *Daily Yomiuri,* May 12, 1994; Shoichi Oikawa, "'Complaint Box' Becomes Hata's Pet Project and Tool," *Daily Yomiuri,* May 24, 1994. *Yomuiri* database.

151. Joel Brinkley, "Cultivating the Grass Roots to Reap Legislative Benefits," *New York Times,* Nov. 1, 1993, A1, B8; Dale Russakoff, "No-Name Movement Fed by Fax Expands," *Washington Post,* Aug. 20, 1995, A1, 22; Jennifer Halperin, "A Funny Thing Happened on the Way to a Conference of the States," *The State of the State,* Sept. 8, 1995, http://www.lib.niu.edu/ipo/1995/ii950908.html. See also David Montgomery and Paul W. Valentine, "Government Haters Fight a Button-Down Battle," *Washington Post,* May 14, 1994, B1, 5.

152. Robert Pear, "A Million Faxes Later, a Little-Known Group Claims a Victory on Immigration," *New York Times,* July 15, 2007, A13.

153. Fleming Meeks, "We Want to Be Intrusive," *Forbes,* Dec. 18, 1995, 170.

154. E.g., Tim Craig, "Community Leaders Decry Lobby Firm's Fax," *Baltimore Sun,* March 9, 2002, www.commondreams.org/headlines02/0309-04.htm; Peter Baker, "With a Private Fax Pipeline, Friends Help Clinton Keep in Touch," *Washington Post,* Jan. 20, 1997, A15.

155. "*EMMS* Goes Shopping for a New Fax Machine," *EMMS,* Sept. 1, 1992, 10–14.

156. Susan Benesch, "Getting the Fax into Panama," *Columbia Journalism Review* 26, 5 (Jan/Feb 1988), 6–8; Bill Keller, "By Pluck and Fax, Tiny Free Press," *New York Times,* March 1, 1993, C6; Paul Fauvet and Marcelo Mosse, *Carlos Cardoso: Telling the Truth in Mozambique* (Cape Town: Double Story, 2003), 231–33, 236, 244, 274–78, 332–33.

157. Carl Bernstein, "The Holy Alliance," *Time,* Feb. 24, 1992, 28; Lech Walesa, *The Struggle and The Triumph: An Autobiography* (New York: Arcade Publishing, 1992), 110; Philip Merrill, "Make Them More Like Us," *Washingtonian,* Sept. 1989, 97–99.

158. Gladys D. Ganley, *Unglued Empire: The Soviet Experience with Communications Technologies* (Norwood, NJ: Ablex, 1996), 23–26, 60–61; Scott Shane, *Dismantling Utopia: How Information Ended the Soviet Union* (Chicago: Ivan R. Dee, 1994), 205.

159. *Dismantling Utopia,* 261–72; Ganley, *Unglued Empire,* 127–95.

160. "Strong Brew," *The Economist,* July 13, 1996, 34–36.

161. "Political Intelligence," *Texas Observer,* May 8, 1992, 24; Barbara Ferry, "The Truth in Guatemala," *Texas Observer,* June 4, 1993, 17, 21; "Suspect Planned Bomb Threats," *Dallas Morning News,* Dec. 15, 2001, A9.

162. "Politics, Kuwaiti-style," *U.S. News & World Report,* March 2, 1992, 46–47; John Lancaster, "Dissidents Take the High-Tech Road," *Washington Post,* Dec. 18, 1994, A40; Louise Leff, "Waging War by Fax Modem," *U.S. News & World Report,* Nov. 27, 1995, 51; Michael Lewis, "The Satellite Subversives," *New York Times Magazine,* Feb. 24, 2002, 33.

163. "Fax and Opinions," *The Economist,* June 15, 1991, 40–41; "Closing in on King Fahd," *The Economist,* Dec. 19, 1992, 40.

164. Richard Baum, *China Watcher: Confessions of a Peking Tom* (Seattle: University of Washington Press, 2010), 172.

165. Michael Gartner, "Up Freedom! Faxes to the Rebels, Gunfire Via Cellular

Phone," *Wall Street Journal*, June 8, 1989, A1; Jonathan Kaufman, "Turmoil in China," *Boston Globe*, June 7, 1989, 3.

166. Frank Viviano, "Foreign Correspondents Set Free in Laptop Era," *San Francisco Chronicle*, Feb. 7, 1995, D16.

167. John Pallatto, "The Fax Machine Is Mightier than the Sword," *PC Week*, July 3, 1989, 63; William Mathewson, "World Wire," *Wall Street Journal*, Nov. 16, 1989, A15. The concept spread: abortion rights supporters in the early 1990s faxed lists of British abortion clinics to Irish fax numbers to evade an Irish court decision banning any information about abortion (Nuala O'Faolain, "Information Ban," *Irish Times* in *World Press Review*, Oct. 1992, 24).

168. Susan R. Koenig, *Survey of Fax Use by Attorneys in Four Judicial Districts* (San Francisco: National Center for State Courts, Aug. 1990), 2–3, 9–20, 23–24.

169. Don J. DeBenedictis, "Idaho Courts OK Fax," *ABA Journal*, May 1990, 19–20; Susan R. Koenig, *Courts in the Fax Lane: The Use of Facsimile Technology by State Courts* (San Francisco: National Center for State Courts, Aug. 1990), 3–6.

170. Susan R. Koenig, Frederick G. Miller, and Frederick M. Russillo, *Model Court Rules for the Use of Facsimile Technology by State Courts* (Williamsburg, VA: National Center for State Courts, 1992).

171. "Fax Ruling Causes a Legal Stir," *Wall Street Journal*, Nov. 22, 1988, B1; Ben Calabrese v. Springer Personnel of New York, Inc. 141 Misc. 2d 566; 534 N.Y.S.2d 83; 1988 N.Y. Misc. LEXIS 649, Oct. 24, 1988; Rorie Sherman, "Filing by Telephone," *National Law Journal*, March 6, 1989, 1, 45.

172. Paul Marcotte, "Fax of (Court) Life," *ABA Journal*, Feb. 1989, 29; Monica Lee, *Facsimile Transmission of Court Documents: A Feasibility Study. Fifty-State Survey of Fax Use by State Courts* (San Francisco: National Center for State Courts, Aug. 1990), 48–50.

173. Katherine Schweit and Janan Hanna, "Federal Courts Say Fax Filing a No-No," *Chicago Daily Law Bulletin*, June 7, 1989, 5, 14; Ken Myers, "Despite Some Doubts, Fax Filing Gains," *National Law Journal*, Jan. 22, 1990, 9.

174. "Minutes. Committee on Rules of Practice and Procedure," Jan. 19–20, 1989, 9, www.uscourts.gov/rules/ST01-1989-min.pdf; "Summary of the Report of the Judicial Conference Committee on the Rules of Practice and Procedure," Sept. 1993, 13, www.uscourts.gov/rules/Reports/ST9-1993.pdf; "Summary of the Report of the Judicial Conference Committee on Rules of Practice and Procedure," Sept. 1994, 1, 18–20, www.uscourts.gov/rules/Reports/ST9-1994.pdf; "Minutes. Advisory Committee on Civil Rules," Oct. 20–21, 1994, 2–3, www.uscourts.gov/rules/Minutes/CV10-1994.pdf.

175. "Minutes. Committee on Rules of Practice and Procedure," June 17–19, 1993, 3–4, www.uscourts.gov/rules/Minutes/june1993.pdf; "Minutes. Advisory Committee on Civil Rules," Oct. 21–23,1993, 2–3, www.uscourts.gov/rules/CV10-1993-min.pdf.

176. Talmadge, "The Fax of Life," 4.

177. "Tax Court Refuses Petition Transmitted by 'Zapmail,'" *Journal of Taxation*, Aug. 1986, 120; "State Court Roundup," *National Law Journal*, April 24, 1989, 36; Benjamin Wright, *The Law of Electronic Commerce, EDI, Fax, and E-mail: Technology, Proof, and Liability* (Boston: Little, Brown, and Company, 1991), 52, 288, 387; Catherine Rubio Kuffner, "Legal Issues in Facsimile Use," *Media Law & Policy* 5, 1 (Fall 1996), 13–14.

178. Connotech Experts-conseils, Inc., "Frequently Asked Questions about Fax Security," July 1999 www.connotech.com/FAXBIBL.HTM.

179. Keyes, "Do You Have Fax Appeal?" 138–40; Carol Kleiman, "Pursuing Participatory Democracy around the Old Office Fax Machine," *Chicago Tribune*, June 17, 1991, 3.

180. S. L. Berry, "Faxpionage!" *Security Management* 35, 2 (April 1991), 60.

181. M. L. Stein, "Fax Fluke Puts Pol's Poll in Hands of Press," *Editor & Publisher*, Aug. 6, 1988, 16, 29; Howard Kurtz, "'Talking Points' Spin Out of Control in Fax Fiasco," *Washington Post*, Oct. 17, 1992, A12; Fred Barbash, "Arms Deal Scandal Roils London Politics," *Washington Post*, June 20, 1995, A16.

182. "Beware! Fax Attacks!" *ABA Banking Journal*, June 1990, 52; "Police Communications in the Information Age," *FBI Law Enforcement Bulletin*, Dec. 1991, 12.

183. Mark Ilansen, "Misfaxed Papers Stop Asbestos Trial," *ABA Journal*, Aug. 1991, 22; James P. Ulwick, "Producing by Mistake," in *The Litigation Manual: A Primer for Trial Lawyers*, ed. John G. Koelth (Washington, DC: American Bar Association, 1999), 298–307.

184. Anne G. Bruckner-Harvey, "Inadvertent Disclosure in the Age of Fax Machines: Is the Cat Really Out of the Bag?" *Baylor Law Review* 46, 2 (June 1994), 385–98; Thomas L. Browne, "Avoiding Malpractice," *Chicago Lawyer*, June 1999, 10.

185. "Fax Machine Interception," *Full Disclosure* 23 (1991), reprinted in www.totse .com/en/phreak/phone_phun/faxint.html; Philip A. Godwin, "Interception and Interpretation of Information from a Subscriber Loop Phone Line," *Proceedings of the 1992 International Carnahan Conference on Security Technology: Crime Countermeasures, October 14–16 1992* (Piscataway, NJ: IEEE, 1992), 126–31; Terry Metzgar, "Hostile Intercepts Aimed at Information Systems," *National Defense*, May/June 1993, 24–26; William J. Orvis and Allan L. Van Lehn, *Data Security Vulnerabilities of Facsimile Machines and Digital Copiers* (Livermore, CA: Lawrence Livermore National Laboratory, Jan. 1995); G-COM Monitoring Technology, "Technical Report DAMS-MT/2" [1996?].

186. Kevin Braun, "Group Vulnerabilities Demand Methods to Practice Safe Fax," *Signal*, Aug. 1994, 57. Advertisements reinforced this message: [Cylink advertisement], "Would you be laughing if your faxes fell into the wrong hands?" *Signal*, Aug. 1994, 55.

187. David Clark Scott, "Cold War May Wane, but Soviet Spying on Faxes Is Up," *Christian Science Monitor*, Nov. 21, 1989, 6; Bruce Livesay, "Trolling for Secrets," *Financial Post*, Feb. 28, 1999, R1; Jeffrey Richelson, "Desperately Seeking Signals," *Bulletin of the Atomic Scientists* 56, 2 (March/April 2000), 47–51. And, of course, "The French are notorious for national economic espionage endeavors, such as breaking into hotel rooms, rifling briefcases, stealing laptop computers, eavesdropping on international business telephone calls, and intercepting faxes and telexes" (Winn Schwartau, *Information Warfare: Chaos on the Electronic Superhighway* [New York: Thunder's Mouth Press, 1994], 277).

188. Paul Taylor, "PLO Chiefs Use Fax-tech to Micro-manage Uprising," *Washington Times*, May 9, 1989, A8.

189. Daniel J. McAuliffe, "Command, Control, Communications, and Intelligence," *Aerospace America*, Sept. 1993, 45; Schwartau, *Information Warfare*, 126.

190. ANC Daily News Briefing, "Attempts Being Made to Sabotage NIA: Committee," Oct. 10, 1997, www.e-tools.co.za/newsbrief/1997/news1010.

191. Doreen Carvajal, "Swiss Investigate Leak to Paper on C.I.A. Prisons," *New York Times*, Jan. 12, 2006, A8; "Swiss Try to Calm Leaked Fax's Uproar," *New York Times*, Jan. 18, 2006, A6.

192. In 1999, 512 firms sold 805 products worldwide to encrypt data, voice, and fax transmissions (Lance J. Hoffman et al., *Growing Development of Foreign Encryption Products in the Face of U.S. Export Regulations*, Report GWU-CIPI-1999-02 [Washington, DC: George Washington University, June 10, 1999], i–ii, 37–45); Anne E. Nichols, "General Kinetics Expects Growth Boost from Line of Secure Facsimile Machines," *Wall Street Journal*, Nov. 2, 1990, A9D; National Intelligence Academy, "Advanced Installation of Telephone Intercepts" (Jan. 20, 2002), 11.

193. "Encryption and Network Fax Issues Surface at TCA," *EMMS*, Oct. 15, 1991, 7–8; John R. Wenek, "Fax Attack: Who's Getting the Message?" *Security Management*, May 1, 1994, 5; National Communications System, "Security and Facsimile" (Arlington, VA: National Communications System, Feb. 1995); "New Gameplan for Fax Security," *Human Communications Digest* 4, 2 (Oct. 1995), 5.

194. Department of the Treasury, "Foreign and Commonwealth Office," *National Asset Register* (London, 1998) http://hm-treasury.gov.uk/media/2/2/193.pdf.

195. Daniel M. Costigan, *Fax: The Principles and Practice of Facsimile Communication* (Philadelphia: Chilton Book Co., 1971), 23; "Magnovox Offers Facsimile Scrambler," *Electronic News*, Sept. 18, 1972, 27; "Fax Units Replace 'Copters as Communicators," *The Office*, Nov. 1982, 121.

196. Berry, "Faxpionage!" 60; John W. Verity, "Scrambled Faxes That Free You from Standing Guard," *Business Week*, July 4, 1994, 84E; Bob Wade and Thomas J. Rohr Sr., "A Snapshot of Communications Security at Kodak," *Security Management*, Nov. 1, 1998, 76.

197. National Counterintelligence Center, *Annual Report to Congress on Foreign Economic Collection and Industrial Espionage: 1998* (Washington, DC: National Counterintelligence Center, 1998), 8–9, 15.

198. Brock N. Meeks, "Fax Board Faire," *Byte* 13, 9 (Sept. 1988), 207.

199. Jeffery T. Walker, "Fax Machines and Social Surveys: Teaching an Old Dog New Tricks," *Journal of Quantitative Criminology* 10, 2 (1994), 181–88.

200. Geraldine Brooks, "How a Recurring Scam Cost an Accountant and His Wife $54,000," *Wall Street Journal*, June 24, 1994, A1, 4; "The Times City Diary," *London Times*, Jan. 21, 1994, 27.

201. Jonathan Kapstein, "Beware of the Fax Directory Hustle," *Business Week*, Oct. 8, 1990, 156; "World Wire," *Wall Street Journal*, March 5, 1992, A10.

202. Bill Lodge, "Jailed Businessman May Have to Incriminate Self to Get Out," *Dallas Morning News*, April 8, 2000, A32.

203. Celia F. McAllister, "Piercing the Chinese Wall: All the News That's Fit to Fax," *Business Week*, Dec. 4, 1989, 114G; M. L. Stein, "Alternative Weekly Participates in 'Fax-in,'" *Editor & Publisher*, Dec. 30, 1989, 23.

204. "AACRAO Fax Guidelines," http://aacrao.org/About-AACRAO/governance -and-leadership/policies/fax-guidelines.aspx#.UR1Tjo52ufR. I am grateful to Lisa Rosenberg for the citation.

205. Bruce Schneier, "Fax Signatures," in "Schneier on Security," June 3, 2008, www .schneier.com/blog/archives/2008/06/fax_signatures_1.html.

206. David B. Pearson and Douglas P. Sauter, "For the Practicing Editor," *Journal of Accountancy*, March 1990, 75–79; "Beware! Fax Attacks!" 53.

207. Nancy Wartik, "Life in the Fax Lane," *Ms.*, Nov. 1989, 42.

208. Andrew Murr and John Schwartz, "A Mounting Pile of 'Junk' Fax," *Newsweek*, July 25, 1988, 54.

209. Hearing on "Telemarketing Practices," Subcommittee on Telecommunications and Finance, House Committee on Energy and Commerce, May 24, 1989, 54 (hereafter, "Telemarketing Practices"); Ronald Alsop, "Marketing," *Wall Street Journal*, Dec. 27, 1988, B1; Blum, "Fax Mania," 44.

210. "Telemarketing Practices," 47; Alyson Pytte, "Avalanche of Fax Messages Lands Close to Home," *Congressional Quarterly*, Sept. 30, 1989, 2553–54.

211. "Telemarketing Practices," 54.

212. Alison Fahry, "States Ponder 'Junk Fax' Rules," *Advertising Age*, Jan. 16, 1989, 32; Richard A. Barton, DMA Senior Vice President of Government Affairs, "Telemarketing Practices," 27–32; Hearing on "Telemarketing/Privacy Issues," Subcommittee on Telecommunications and Finance, House Committee on Energy and Commerce, April 24, 1991, 105; Hearings on S. 1410 and S. 1462 Before the Subcommittee on Communications, Senate Committee on Commerce, Science, and Transportation, July 24, 1991, 35–36.

213. Pytte, "Avalanche of Fax Messages."

214. "House Bill Curtails Unsolicited Ads," *Congressional Quarterly*, Aug. 4, 1990, 2508; "FCC Bans Fax Ads, Prerecorded Telemarketing," *Home Office Computing*, March 1993, 10; Bob Donath, "Don't Try Relationship Marketing in Reverse," *Marketing News*, Oct. 21, 1996, 14.

215. "Businesses Tired of Faxed Ads Sue the Senders," *Wall Street Journal*, May 9, 1995, B1, 12; "Junk Faxes Are Illegal, So Why Are You Getting Them?" *Consumer Reports*, March 2004, 47; Government Accountability Office, "Telecommunications: Weaknesses in Procedures and Performance Management Hinder Junk Fax Enforcement," April 5, 2006; Gwen Fontenot, Raj Srivastava, and Anne Keaty, "Marketing and the Junk Fax Prevention Act: What Now?" *Journal of Database Marketing & Customer Strategy Management* 15, 1 (Oct. 2007): 49–55.

216. James Gleick, *The Information: A History, a Theory, a Flood* (New York: Pantheon Books, 2011), 319–20.

217. CIAJ data.

CHAPTER 6: **The Fax and the Computer**

Epigraph. "Ford Motor Company Chooses E-Sync Networks for Global On-Line Fax Services," *Business Wire*, Sept. 20, 1999.

1. Nicholas Negroponte, *Being Digital* (New York: Alfred A. Knopf, 1995), 187.

2. "Approximately 10% of Business Phone Lines Now Tied to Fax Machines," *EMMS*, March 15, 1991, 10–11; "Reports of the Demise of Fax Are Greatly Exaggerated," *EMMS*, Oct. 16, 1995, 7–8; LobsenzStevens, "Gallup/Pitney Bowes Fortune 500 Telecommunications Managers Fax Management and Cost Study," Executive Summary (New York, 1995); Frost & Sullivan, *World Internet Protocol Faxing Markets* (Mountain View, CA: Frost & Sullivan, 1998), 1–35.

3. Kara Blond, "Firms look for ways to combat the fax of life," *Dallas Morning News*, July 8, 1996, D1–2.

4. Peter J. Davidson, "100 Million Potential On-Ramps: The Worldwide Fax Machine Installed Base and Market Forecast, 1997–2002" (Burbank, CA: Dec. 1998), 6;

CIAJ, *Tsushin kiki chuki juyo yosoku* [Communications Equipment Demand Forecast] (Tokyo: CIAJ) data.

5. Paul Saffo, "Worlds in Collision," *Personal Computing*, July 27, 1990, 53–55.

6. E.g., Brooktrout, before being absorbed by Cantata, which in turn was purchased by Dialogic, had also acquired several smaller firms.

7. "Delrina Fax Strategy Acquires a Broader Form," *EMMS*, Jan. 15, 1992, 4–8.

8. "World Bank Beats Blitz," *Communications News* 27, 4 (April 1990), 47; Paula Klein, "Gains Aren't Just on Paper," *Information Week*, May 24, 1993, 51–52; Blond, "Firms Look for Ways."

9. "Busy Fax Receivers: Is the Answer Receive-Only or Receive-Optimized?" *EMMS*, Feb. 15, 1991, 10–13; "Add a Second and Third Fax Machine—or a Multifax Internet fax?" *EMMS*, June 3, 1991, 6–9.

10. PC fax software, PC fax modems, intelligent fax boards, LAN fax servers, fax-email gateways, wireless fax, fax-on-demand, and unified messaging ("Letter from the President," *Human Communications Digest* 5, 1 [July, 1996], 2).

11. Marketfinders, *Directory* (Austin: Marketfinders, 1993), 4–5; Aisha M. Williams, "The Fax Advantage," *InformationWeek*, Aug. 3, 1998, www.informationweek.com /694/94iufax.htm.

12. "Can LAN Fax Become a $10 Billion Market?" *EMMS*, Aug. 15, 1991, 1; Tom Brennan, "Beware the Fax Beast," *Network World*, Nov. 27, 1995, 51; Mary E. Thyfault, "Small Word, Big Savings," *InformationWeek*, April 22, 1996, 89.

13. John T. Mulqueen, "Better to Switch than Fight," *Data Communications*, March 1991, 58; Mitch Betts, "Catch a Ride on the Fax Wave," *Computerworld*, April 8, 1991, 53.

14. International Resource Development, Inc., *Facsimile and PC-Based Image Transmission Markets* (New Canaan, CT: IRDI, Feb. 1989), 19; "The Year in Messaging," *EMMS*, Jan. 4, 1988, 2; "It Was a Very Good Year," *EMMS*, Dec. 15, 1988, 3.

15. William M. Bulkeley, "Technology," *Wall Street Journal*, Nov. 24, 1993, B1; Nina Munk, "Fax Killer," *Forbes*, April 24, 1995, 74.

16. Don Dunn, "Let Your PC Do the Faxing," *Business Week*, June 3, 1991, 140.

17. Hank G. Magnuski interview, July 6, 1992.

18. International Resource Development Inc., *Consumables and Paper for Fax and E-Mail* (New Canaan, CT: IRDI, May 1989), 41.

19. Ned Snell, "Users Want the Bare Fax," *Datamation*, Sept. 1, 1992, 117; Dayna Delmonico, "Buyer's Guide," *Windows Magazine Shopper*, June 1993, 242–60.

20. "Has Alcom Reached the Top of the LAN/Fax Hierarchy?" *EMMS*, April 15, 1993, 5.

21. Saffo, "Worlds in Collision," 53. Another early problem was the inability to adjust the received image. If sent upside down, the only way to read a fax was to turn the monitor or the person.

22. Brock N. Meeks, "Fax Board Faire," *Byte*, Sept. 1988, 208.

23. Robert A Myers, "Japanese I/O Technology—Where It Is and How It Got There," *VLSI and Computer Peripherals CompEuro '89*, 3rd Annual European Computer Conference May 8–12 1989, P/2-4; "Are Computer/Fax Standards Failing the Group 3 Test?" *EMMS*, March 15, 1993, 5–6; "Can the Fax Industry Conform to Its Group 3 Tradition?" *EMMS*, Dec. 15, 1993, 6.

24. "Other TR-29 News," *Human Communications Digest* 1, 4 (April 1993), 10–11.

25. Shawn Willett and Kate Orrange, "Delrina continues to nurse WinFax Pro's growing pains," *InfoWorld*, Dec. 27, 1993, 6.

26. Eric Giler and David Duehren interview, Feb. 19, 2002; Magnuski interview.

27. "Did We Mis-CAS(T) GammaLink's Position on Fax Development Tools?" *EMMS*, April 15, 1991, 6; "LAN Fax Market Gets Significant New Products," *EMMS*, Oct. 1, 1991, 8.

28. "Intel Advances PC Fax Board Standard," *EMMS*, Sept. 1, 1988, 4–6; "Consensus Evaporating on New Fax Modem Standard," *EMMS*, Jan. 15, 1991, 1–5; "de/FAX: The Missing Link or Just Symptomatic of the Missing Link," *EMMS*, March 15, 1991, 8; Magnuski interview.

29. "Athena Springs Forth as FaxBios," *EMMS*, Sept. 3, 1991, 8–9; "Can Computer/ Fax Fulfill the Great Expectations Created For It?" *EMMS*, Jan. 1, 1993, 6; "Facsimile API News," *Human Communications Digest* 1, 5 (July 1993), 14–15.

30. Not until Intel's 1991 support of a French standard did some American firms realize standardization efforts existed outside the United States ("The Fax API Wars: A Last-Minute French Kiss," *EMMS*, Dec. 16, 1991, 12).

31. "Consensus Evaporating on New Fax Modem Standard," *EMMS*, Jan. 15, 1991, 3; "New Version of T.611 Program Interface Approved," *Human Communications Digest* 3, 1 (July 1994), 3–4; "International Standards Passed for Class 1 and 2 Fax Interfaces," *Human Communications Digest* 3, 4 (April 1995), 1–3; Dennis Fowler, "Fax Forward on the Network," *PC Computing*, June 1995, 113.

32. "Seven Dwarves and One Giant Chase $200 Million in Windows Fax Software," *EMMS*, Dec. 1, 1992, 8–11; Sharon Massey, "Microsoft Offers Software to Link Office Machines," *Wall Street Journal*, June 9, 1993, B1, 8; "Microsoft Announces 'At Work' Architecture That Permits Messaging Between Copiers, Printers, Computers, and Faxes," *EMMS*, June 15, 1993, 1–4; "Has MAW Been Laid Off?" *EMMS*, Jan. 9, 1995, 6–9; Nina Munk, "Fax Killer," *Forbes*, April 24, 1995, 74; Nick Baran, "Whatever Happened to . . . Microsoft at Work," *Byte*, July 1995, 30. http://www.byte.com/art/9507/sec3/art5 .htm.

33. Craig Stinson, *Running Microsoft Windows 95* (Redmond, WA: Microsoft Press, 1995), 641–74; "Midnight Madness Ushers Fax into a Windows 95 Dawn," *EMMS*, Sept. 4, 1995, 3–7.

34. Gerry Blackwell, "Exploring the LAN Fax Option," *Canadian Datasystems* 24, 7 (July 1992), 41; F&S, *U.S. Computer Facsimile*, 9-6.

35. "Fax Machines: If You Can't Beat 'em, Join 'em," *EMMS*, Aug. 15, 1992, 8–10; "Office Staffers Cling," *Wall Street Journal*, April 23, 1996, A1.

36. "Can LAN Fax Overcome Its Poor Inbound-Routing Image?" *EMMS*, Dec. 2, 1991, 6–13.

37. "CCITT to Adopt Changes to Group 3 Fax Standard," *Human Communications Digest* 1, 1 (Aug. 1992), 2; Tom McCusker, "Easing the LAN Fax Burden," *Datamation*, July 15, 1993, 76; Peter Davidson, *Computer-Based Fax, 2004–2009* (Burbank, CA: Davidson Consulting, 2004), 4.

38. Other names included interactive fax, fax retrieval, fax databasing, fax bulletin boards, automated fax processing, fax response, and automated integrated voice/fax processing ("Fax-on-Demand Seeks Definitions in the Market and the Glossary," *EMMS*, Jan. 15, 1993, 9).

39. Theresa Henry, "Fax Technology Allows Retrieval of Cancer Research," *Computing Canada*, Aug. 4, 1992, 45–47; William M. Bulkeley, "Fax-on-Demand Provides Ready Answers," *Wall Street Journal*, Oct. 20, 1992, B3; "Rats on Demand," *InformationWeek*, Jan. 9, 1995, 7; Maury S. Kauffman interview, Jan. 17, 2008.

40. Philip C. W. Sih, *Fax Power: High Leverage Business Communications* (New York: Van Nostrand Reinhold, 1993), 109–33; Maury S. Kauffman, "'FODBO' Is the Way to Go," *Voice Processing Magazine*, Nov. 1993, 24–26; Jerry Lazar, "The Truth about Fax," *Information Week*, Feb. 19, 1996, 56.

41. Richard Cross, "Fax-on-Demand: A Tool for Instant Interactive Marketing," *Direct Marketing*, May 1, 1994, www.allbusiness.com/technology/telecommunications -phone-systems/463422-1.html; Maury Kauffman, "Instant Gratification," *Marketing Tools*, Jan/Feb 1995, 64–66; Eric Carr, "Fax-on-Demand for Everyone," *Network Computing*, May 1, 1995, 56.

42. Seth Godin, *eMarketing: Reaping Profits on the Information Highway* (New York: Perigee Books, 1995), 55; Cecil C. Hoge Sr., *The Electronic Marketing Manual* (New York: McGraw-Hill, 1993), 175–85.

43. Betsy Wiesendanger, "Electronic Delivery and Feedback Systems Come of Age," *Public Relations Journal*, Jan. 1993, 11.

44. Mitch Betts, "Catch a Ride on the Fax Wave," *Computerworld*, April 8, 1991, 53; "Information Processing," *Business Week*, Oct. 9, 1989, 148A.

45. Tinkelman Enterprises hosted New York City and faxed thousands of pages monthly in 1994 for only $100 in local telephone charges ("Experts Demonstrate Fax Deliveries over the Internet," *EMMS*, Sept. 15, 1993, 6–8; "After 1 Year, Internet-to-Fax Experiment Continues," *EMMS*, July 11, 1994, 1–5).

46. In 1996, FaxSav charged 45 cents, a quarter of regular rates, to fax one page as email from America to Japan (Mary E. Thyfault, "Small Word, Big Savings," *InformationWeek*, April 22, 1996, 89; Liz Simoneau, "Is It Time to Unplug Your Fax Machine?" *Harvard Business Review* 76, 5 [Sept/Oct 1998], 14–15).

47. "Internet Fax Standards Work Underway," *Human Communications Digest* 5, 3 (Jan. 1997), 7; William Dutcher, "CTI Message-Handling Utopia Just Around Corner," *PC Week*, April 21, 1997, 86; Erich Almasy and Catherine Kunkmueller, "Right Around the Corner?" *Telephony Online*, Jan. 12, 1998, http://telephonyonline.com/mag /telecom_right_around_corner/index.html; Richard Sewell, "The Time Is Now—Swim and Swim Fast," *Telephony Online*, Feb. 16, 1998, http://connectedplanetonline.com /mag/telecom_time_nowswim_swim.

48. Fred Hapgood, "Just the (Digital) Fax," *CIO WebBusiness Magazine*, Aug. 1998, www.cio.com/archive/webbusiness/080198_power.html.

49. Gartner Dataquest survey, "Fax Attitudes and Buying Behavior: A Study of Ownership, Attitudes, and Buying Processes in U.S. Businesses" (FMFP-NA-UW-0001), 2001, 5.

50. "Global Fax, Local Call," *Data Communications*, Nov. 21, 1995, 16; "Beyond the Start-up Nation," *The Economist*, Jan. 1, 2011, 60; F&S, *World Internet Protocol Faxing Markets*, 1-52-54.

51. Denise Pappalardo, "GTE and ATT WorldNet Add IP Faxing Services," *Network World*, March 23, 1998, www.networkworld.com/news/0323ipfax.html; Vince Vittore, "IP Fax Picks Up Steam: Vendor Alliances, Real-time Capabilities Add Fuel to

Red-hot Market," *Telephony Online*, Sept. 14, 1998, http://telephonyonline.com/mag
/telecom_ip_fax_picks/index.html; Kim Shiffman, "Size Matters," *Profit*, Dec. 2004, 16.

52. "WorldCom, Inc.," *Wall Street Journal*, July 9, 1997, B4; Sewell, "The Time Is
Now"; http://connectedplanetonline.com/mag/telecom_time_nowswim_swim/Brian
Morrissey, "Will WorldCom's Woes Engulf UUNet?" *internetnews.com*, July 3, 2002,
www.internetnews.com/bus-news/article.php/1380561.

53. Names included email fax, messaging fax, store-and-forward internet faxing,
node-to-nearest-node routing, iFax, least-cost routing, bypass point-point network faxing,
and IP faxing (James E. Gaskin, "Faxing across the Net," *Information Week*, June 23,
1997, 54–59; Kiyoshi Toyoda, Shiro Tamoto, and David H. Crocker, "Internet Facsimile
as an Internet Office Appliance," *IEEE Communications Magazine*, Oct. 2001, 60).

54. "TR-29 Roundup," *Human Communications Digest* 5, 2 (Oct. 1996), 11.

55. "Internet Fax Standards Discussions Begin," *Human Communications Digest* 5, 2
(Oct. 1996), 3; Chris Bucholtz, "Fax Fees Get the Ax: ISPs Move in on Fax Market with
Services that Cut Fax Costs," *Telephony Online*, Feb. 16, 1998, http://telephonyonline
.com/mag/telecom_fax_fees_ax/index.html.

56. Sewell, "The Time Is Now." http://connectedplanetonline.com/mag/telecom
_time_nowswim_swim/

57. IETF, "A Mission Statement for the IETF," RFC 3935, Oct. 2004, www.ietf.org
/rfc/rfc3935.txt; Harald Alvestrand, "The Role of the Standards Process in Shaping the
Internet," *Proceedings of the IEEE* 92, 9 (Sept. 2004), 1371–74.

58. A major conduit was the first large Japanese internet effort, the Widely Integrated
Distributed Environment ("TR-29 Roundup," *Human Communications Digest* 7, 3 [Jan.
1999], 27; "Activities by Ohno Laboratory [WIDE Project] for Internet FAX," www.ohno
lab.org/researches/ifax/ifax_with_ohnolab-e.html; Ken Coates and Carin Holroyd, *Japan and the Internet Revolution* [London: Palgrave Macmillan, 2003], 47).

59. Including the Electronic Messaging Association, Multi-Function Peripheral Association, Enterprise Computer Telephony Forum, and Group 5 Forum ("Internet Fax
Standards Work Underway," *Human Communications Digest* 5, 3 [Jan. 1997], 4).

60. "Internet Fax Priorities Set," *Human Communications Digest* 5, 4 (April 1997),
16–21; Bridget Mintz Testa, "Special Delivery," *Telephony Online*, Sept. 20, 1999, http://
telephonyonline.com/mag/telecom_special_delivery/index.html; "SIP Forum Fax over
IP Interop Group Reaches Milestone," www.phoneplusmag.com/news/2009/09/sip
-forum-fax-over-ip-interop-group-reaches-miles.aspx.

61. Anne Zieger, "Fax Plods Along," *InformationWeek*, Sept. 27, 1999, www.informa
tionweek.com/754/fax.htm; "Clarity Integrates the Office Fax Machine with the Internet," *Business Wire*, June 3, 1997.

62. Dataquest, "Dataquest Internet Fax Study Preliminary Survey Analysis," July 25,
1997 [Powerpoint presentation]; Center for Telecommunications Management, "The
Telecom Outlook Report, 1999–2010. White Paper" (Chicago: International Engineering Consortium, 1999).

63. Dave Trowbridge, "IP Faxing Could Save Millions," *Computer Technology Review* 16, 12 (Dec. 1996), 1; "US West Calls on Faxstorm," *Telephony Online*, Oct. 18,
1999, http://telephonyonline.com/mag/telecom_west_calls_faxstorm/index.html; "With
Emphasis on Text Messaging, Fax Over Internet Almost Overlooked," *EMMS*, April 14,
2000, 1.

64. Dataquest, "Dataquest Internet Fax Study"; Bernard A. Gelb, "RS20319—Telecommunications Services Trade and the WTO Agreement," Congressional Research Service, Jan. 10, 2001.

65. Federal Communications Commission, *Trends in Telephone Service* (Washington, DC: GPO, Feb. 2007), table 6.1, http://hraunfoss.fcc.gov/edocs_public/attach match/DOC-270407A1.pdf.

66. A similar venture earlier that decade failed. HomeFax experimented with the concept of receiving free equipment or services in exchange for personal information but could not attract the venture capital to create an "advertiser-supported home fax network" (Esther Dyson, "Who Pays for Data," *Forbes*, Feb. 3, 1992, 96).

67. "Free Faxes Online," *Washington Post*, "Washington Business," May 10, 1999, 3; Charles Whaley, "Fax Is Still a Favourite, Despite the Alternatives," *Computing Canada*, June 25, 1999, 9; Saul Hansell, "In the Wired World, Much Is Now Free at Click of a Mouse," *New York Times*, Oct. 14, 1999, A1, C21; Rob Turner, "Fax Service Without a Fax Machine," *Money*, June 1, 2000, 181.

68. Seth Schiesel, "Voice Mail and Faxes Add Intrigue to E-Mail Offer," *New York Times*, June 7, 1999, C4; Lewis Perdue, "Well-Financed onebox.com Enters Crowded Unified Messaging Market," *internetnews.com*, July 27, 1999, www.internetnews.com /bus-news/article.php/16947; Mary E. Thyfault, "Small Word, Big Savings," *informationweek*, April 22, 1996, 89; Todd Wallack, "Phone.com to Acquire Onebox.com," *San Francisco Chronicle*, Feb. 15, 2000, C1.

69. Maury Kauffman, "Fax over IP Profits (Part I)," *Computer Telephony*, Jan. 1999, 143; Heather Harreld, "Fax over IP: Great Expectations," *FCW.com*, Nov. 1, 1999, www .fcw.com/fcwitharticles/1999/FCW_110199_70.asp; Davidson, *Computer-Based Fax*, 2004–2009.

70. Kevin Savetz, "FAQ: How Can I Send a Fax from the Internet?" May 9, 2006, www.savetz.com/fax; j2 Global Communications, SEC Form 10-K for 2000, 27, 33; 2002, 3, 26; 2010, 3–4, 22.

71. Don J. DeBenedictis, "Idaho Courts OK Fax," *ABA Journal*, May 1990, 19–20.

72. Geoffrey Wheelwright, "New Front Opens in the Net versus Fax Battle: Internet Faxing," *Financial Times*, March 21, 2001, 2.

73. Michael Gianturco, "The $188 fax machine," *Forbes*, March 2, 1992, 106.

74. International Resource Development, Inc., *Facsimile Markets* (Norwalk, CT: IRDI, Oct. 1985), 133. The reference was to *The Techno/peasant survival manual* (New York: Bantam Books, 1980).

75. E.g., fax services in 1969 were "what amounts to electronic message delivery services" ("Courts and Commissions," *Telephony*, July 5, 1969, 15, 34) and in 1978 faxing was "becoming one of the primary vehicles of so-called electronic mail" ("Burroughs Offering High-Speed Machine for Sending of Data," *Wall Street Journal*, Jan. 25, 1978, 7); Peter Vervest, *Electronic Mail and Message Handling* (Westport, CT: Quorum Books, 1985), 12.

76. "What to Do with the X.400 Mess?" *EMMS*, Nov. 15, 1991, 8; B. Plattner et al., *X.400 Message Handling: Standards, Interworking, Applications*, Stephen S. Wilson, trans. (Wokingham, UK: Addison-Wesley, 1991), 163–64; John Rhoton, *X.400 and SMTP: Battle of the E-mail Protocols* (Boston: Digital Press, 1997), 4, 67–70.

77. Margaret A. Emmelhainz, *Electronic Data Interchange: A Total Management Guide* (New York: Van Nostrand Reinhold, 1990), 63.

78. Gerry Blackwell, "Exploring the LAN Fax Option," *Canadian Datasystems* 24, 7 (July 1992), 41; Joe Dysart, "Shippers Send EDI messages via FAX to Save Time, Money," *Traffic World*, April 9, 1990, 39–41; Richard H. Baker, *EDI: What Managers Need to Know about the Revolution in Business Communications* (Blue Ridge Summit, PA: TAB Books, 1991), 6–13; Max Anhoury and Roger D. Brown, "Fax Can Surmount EDI Format Snafus," *Business Credit* 94, 5 (May 1992), 18–19.

79. John Tydeman and Laurence Zwimpfer, "Videotex in the United States—Toward Information Diversity," *Videotex '81* (Middlesex, UK: Online Conferences, Ltd., 1981), 348–50; Vincent Mosco, *Pushbutton Fantasies: Critical Perspectives on Videotex and Information Technology* (Norwood, NJ: Ablex Publishing, 1982), 67–84.

80. James Martin, *Viewdata and the Information Society* (Englewood Cliffs, NJ: Prentice-Hall, 1982), 3; "Teletex and Computer-Based Message Systems," *EMMS*, May 3, 1982, 11.

81. Svend Erik Jeppesen and Knud Brunn Poulsen, "The Text Communications Battlefield," *Telecommunications Policy* 18, 1 (1994), 66–77; Christopher H. Sterling, "Pioneering Risk: Lessons from the US Teletext/Videotex Failure," *IEEE Annals of the History of Computing* 28, 3 (July-Sept. 2006), 41–47.

82. Vervest, *Electronic Mail*, 69; "WU Easylink Service Now Available," *PR Newswire*, June 29, 1987; "Electronic Messaging—A $10 Billion Franchise Up for Grabs," *EMMS*, June 17, 1991, 3; Daniel J. Blum and David M. Litwack, *The E-Mail Frontier. Emerging Markets and Evolving Technologies* (Reading, MA: Addison-Wesley, 1995), 8–11.

83. L. Brett Glass, "Fax Facts," *Byte*, Feb. 1991, 301-08; "Ballot," *EMMS*, Jan. 15, 1994, 12.

84. Andy Reinhardt, "Smarter Copiers, Printers, and Fax Devices Are Coming," *Byte*, Nov. 1994, 81; Kate Orrange, "Binary file transfer to transform faxing," *InfoWorld*, Dec. 6, 1993, 6.

85. "U.S. Standard for Binary File Transfer Approved," *Human Communications Digest* 1, 5 (July 1993), 11; "Microsoft Announces 'At Work' Architecture That Permits Messaging Between Copiers, Printers, Computers, and Faxes," *EMMS*, June 15, 1993, 2.

86. Rhoton, *X.400 and SMTP*, 88, 97.

87. Sanjiv P. Patel, Grant Henderson, and Nicolas D. Georganas, "The multimedia fax-MIME gateway," *Multimedia* 1, 4 (Winter 1994), 64–70; Marshall T. Rose and David Strom, *Internet Messaging. From the Desktop to the Enterprise* (Englewood Cliffs, NJ: Prentice Hall, 1998), 290; Wheelwright, "New front"; Michael Osterman, "Moving really large files efficiently," *Network World* Nov.30, 2006, www.networkworld.com/news letters/gwm/2006/1127msg2.html.

88. Barbara Gengler, "The Death of the Fax?" *Communications International* 26, 9 (Sept. 1999): 74–76; Printer Working Group, "Frequently Asked Questions," www.pwg .org/ipp/faq.html.

89. Richard Brandt, "Does Adobe Have a Paper Cutter?" *Business Week*, Nov. 16, 1992, 83–85; Jim Carlton, "Adobe Systems to Unveil Program to Let Computers Ship Complex Documents," *Wall Street Journal*, June 14, 1993, 4; Jeanette Borzo, "Tools Resurrect Hope for Paperless Office Concept," *InfoWorld*, June 14, 1993, 1, 24.

90. P. J. Connolly, "Acrobat Updates Put a Shine on PDFs," *InfoWorld*, Aug. 4, 2003, 34; Lee Thé, "Paper-free Publishing Gets Real" *Datamation*, Sept. 1, 1993, 39–41; "Driving Adobe: Co-founder Charles Geschke on Challenges, Change, and Values," *Knowledge@Wharton*, Sept. 3, 2008, http://knowledge.wharton.upenn.edu/article.cfm.

91. Matt Bradley, "Whatever Happened to the Paperless Office?" *Christian Science Monitor*, Dec. 12, 2005, www.csmonitor.com/2005/1212/p13s01-wmgn.html; Gregory Deatz, "The Ubiquitous Fax Goes Digital," *Infonomics* 21, 1 (Jan/Feb 2007), 55–57.

92. Janice Maloney, "Goliath.com Still Winning, but David Has an On-Line Sling," *New York Times*, Sept. 22, 1999, 34.

93. ITU, *World Telecommunication Development Report* (Geneva: ITU, 1994), 43; ITU and TeleGeography, Inc., *Direction of Traffic: International Telephone Traffic, 1983–1992* (Geneva: ITU, 1994), 3; ITU, "Key Global Telecom Indicators for the World Telecommunication Service Sector," 2006, www.itu.int/ITU-D/ict/statistics/at_glance/KeyTelecom99.html.

94. Matt Kramer, "Within Limits, Fax Shines as On-line Services Option," *PC Week*, Oct. 23, 1989, 76.

95. Netcraft, "November 2006 Web Server Survey," http://news.netcraft.com/archives/2006/11/01/november_2006_web_server_survey.html; "Internet 2012 in numbers," Jan. 16, 2013, http://royal.pingdom.com/2013/01/16/internet-2012-in-numbers.

96. Thomas Haigh, "The Web's Missing Links: Search Engines and Portals," in *The Internet and American Business*, ed. William Aspray and Paul E. Ceruzzi (Cambridge, MA: MIT Press, 2008), 159–99.

97. "What Has Happened to On-Demand Fax Systems?" *Digital Publisher* 1, 2 (1996), 22.

98. Alison L. Sprout, "The Internet Inside Your Company," *Fortune*, Nov. 27, 1995, 168; "Economists Switch to Efficient E-Mail, Shelving Costly Faxes," *Wall Street Journal*, June 8, 1999, B4; "Times Fax Service Expands," *New York Times*, Feb. 16, 2001, C4; Tom Colbert interview, Sept. 13, 2010.

99. Michael W. Von Orelli, *Bloomberg Basic Manual* (Bethlehem, PA: Lehigh University, 2001), http://firestone.princeton.edu/econlib/blp/docs/bloombergmanuallehighuniversity.pdf; "Bloomberg," http://ketupa.net/bloomberg.htm.

100. Claudia H. Deutsch, "Pitney Bowes Is Casting Off Copier and Fax Units Today," *New York Times*, Dec. 3, 2001, C9.

101. Lisa Nadile, "Fax on Demand Hits Web," *PC Week*, Sept. 4, 1995, 16; Hillary Rettig, "Fax the Web! New Technologies Let you Hook Up Your Fax Solution to the Internet," *VARbusiness*, July 15, 1996, 47; Barbara DePompa, "To Send a Fax, Click Here," *Information Week*, Oct. 14, 1996, 68. In Japan, customers faxed a web address to Nissin Electric, which faxed the webpage's contents back ("Anything to Avoid Using a Keyboard!" *Japan Internet Report* 23 (Jan.1998), http://japaninternetreport.com/jir1_98.html).

102. Jun Fukami and Takayuki Naitoh interview, July 15, 2009; Castelle, SEC Form 10-K for 2004, 3.

103. Maury Kauffman, "There's No Denying the Fax," *Information Week*, Feb. 20, 1995, http://informationweek.com/515/15uwfw.htm.

104. LobsenzStevens, "1998 Pitney Bowes Fax Usage and Technology Study," Execu-

tive Summary; Scott Cullen, "The year in fax," *Office World News* 26, 12 (Dec. 1998), 1.

105. E.g., Teresa Leung, "Castelle Champions Fax Renaissance," *Asia Computer Weekly*, April 11, 2005, 1; Erica Christoffer, "Point, Shoot, and Fax, *Realtor Magazine* 44, 1 (Jan. 2011), 16.

106. David Haueter, "The 2000 U.S. Fax Market: Unit Shipments Up, Spending Down," Gartner Dataquest Market Analysis, April 3, 2001, 2–5.

107. j2 Global Communications, SEC Form 10-K for 2002, 6, 13.

108. Castelle, SEC Form 10-K for 2004, 4; Captaris, SEC Form 10-K for 2007, 1; www .networkworld.com/news/2008/090408-open-text-to-buy.html.

109. Michael Kinsley, "The Morality and Metaphysics of Email," *Forbes*, Dec. 2, 1996, S113.

110. International Data Corporation, *Worldwide Facsimile Services Market Review and Forecast, 1997–2002* (Framingham, MA: IDC, Aug. 1998), 17–18, 22; Frost & Sullivan, *U.S. Facsimile Machine Markets* (Sunnyvale, CA: Frost & Sullivan, 1998), www .highbeam.com/doc/1G1-53017030.html.

111. Hiro Hiranandani interview, Feb. 14, 2002; Ryan Ginstrom, "End of an Era: Goodbye, Fax-to-Email Service," Jan. 15, 2009, http://ginstrom.com/scribbles/2009/01/15 /end-of-an-era-goodbye-fax-to-email-service.

112. Excepting the dedicated fax circuits of Hong Kong (D. C. A. Connolly, "Development of Data and Fax Communications in Hong Kong," *IEEE Region 10 Conference on Computer and Communication Systems* [Hong Kong: IEEE, Sept. 1990], 793–96).

113. E.g., Peter Davidson, *Worldwide Fax Machine Report, 2001* (Burbank, CA: Davidson Consulting, 2002) and "Computer-Based Fax, 2004–2009."

114. Sadayuki Ito and Shinitsu Narazaki interview, April 21, 2009.

115. Claudio Allocchio email communication, Sept. 2, 2010.

116. Communications and Information Network Association of Japan, *Yearbook 2002* (Tokyo: CINAJ, 2002), 2.

117. E.g., Anna Jane Grossman, *Obsolete* (New York: Abrams Image, 2009), 74–76.

118. CIAJ annual data, 1980–2009.

119. Communications and Information Network Association of Japan, *Yearbook 2002* (Tokyo: CIAJ, 2002) 14; Eric A. Taub, "Ease of Paperless E-Mail Sidelines the Forlorn Fax," *New York Times*, March 13, 2003, E7.

120. Robert Johnson, "The Fax Machine: Technology That Refuses to Die," *New York Times*, March 27, 2005, Business 8.

121. General Accounting Office, "Diffuse Security Threats: Technologies for Mail Sanitization Exist, but Challenges Remain" (Washington, DC: GAO, April 2002); John W. Miller and Maarten van Tartwijk, "Hacking Scandal Roils Dutch Public," *Wall Street Journal*, Sept. 7, 2011, A11.

122. "Comdex Fall: Have Multifunctionals Come of Age?" *EMMS*, Nov. 27, 1995, 3–5; "Is It a Copier, Fax Machine, or Printer?" *PC Week*, Jan. 20, 1997, 46.

123. CIAJ, 2007, 62–63; 2009, 63.

124. Haueter, "The 2000 U.S. Fax Market," 2–5; CIAJ, 2001, 51; 2004, 19–20.

125. Jerry Lazar, "The Truth about Fax," *Information Week*, Feb. 19, 1996, 54; Lenny Liebmann, "The Fax of Life," *Communications News* 39, 3 (March 2002), 58; Susan Avery, "Fax Continues to Battle Competition from E-mail," *Purchasing*, May 20, 2004,

49–50; James E. Gaskin, "Undying Facsimile," *Network World*, May 4, 2006, www.net workworld.com/newsletters/sbt/2006/0501networker3.html.

126. "But Which Way Is Up?" *The Economist*, Jan. 25, 2003, 68.

127. Aisha M. Williams, "The Fax Advantage," *InformationWeek*, Aug. 3, 1998, www .informationweek.com/694/94iufax.htm.

128. Davidson, "Fax Services, 2004–2009," 15-16; Robert Lowes, "Tech Talk: Facts about Paperless Faxing," *Medical Economics* 84, 3 (Feb. 2, 2007), 24–25; "Plumbers in Suits," *The Economist*, July 6, 2013, 63.

129. Peter Kaufman, "Doctors May Now Electronically Prescribe Schedule II Drugs," July 1, 2010, www.practicalpainmanagement.com/treatments/pharmacological /doctors-may-now-electronically-prescribe-schedule-ii-drugs; "Pharmacist's Manual," www.deadiversion.usdoj.gov/pubs/manuals/pharm2/pharm_content.htm; The Little Bluebook, 2012 *National Physicians Survey*, http://static.sharecare.com/promo/tlbb /tlbb_nps_2012_24.pdf; Erin McCann, "Getting the Fax Straight," Aug. 3, 2012, www .healthcareitnews.com/news/getting-fax-straight; Sachin H. Jain, "Care Resolution for 2014: Let's Retire the Fax Machine," *Huffington Post*, Feb. 10, 2014, www.huffingtonpost .com/sachin-h-jain-md/a-late-resolution-for-the_b_4733235.html.

130. Jaikumar Vijayan, "E-signatures Slow to Gain Ground," *Computerworld*, Oct. 22, 2001, 1; Coates and Holroyd, *Japan and the Internet Revolution*, 58; "Summary of the Law Governing the Use of Information and Communications Technology in the Preservation of Documents that Private Businesses Perform," www.kantei.go.jp/foreign /policy/it/051031/summary.pdf.

131. Nanette Gottlieb, *Word-Processing Technology in Japan: Kanji and the Keyboard* (London: Curzon, 2000), 64.

132. Miwako Doi and Haitao Lei, "Word Processing for the Japanese Language," IEEE Global History Network, http://ieeeghn.org/wiki/index.php/STARS:Word_Pro cessing_for_the_Japanese_Language.

133. Ministry of Public Management, Home Affairs, Posts and Telecommunications, *Information and Communications in Japan: Building a Ubiquitous Network Society That Spreads Throughout the World. White Paper 2004* (Tokyo: MPMHAPT, 2004), 49.

134. Doi and Lei, "Word Processing."

135. Gottlieb, *Word-Processing Technology*, 5; IPSJ Computer Museum, "Japanese Word Processors," http://museum.ipsj.or.jp/en/computer/word/history.html.

136. Yasunori Baba, Shinji Takai, and Yuji Mizuta, "The User-Driven Evolution of the Japanese Software Industry: The Case of Customized Software for Mainframes," in *Evolution of the Japanese Software Industry*, ed. David Mowry (Oxford: Oxford University Press, 1995), 104–30; Hisao Fujimoto, "Evolution of Marketing Channel of Distribution in Japan," *Osaka Keidai Ronshu* 57, 5 (Jan. 2007), 54.

137. Joel West and Jason Dedrick, "Competing through Standards: DOS/V and Japan's PC Market" (Berkeley, CA: Center for Research on Information Technology and Organizations, Nov. 30, 1996), 16; Andrew Pollack, "U.S. Leads Japanese into Information Age," *New York Times*, Sept. 4, 1995, 35.

138. Reiji Yoshida, "Will Windows 98 Spur PC Revival?" *Japan Times*, July 27, 1998, http://search.japantimes.co.jp/cgi-bin/nn19980727b1.html. See also John McGlynn, "Windows 95J: Marketplace Evolution, or Revolution?" *Computing Japan*, Dec. 1995,

www.japaninc.com/cpj/magazine/issues/1995/dec95/1295winevolu.html; Joel West, "Moderators of the Diffusion of Technological Innovation: Growth of the Japanese PC Industry" (Berkeley, CA: Center for Research on Information Technology and Organizations, Jan. 1, 1996), 10–12.

139. Bruce Hahne, "Japan Slow on Internetworking," *Computer-Mediated Communication Magazine* 1, 2 (June 1, 1994), 3, www.ibiblio.org/cmc/mag/1994/jun/japan.html; Michael Zielenziger, *Shutting Out the Sun: How Japan Created Its Own Lost Generation* (New York: Vintage, 2007), 113-14; American Chamber of Commerce in Japan, *Internet Economy White Paper* (Tokyo: ACCJ, 2009), 40.

140. "Industry News Briefs," *Japan Internet Review* 17 (July 1997), http://japaninter netreport.com/jir7_97.html; Coates and Holroyd, *Japan and the Internet Revolution,* 48–50, 57, 68.

141. Sven Lindmark, Erik Bohlin, and Erik Andersson, "Japan's Mobile Internet Success Story—Facts, Myths, Lessons and Implications," *Info: the Journal of Policy, Regulation and Strategy for Telecommunications* 6, 6 (2004), 350–52.

142. Lara Srivastava, "Ubiquitous Network Societies: The Case of Japan" (Geneva: ITU, April 2005), 19.

143. A similar variation in use by age occurred with word processors in the 1980s (Gottlieb, *Word Processing Technology,* 62–63).

144. Fukami and Naitoh interview.

145. "Fax Machines Still Going Strong," *Daily Yomiuri,* Oct. 20, 2003, https://data base.yomiuri.co.jp/rekishikan; CIAJ, 2000, 6.

146. Matthew Edmund interview, July 28, 2009.

147. Eric Gan interview, Feb. 24, 2009; Ko Fujii personal communication, April 13, 2009.

148. CIAJ, 1999, 43; 2000, 26; 2001, 52; 2006, 56; "Telephone Service Subscribers," *Japan Statistical Handbook* 2010, www.stat.go.jp/english/datga/hankdbook/co8cont.htm.

149. "Telephone Service Subscribers."

150. *NTT Annual Reports* 2003, 27, 33; 2005, 39, 45; 2008, 59; NTT, SEC Form 20-F for 2009, 24.

151. Ryoji Sakai, executive officer, 7-Eleven Japan, personal communication, June 15, 2009.

152. Chico Harlan, "In Japan, Fax Machines Remain Important Because of Language and Culture," *Washington Post,* June 7, 2012, www.washingtonpost.com/world /asia_pacific/in-japan-fax-machines-find-a-final-place-to-thrive/2012/06/07/gJQAsh FPMV_story_1.html; Martin Fackler, "In High-Tech Japan, the Fax Machines Roll On," *New York Times,* Feb. 14, 2013, A1, 10.

Conclusion

Epigraph. Thomas Edison to John Clark Van Duzer, December 6, 1868, *Papers of Thomas A. Edison* (Baltimore: Johns Hopkins University Press, 1989), v. 1, 95 [emphasis in original].

1. Joseph N. Pelton, *Global Talk: The Marriage of the Computer, World Communications and Man* (Brighton UK: Harvester Press, 1981), 25.

2. "The 'Daily Mirror's' Fourth Birthday Celebrated by the Installation of the Korn

Machine for Telegraphing Photographs," *Daily Mirror*, Nov. 2, 1907, 1; George Seldes, *The Vatican: Yesterday, Today, Tomorrow* (New York: Harper & Brothers, 1934), 123; Molesworth Associates, "Alden Weather-recording Electronic Device Chosen by U.S. for Tunis International Fair," Oct. 13, 1960, "AEIREC" press releases & clippings, April 1–Oct. 15, 1960. Alden files.

3. Naomi S. Baron, *Alphabet to Email: How Written English Evolved and Where It's Heading* (London: Routledge, 2000), 189–193.

4. Clayton M. Christensen, *The Innovator's Dilemma: When New Technologies Cause Great Firms to Fail* (Boston: Harvard Business School Press, 1997), 77–84.

5. Susanne K. Schmidt and Raymund Werle, *Coordinating Technology: Studies in the International Standardization of Telecommunications* (Cambridge, MA: MIT Press, 1998), 41.

6. Amy McRary, "Artistic Faxes: New Knoxville Museum of Art Exhibit Faxed Art," *knoxnews.com*, Aug. 20, 2011, www.knoxnews.com/news/2011/aug/20/artistic-faxes-new-knoxville-museum-of-art-faxed.

7. "Cisco Fax Conversion Program," Nov. 7, 2012, www.instantinfosystems.com/2012/11/07/cisco-fax-conversion-program.

Essay on Sources

Perspectives

Ranging around the world and over the decades, fax's history intertwines with many threads in economic and technological history. Seas, if not oceans, of ink have been written about the information revolution, information technology (IT), and information and communications technologies (ICT). Indeed, the changing names are worthy of a study in themselves. Offering broad interpretative frameworks for thinking about the information age are Ithiel de Sola Pool, *Technologies without Boundaries: On Telecommunications in a Global Age*, edited by Eli M. Noam (Cambridge, MA: Harvard University Press, 1990); Francis Cairncross, *The Death of Distance: How the Communications Revolution Will Change Our Lives* (Boston: Harvard Business School Press, 1997); Manuel Castells' trilogy, *The Rise of the Network Society, The Power of Identity*, and *End of Millennium* (Oxford: Blackwell's, 1996, 1997, and 1998), and *Communication Power* (Oxford: Oxford University Press, 2009); and Gerald W. Brock, *The Second Information Revolution* (Cambridge, MA: Harvard University Press, 2003). Carl Shapiro and Hal R. Varian, *Information Rules: A Strategic Guide to the Network Economy* (Boston: Harvard Business School Press, 1999) provides a stimulating and easily accessible economic perspective.

Broader, more critical perspectives on the history of communications come from Brian Winston, *Media Technology and Society, A History: From the Telegraph to the Internet* (London: Routledge, 1998), and Marshall T. Poe, *A History of Communications* (Cambridge: Cambridge University Press, 2011); while Anton A. Huurdeman, *The Worldwide History of Telecommunications* (Hoboken, NJ: John Wiley & Sons, 2003) offers one of the better overviews.

Part of the fun of studying faxing was that its post-1980 success occurred as an analog-digital hybrid in a technical world that increasingly considered digital as the only true technology. Much literature has focused on electronics, the Internet, the Web and the larger concept of digitalization. Four books tracing the evolution of those technologies and concomitant academic interest are Tom Forester, ed., *The Microelectronics Revolution: The Complete Guide to the New Technology and Its Impact on Society* (Cambridge, MA: MIT Press, 1983); Tom Forester, *High-Tech Society. The Story of the Information Technology Revolution* (Cambridge, MA: MIT Press, 1987); Janet Abbate, *Inventing the Internet* (Cambridge, MA: MIT Press, 1999), and William Aspray and Paul E. Ceruzzi, eds. *The Internet and American Business* (Cambridge, MA: MIT Press, 2008). To understand the failure of the Open Systems Interconnection (OSI), see Andrew L. Russell,

Open Standards and the Digital Age: History, Ideology, and Networks (New York: Cambridge University Press, 2014).

Telecommunications deregulation in the last third of the twentieth century drastically changed the legal and economic environments for facsimile. Good guides are Peter Temin with Louis Galambos, *The Fall of the Bell System* (Cambridge: Cambridge University Press, 1987), and Robert B. Horwitz, *The Irony of Regulatory Reform: The Deregulation of American Telecommunications* (New York: Oxford University Press, 1989). Articles in journals like *Telecommunications Policy* as well as that excellent business observer, *The Wall Street Journal*, provided contemporary coverage.

The literature on the history of technology continues to grow faster than I can read it, the sign of a healthy field (and slow reader). The single most important volume for my generation of graduate students was Wiebe E. Bijker, Thomas P. Hughes, and Trevor J. Pinch, eds. *The Social Construction of Technological Systems* (Cambridge, MA: MIT Press 1987); while economic historians like Nathan Rosenberg reminded us of the importance of economics in technological evolution in *Perspectives on Technology* (Cambridge: Cambridge University Press, 1976). More recent approaches to the history of technology can be found in Patrice Flichy, *Understanding Technological Innovation: A Socio-Technical Approach* (Cheltenham: Edward Elgar, 2007), and Michael B. Schiffer, *Power Struggles: Scientific Authority and the Creation of Practical Electricity before Edison* (Cambridge, MA: MIT Press, 2008).

Understanding the diffusion of innovation owes much to Everett M. Rogers, *Diffusion of Innovations*, 5th ed. (New York: Free Press, 2003). The do-it-yourself movement of innovators, tinkerers, and entrepreneurs has long existed in technology, but blackboxing of complex technologies as demonstrated in modern faxing greatly expanded the potential audiences of creators and users. This concept of innovation from below is explored in Eric von Hippel, *Democratizing Innovation* (Cambridge, MA: MIT Press, 2005); Kathleen Franz, *Tinkering: Consumers Reinvent the Early Automobile* (Philadelphia: University of Pennsylvania Press, 2005); David Edgerton *The Shock of the Old: Technology and Global History Since 1900* (Oxford: Oxford University Press 2007); and Alan L. Olmstead and Paul W. Rhode, *Creating Abundance: Biological Innovation and American Agricultural Development* (Cambridge: Cambridge University Press, 2008).

Historians are becoming increasingly interested in the world of the user, both as consumer and as actual adapter and constructor of a technology's applications and meanings. The volumes edited by Nelly Oudshoorn and Trevor Pinch, *How Users Matter: The Co-construction of Users and Technology* (Cambridge, MA: MIT Press, 2003), and Lisa Bud-Frierman, *Information Acumen: The Understanding and Use of Knowledge in Modern Business* (London: Routledge, 1994), provide guides to this challenging area. Closely related is adapting to new technologies, a challenge well elucidated by Edward Tenner in *Why Things Bite Back: Technology and the Revenge of Unintended Consequences* (New York: Alfred A. Knopf, 1996), and Joe Corn, *User Unfriendly: Consumer Struggles with Personal Technologies, from Clocks and Sewing Machines to Cars and Computers* (Baltimore: Johns Hopkins University Press, 2011).

As I researched more, I began to appreciate the importance of expectations and predictions about the future in shaping the evolution of faxing. To place the enthusiasm by advocates for faxing in perspective, I consulted Joseph J. Corn, ed., *Imagining Tomorrow: History, Technology, and the American Future* (Cambridge, MA: Harvard University

Press, 1986); Eric Schatzberg, *Wings of Wood, Wings of Metal: Culture and Technical Choice in American Airplane Materials, 1914–1945* (Princeton, NJ: Princeton University Press, 1995); and Marita Sturken, Douglas Thomas, and Sandra J. Ball-Rokeach, eds., *Technological Visions: The Hopes and Fears That Shape New Technologies* (Philadelphia: Temple University Press, 2004). For the importance of selling promises, see Sylvie Blanco, "How Techno-Entrepreneurs Build a Potentially Exciting Future," in Francois Therin, ed., *Handbook of Research on Techno-Entrepreneurship* (Cheltenham, UK: Edward Elgar, 2007), and the "Gartner hype cycle" http://www.gartner.com/technology /research/methodologies/hype-cycle.jsp.

One downside of overenthusiasm but a normal part of technology is failure, an understudied part of the history of technology. This neglect is changing with articles like Kenneth Lipartito's "Picturephone and the Information Age: The Social Meaning of Failure" *(Technology & Culture, 2003)*, and the works of Martin Bauer, ed., *Resistance to New Technology: Nuclear Power, Information Technology, and Biotechnology* (Cambridge: Cambridge University Press, 1995); Henry Petroski, *Success through Failure: The Paradox of Design* (Princeton, NJ: Princeton University Press, 2006): Tim Harford, *Adapt: Why Success Always Starts with Failure* (New York: Farrar, Straus and Giroux, 2011); and, at the risk of tooting my own horn, my "Failure & Technology" *(Japan Journal for Science, Technology & Society, 2009)*. Charles R. Acland edited a splendid volume, *Residual Media* (Minneapolis: University of Minnesota Press, 2007), exploring what happens when communications technologies become obsolescent. More polemic than analytical, Craig Brod's *Technostress: The Human Cost of the Computer Revolution* (Reading, MA: Addison-Wesley, 1984) displays the personal costs of new office technologies.

Academics, especially economists, have justly devoted much attention to standards and their importance. Among the most important contributions are Charles P. Kindleberger, "Standards as Public, Collective and Private Goods" *(Kyklos, 1983)*; Paul A. David and Shane Greenstein, "The Economics of Compatibility Standards: An Introduction to Recent Research" *(Economics of Innovation and New Technology, 1990)*; and Susanne K. Schmidt and Raymund Werle, *Coordinating Technology: Studies in the International Standardization of Telecommunications* (Cambridge, MA: MIT Press, 1998). The classic cases of clashing standards are the QWERTY keyboard, Paul A. David, "Clio and the Economics of QWERTY" *(American Economic Review, 1985)*, and the VHS-Beta videocassette, Michael A. Cusumano, Yiorgos Mylonadis, and Richard S. Rosenbloom, "Strategic Maneuvering and Mass-Market Dynamics: The Triumph of VHS over Beta" *(Business History Review, 1992)*. See also Jonathan Coopersmith, "Creating Fax Standards: Technology Red in Tooth and Claw?"*(Kagakugijutsushi. The Japanese Journal for the History of Science and Technology, 2010)*.

My appreciation of technology push and market pull derives from David Mowery and Nathan Rosenberg, "The Influence of Market Demand upon Innovation: A Critical Review of Some Recent Empirical Studies" *(Research Policy, 1979)*; C. A. Voss, "Technology Push and Need Pull: A New Perspective" *(R&D Management, 1984)*; and John Howells, "Rethinking the Market-Technology Relationship for Innovation" *(Research Policy, 1997)*. Nicholas Economides and Charles Himmelberg, "Critical Mass and Network Size with Application to the US FAX Market" (New York University, 1995) provide a formal analysis of network externalities.

In a capitalist society that champions the concept of competition, the idea of too much competition seems counterintuitive. Nonetheless, over-competition can adversely affect the evolution of a market by providing buyers with too many choices, fracturing the market and depriving suppliers of necessary profits, as shown by N. Gregory Mankiw and Michael D. Whinston, "Free Entry and Social Inefficiency" *(RAND Journal of Economics*, 1986); Venkatesh Bala and Sanjev Goyal, "The Birth of a New Market" *(Economic Journal*, 1994), and Anthony Creane, "Note on Uncertainty and Socially Excessive Entry" *(International Journal of Economic Theory*, 2007).

The state has played a major role in shaping the development of technologies, as demonstrated by Nathan Rosenberg, Ralph Landau, and David C. Mowery, eds., *Technology and the Wealth of Nations* (Stanford, CA: Stanford University Press, 1992); Kenneth Flamm, *Creating the Computer: Government, Industry, and High Technology* (Washington, DC: Brookings, 1988); Office of Technology Assessment, *Innovation and Commercialization of Emerging Technologies* (Washington, DC: GPO, 1995); and Tessa-Suzuki Morris, *The Technological Transformation of Japan: From the Seventeenth to the Twenty-First Century* (Cambridge: Cambridge University Press, 1994). For an appreciation of the importance of the military in shaping technologies, see Merritt Roe Smith, ed., *Military Enterprise and Technological Change* (Cambridge, MA: MIT Press, 1982), and Barton C. Hacker, "Military Institutions, Weapons, and Social Change: Toward a New History of Military Technology" *(Technology & Culture*, 1994).

Viewing replication, repetition, and reproduction—all essential components of faxing—through the prism of postmodernism provided a different perspective on communications technologies. Good guides for exploring concepts of visual culture are Andrew Darley, *Visual Digital Culture: Surface Play and Spectacle in New Media Games* (London: Routledge, 2000); Nicholas Mirzoeff, *An Introduction to Visual Culture* (London: Routledge, 1999); and Marita Sturken and Lisa Cartwright, *Practices of Looking: An Introduction to Visual Culture* (Oxford: Oxford University Press, 2001). For deeper diving, Fredric Jameson, *Postmodernism, or, the Cultural Logic of Late Capitalism* (Chapel Hill, NC: Duke University Press, 1991), and Jean Baudrillard, *Simulacra and Simulation*, trans. Sheila Faria Glaser (Ann Arbor: University of Michigan Press, 1994) will stimulate.

Primary Sources

Unlike my earlier work on Russian electrification, which was shaped by limited information, the history of fax was awash in data. This book is grounded in the holdings of government, corporate, public, and private archives as well as a sea of published material.

For nineteenth-century and prewar European efforts, the France Telecom Bibliotheque historique des postes et des telecommunications in Paris, and the Institution of Engineering and Technology library, the *Daily Mirror* microfilm collection, and the British Telecom archives in London were invaluable. The Library of Congress offered the papers of Alexander Graham Bell and the Smithsonian Institution Archives, the William J. Hammer papers. Beyond the inventor's work, the Thomas A. Edison Papers and website contained material on the state of contemporary facsimile. The Samuel C. Williams Library, Stevens Institute of Technology, provided manuscripts of Arthur Korn.

The Bain-Morse litigation generated partisans on both sides. For a pro-Bain account, see John Finlaison, *An Account of some Remarkable Applications of the Electric Fluid to the Useful Arts, by Alexander Bain* (London: Chapman and Hall, 1843); while Arthur J. Stansbury, *Report of the Case of Alexander Bain, Appellant, vs. Samuel F. B. Morse,*

Respondent, before the Hon. William Cranch, Chief Justice of the District of Columbia on Appeal from the Decision of the Commissioner of Patents (Washington, DC: Gideon & Co., 1849) provided the American justification.

Understanding the explosive burst of interwar interest in faxing would have been impossible without access to corporate archives. Reflecting its importance in twentieth-century telephony, the AT&T archives, originally in Warrington, New Jersey, and now in San Antonio, Texas, contained over half a century of information about faxing, including potential competitors. More chronologically focused, the RCA holdings of the Hagley Library in Hagley, Delaware, and the papers of Charles J. Young preserved by his son, Neils Young, in Boise, Idaho, revealed RCA's extensive efforts. At the Smithsonian Institution Archives in Washington, D.C., the Clark Radioana and Western Union collections covered Western Union fax efforts through the 1960s as well as the interwar emergence of radio-based fax. In London, the British Telecom archives provided records about the British Post Office's venture into Picture Telegraphy. The California State Railroad Museum Library provided train timetables to compare the cost of sending a person across the country with a telephotograph.

The Eleanor McClatchy Collection of the Sacramento Archives & Museum Collection Center provided insight into California newspaper interest in fax broadcasting as well as the normally elusive newspaper reader. The struggle over wirephoto unfolded in *Editor & Publisher*, a struggle also covered by Kent Cooper in *Kent Cooper and the Associated Press: An Autobiography* (New York: Random House, 1959). John R. R. Hunt, *Pictorial Journalism* (London: Sir Isaac Pitman & Sons, 1937), and Sammy Schulman, "*Where's Sammy?*" (New York: Random House, 1943) portray the competitiveness in obtaining pictures.

For World War II and the postwar era, I spent many exhilarating days working in the National Archives and Records Administration in Washington, D.C., on military faxing, the 1948 FCC inquiry, and the Post Office's abortive Speed Mail. The Army Signal Corps Museum and the archives at the Signal Corps Engineering Laboratory at Fort Monmouth, New Jersey, chronicled postwar military activities. I also found valuable material at the Naval History and Heritage Command in Washington, D.C.; the Federal Communications Commission library; the *Miami Herald* archive in Miami; the corporate holdings of Alden in Framingham, Massachusetts, and Times Facsimile Corporation at the *New York Times* in New York City; and the records of fax researcher Homer Dana at the Washington State University in Bellingham, Washington.

Researching the age of modern facsimile depended heavily on market research reports and specialized newsletters as well as on the 3M holdings at the Minnesota Historical Society in St. Paul, the wide-ranging materials of the Charles Babbage Institute at the University of Minnesota in Minneapolis, and the archives of the Telautograph Corporation in Los Angeles. The private papers of Robert Wernikoff, maintained by his wife, Denise, in West Newton, Massachusetts, provided an intimate portrayal of a startup and its challenges. The library of the Newspaper Association of America in Arlington, Virginia, had decades of reports on facsimile in newspapers, including ANPA Telecommunications Department, 1991 *Directory: Newspaper Electronic Information Services* (Washington, DC: ANPA, 1991). As well as providing an enriching academic home, the National Museum of American History allowed me to use its files on the acquisition of its first fax machine, which has moved from the library to the museum's collections.

The Smithsonian Institution Archives also provided invaluable information on Federal Express's Zapmail.

Market research firms like Frost & Sullivan and International Research Development, Inc. sold, often for several thousand dollars, reports about the state of the facsimile field. A surprising number of these invaluable studies were publicly accessible. Not as expensive and more topical were the many years of the bimonthly *EMMS* (*Electronic Mail and Messaging Service*) newsletter, which provided ongoing analysis and data about electronic mail, broadly defined.

The annual *Tsushin kiki chuki juyo yosoku* [Communications Equipment Demand Forecast] by the Communications Industry Association of Japan provided invaluable statistical and qualitative material otherwise unobtainable. The government of Japan published a range of very informative white papers, policy statements, and annual reports. The AT&T archives preserved the reports of the 1977 trip by two Bell Telephone Laboratory researchers to Japan, offering a comprehensive firm-by-firm study of that point in time.

The decades of struggles over national and international standards have been preserved in the thousands of contributions and other records at the International Telecommunications Union library in Geneva and the Telecommunications Industry Association (née Electronic Industries Association) in Washington, D.C. Invaluable for the 1990s was the journal *Human Communications Digest* generously provided by its editor, James Rafferty.

Electrical and mechanical engineering journals provided much contemporary technical information. To leaf through *Mechanic's Magazine, Scientific American, Electrical Review*, and similar journals was to realize the intricate beauty of the early fax machines. Overviews of nineteenth-century telegraphy usually mentioned facsimile, almost always at the end. The impression given was that it was an interesting technology on the periphery, like a promising baseball player who has yet to, and may never, leave AAA for the major leagues. George B. Prescott chronicled the changes in faxing in his many versions of his *Electricity and the Electric Telegraph* in the 1860s–80s, while Louis Figuier offered a French perspective in *Les Merveilles de la science* (Paris: Furne, Jouvet, 1867), and Karl E. Zetzsche gave a German perspective in *Handbuch der elektrischen Telegraphie* (Berlin: Verlag von Julius Springer, 1877).

Reflecting the spread of faxing into the business world in the 1960s–70s, journals like *The Office* provided useful information to administrators and managers about the challenges of actually deploying and using fax in the workplace. In the 1980s the spread of articles into more popular journals reflected the explosion of faxing into the public imagination. Newspapers, especially the *New York Times* and the *Wall Street Journal*, proved very useful in tracking business interest in faxing.

Secondary Sources

Surprisingly few books on facsimile exist. Four exceptions are Charles R. Jones *Facsimile* (New York: Murray Hill Books, Inc., 1949); Daniel M. Costigan, *Fax: The Principles and Practice of Facsimile Communication* (Philadelphia: Chilton Book Co., 1971) and *Electronic Delivery of Documents and Graphics* (New York: Van Nostrand Reinhold Company, 1978); and Kenneth McConnell, Dennis Bodson, and Stephen Urban, *Fax: Facsimile Technology and Systems*, 3rd ed. (Norwood, CT: Artech House, 1999). While all have historical sections, none are historical — unlike Jennifer S. Light's "Facsimile: A Forgotten 'New Medium' from the 20th Century" in *new media & society* (2006). M. J. Peterson "The Emergence of a Mass Market for Fax Machines" in *Technology in Society* (1995) provides a contemporary political science analysis of faxing's explosion.

For an appreciation of the design and engineering of the modern fax machine, see Marvin Hobbs, *Servicing Facsimile Machines* (Englewood Cliffs, NJ: Prentice Hall, 1992), and Gordon McComb, *Troubleshooting and Repairing FAX Machines* (Blue Ridge Summit, PA: TAB Books, 1991). Sadly, *Just the Fax, Ma'am* by Leslie O'Kane (Toronto: Worldwide, 1997), while an entertaining mystery, did not contribute to my understanding of faxing but did show how engrained into popular culture faxing had become. While most viewers of *Bullit, Die Hard 2,* and *Rising Sun* probably did not watch the movies for the fax machine scenes, those segments — product placement — provide nice touches of suspense for the plot and a sense of how contemporary audiences understood the technology as exotic or novel.

Although W. James King published "The Development of Electrical Technology in the 19th Century: 2. The Telegraph and the Telephone" (*United States National Museum Bulletin* 1962), and Arno Press reprinted in 1971 Alvin F. Harlow's *Old Wires and New Waves: The History of the Telegraph, Telephone, and Wireless* (New York: D. Appleton-Century Co., 1936), there was no broad history of American telegraphy until David Hochfelder, *The Telegraph in America, 1832–1920* (Baltimore: Johns Hopkins University Press, 2012). Richard R. John, *Network Nation: Inventing American Telecommunications* (Cambridge, MA: Belknap, 2010) provides an excellent political economy of the nineteenth- and early twentieth-century telegraph and telephone, while Christopher McDonald, "Western Union's Failed Reinvention: The Role of Momentum in Resisting Strategic Change, 1965–1993" (*Business History Review,* 2012) does the same for the last years of Western Union.

K. G. Beachamp provides an entry into the important world of electrical exhibitions in *Exhibiting Electricity* (London: Institution of Electrical Engineers Press, 1997). For electronics in the Great War, see Frederik Nebeker, "The Role of the First World War in the Rise of the Electronics Industry" (*Engineering Science and Education Journal,* 2001), and A. E. Kennelly "Advances in Signaling Contributed During the War" in the reprint of the 1920 Robert M. Yerkes, ed. *The New World of Science: Its Development During the War* (Freeport, NY: Books for Libraries Press, 1969).

The history of radio played a significant role in twentieth-century faxing, mostly in roads not taken successfully. Essential background for understanding the complex environment comes from William F. Boddy, "Launching Television: RCA, the FCC, and the Battle for Frequency Allocations, 1940–1947" (*Historical Journal of Film, Radio, and Television,*1981); Erik Barnouw, *The Golden Web: A History of Broadcasting in the United States,* Vol. 2, *1933 to 1953* (New York: Oxford University Press, 1968); Gary L. Frost, *Early FM radio: Incremental Technology in Twentieth-century America* (Baltimore: Johns Hopkins University Press, 2010); Gwenthy L. Jackaway, *Media at War: Radio's Challenge to the Newspapers, 1924–1939* (Westport, CT: Praeger, 1995); Robert W. McChesney, *Telecommunications, Mass Media, and Democracy: The Battle for the Control of U.S. Broadcasting, 1928–1935* (New York, 1993); and Hugh R. Slotten, *Radio and Television Regulation: Broadcast Technology in the United States, 1920–1960* (Baltimore: Johns Hopkins University Press, 2000). For amateurs in radio, see Susan J. Douglas, *Listening*

In: Radio and the American Imagination . . . from Amos 'n Andy and Edward R. Murrow to Wolfman Jack and Howard Stern (New York: Random House, 1999), and Gordon Bussey, *Wireless: The Crucial Decade: History of the British Wireless Industry, 1924–34* (London: Peter Peregrinus, 1990).

For evolution of photography in newspapers, see Oliver Gramling, *AP: The Story of News* (New York: Farrar and Rinehart, 1940); Ken Baynes et al., *Scoop, Scandal, and Strife: A Study of Photography in Newspapers* (London: Hastings House, 1971); Barbie Zelizer, "Words against Images: Positioning Newswork in the Age of Photography," in Hanno Hardt and Bonnie Brennen, eds., *Newsworkers: Toward a History of the Rank and File* (Minneapolis: University of Minnesota Press, 1995); and John Nerone and Kevin G. Barnhurst, *The Form of News: A History* (New York: Guilford Press, 2001).

A historical perspective on the development of microwave technology is provided by Harold Sobol, "Microwave Communications—An Historical Perspective" (*IEEE Transactions on Microwave Theory and Techniques*, 1984), and Philip L. Cantelon, "The Origins of Microwave Telephony—Waves of Change" (*Technology and Culture*, 1995).

The growth of a copying culture in offices began long before the Xerox 914, providing inspiration, competition, and support for faxing. This pre-Xerox world is well portrayed by H. R. Verry, *Document Copying and Reproduction Processes* (London: Fountain Press, 1958); Ian Batterham, *The Office Copying Revolution: History, Identification, and Preservation* (Canberra: National Archives of Australia, 2008); Barbara J. Rhodes and William Wells Streeter, *Before Photocopying: The Art and History of Mechanical Copying, 1780–1938* (New Castle, DE: Oak Knoll Press, 1999); and JoAnne Yates, *Control through Communication: The Rise of System in American Management* (Baltimore: Johns Hopkins University Press, 1989).

The wealth of information on the Japanese electronics and ICT industries reflects both its considerable accomplishments and late twentieth-century Western fear about this rising economic power. Among the more informative books are Tessa Morris-Suzuki, *Beyond Computopia: Information, Automation, and Democracy in Japan* (London: Kegan Paul, 1988); Marie Anchordoguy, *Computers Inc.: Japan's Challenge to IBM* (Cambridge, MA: Harvard University Press, 1989); Erik Baark, *Towards an Advanced Information Society in Japan: A Preliminary Study of Sociocultural and Technological Driving Forces* (Lund, Sweden: Research Policy Institute, 1985); Martin Fransman, *The Market and Beyond: Cooperation and Competition in Information Technology Development in the Japanese System* (Cambridge: Cambridge University Press, 1990) and *Japan's Computer and Communications Industry: The Evolution of Industrial Giants and Global Competitiveness* (Oxford: Oxford University Press, 1995); Bob Johnstone, *We Were Burning: Japanese Entrepreneurs and the Forging of the Electronic Age* (New York: Basic Books, 1999); Fumio Kodama, *Analyzing Japanese High Technology: The Techno-Paradigm Shift* (London: Pinter Publishers, 1991), and *Emerging Patterns of Innovation: Sources of Japan's Technological Edge* (Cambridge, MA: Harvard Business School Press, 1995); Ken Coates and Carin Holroyd, *Japan and the Internet Revolution* (London: Palgrave Macmillan, 2003); and Mizuko Ito, Diasuke Okabe, and Misa Matsuda, eds., *Personal, Portable, Pedestrian: Mobile Phones in Japanese Life* (Cambridge, MA: MIT Press, 2005). For Japanese interest in quantifying information flows, see Ithiel de Sola Pool et al., *Communications Flows: A Census in the United States and Japan* (New York: Elsevier Science Publishing, 1984).

Index

Page numbers in *italics* refer to figures and tables.

Lightning Source UK Ltd.
Milton Keynes UK
UKHW042006111122
412045UK00004B/277